Gasp!

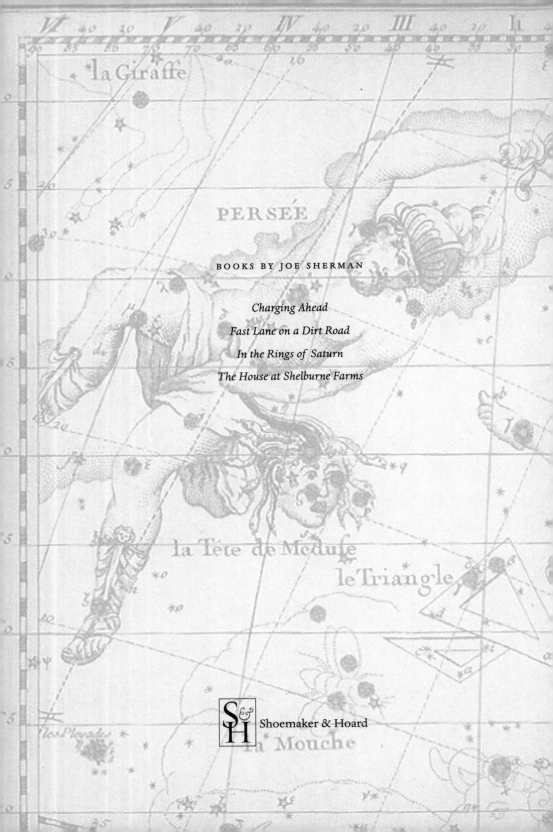

BOOKS BY JOE SHERMAN

Charging Ahead
Fast Lane on a Dirt Road
In the Rings of Saturn
The House at Shelburne Farms

Shoemaker & Hoard

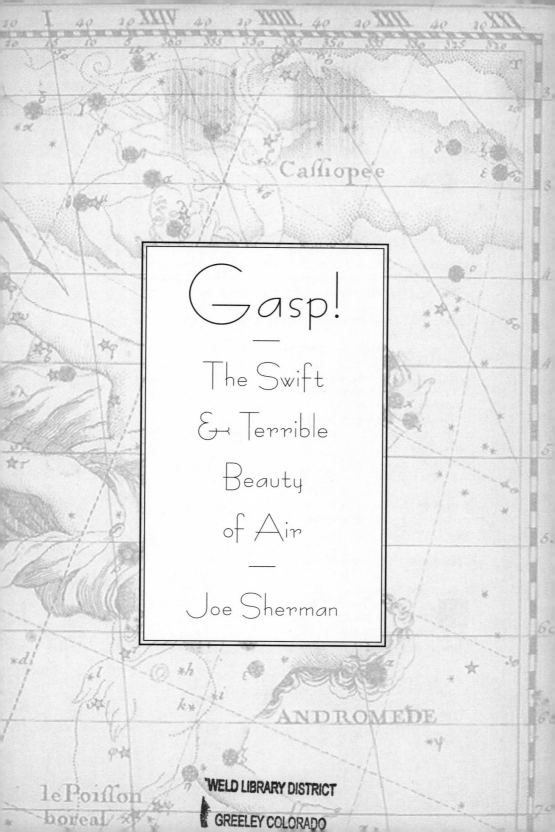

Gasp!

—

The Swift
& Terrible
Beauty
of Air

—

Joe Sherman

Library of Congress Cataloging-in-Publication Data
Sherman, Joe, 1945–
Gasp! : the swift and terrible beauty of air / Joe Sherman.
p. cm.
Paperback: ISBN-10 1-59376-140-6
            ISBN-13 978-1-59376-140-0
Hardcover: ISBN-10 1-59376-025-6
            ISBN-13 978-1-59376-025-0
1. Air. 2. Atmosphere. 3. Respiration. 4. Gases. 5. Air—Pollution. I. Title.
QC161.S48 2004
551.5—dc22
2004011662

Cover design by Kimberly Glyder Design
Interior design by Mark McGarry, Texas Type & Book Works
Set in Dante

Printed in the United States of America

Shoemaker & Hoard
An Imprint of Avalon Publishing Group, Inc.
1400 65th St., Suite 250
AVALON Emeryville, CA 94608
Distributed by Publishers Group West

10 9 8 7 6 5 4 3 2 1

For J. T.

Our souls are air. They hold us together. Listen.

—MICHAEL LONGLEY, *HARMONICA*

Contents

Gasp!

Part 1

Inhale

There is something from every living being in the atmosphere:
close inspection of the air would disclose beggars,
lawyers, bandits, mailmen, seamstresses
and a little of each occupation, a humbled remnant
wants to perform its own work within us.

—PABLO NERUDA, *NOCTURNAL COLLECTION*

1

Our Airy Bodies

Life needs air. It delivers what ancient yogis called *prana*, the vital force. In the womb, *prana* reached you indirectly, through the blood of your mother. Born in a few seconds or minutes, you burst into the shocks of an outside world, into the world of air, and suddenly had to breathe on your own. After the first breath, you could count on about 650 million more over an average life span. As an adult, air washes into your lungs an average of twenty-three thousand times every day. Breathing is involuntary, a reflex action, and rarely given a second thought. Until you can't breathe. Throughout the life of the body, blood absorbs gases, notably oxygen, and transfers them to their destinations, sometimes under great stress. But when you can't breathe, cells begin to die in minutes, setting off complex defense mechanisms. Without air, despite how alive you feel, you are unconscious in four or five minutes and dead in about ten. End of *prana*. End of life.

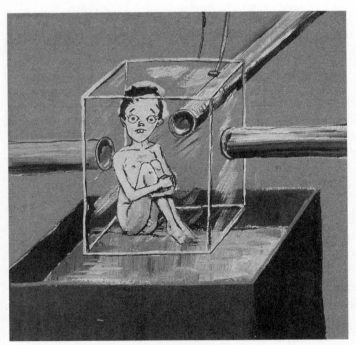

You come into the world of air breathing about thirty-three times a minute at birth. Your lungs double in weight the first eighteen months, then double again by the time you're eight. Here a vulnerable boy sits in an artist's rendition of a body box, a respiratory chamber to measure lung size. His will reach their maximum size when he is around twenty-three.

GIVEN the importance of air, I'm a little embarrassed to admit that I don't remember my son's first breath—and I certainly don't remember mine. I was almost totally ignorant of the enormity of its demands, the marvels it unleashed in his tiny body, and its other unique attributes, including the fact his lungs disgorged water and amniotic fluid, along with blood and vaginal secretions, in order to haul in on air. In the operating room, that blindingly bright theater of cleanliness and anaesthetized hysteria, I was more focused on the Frank's breech (it had brought my wife in there suddenly for a cesarean section) and on listening to her breathing, than I was on my son's first breath. But if I had to do it again, I would shift my attention. I would pay attention to that gasp and first cry, one that a friend recently told me happened with his child almost immediately after the head emerged from the birth canal. When my son did cry, it wasn't very loud. That I do remember. It may have been a sign of his respiratory health. A robust system insists on gulping air as soon as the head emerges, but a less developed one allows the child to be without air for a few seconds, or even a minute, until collective forces within its body manage to communicate the magical word, "Breathe!"

That first breath of air is something we all share. It's a universal. It's an evolutionary crossroads, one that took tens of thousands of years for biology to construct, a passage from the water world to the air one. We zoom through it with a gasp and a wail. It's amazing, but even in this age of medical sophistication, we don't know exactly what triggers this transition, what suddenly makes a baby haul in on air with all its might. "There's a lot of information on why babies take their first breath," says Jay Leiter, an expert on respiration at Dartmouth Medical School, "but no definitive answer." The command seems to be chemical. It appears to be given in the brain, probably by neurotransmitters waiting patiently for the grand exit, or grand entrance, over which they exercise ultimate power. But whatever the ultimate kicker, the mechanical act of taking a first breath and the chemical transactions taking place on the molecular level are the same for every person on the planet. And for every mammal as well, for that matter.

My son, your daughter or son, you and I, your pet dog—in the womb we all had lungs of negligible importance. They were tiny buds that sprouted from the gut. All that the lungs did initially was grow and secrete amniotic fluid. Breathing movements in my wife's womb could have been detected at around twelve weeks, though we never went looking for them. Our son's first breath, which I didn't pay much attention to, was really a continuation of the breathing movements and growth in the womb. Once he came out wet and dripping and not breathing into the bright and airy world of the operating room, there was a moment when he was balanced between the future, when he would grow because of oxygen's power to fuel his cells, and the past, when he had existed for nine months in an aquatic world closer to the Paleozoic than to the world as we know it. The sudden increase in oxygen all around him, the drop in temperature, even the sounds of my wife saying, teary-eyed, "You got your boy," and the doctor adding, "Your son just pissed on me!" as well as the strong grip of the short resident who had hauled him out butt first and passed him along to the doc—all these sudden stimuli probably played roles in awakening the boy to the world of air. And though his cry wasn't loud, his first inhalation, and those of practically all newborns, was as deep as any he would take for months. The energy demands on such a small, rather inanimate thing, which hovered, wrinkled and pink, on the umbilical cord, were extraordinary. This freshly made boy had to expand his lungs, overcome the resistance of fluids in his air passages, and work his respiration muscles, which had practiced some in the womb but not that much, in a powerful display that none of us could see.

"And just think," Leiter says, "he had so little practice."

At birth, the boy's lungs went through what an emergency room physician friend of mine named Louis Dandurand called "an incredible transition." Their surfaces actually foamed a little from the first flush of blood and oxygen. The blood enlivened tissues, worked their elasticity. For a baby's lungs, the first minutes out of the womb and into air are showtime. They take a few breaths. They try to get some kind of rhythm going, but it's usually out of synch. The movement of the ribs doesn't quite jibe with the breath. The first hours of breathing tend to be all over the place: fast to slow to hardly any breathing at all. This is normal, but anxiety producing if you aren't prepared for it. The breathing rhythm usually evens out by the second day, with the baby breathing through its nose. Breathing through the nose is the respiratory default. Breathing air through the mouth is learned behavior that comes later.

Sometimes, though, the wait for the first breath is prolonged and nerve-racking. Stranded between eons, between the aquatic and the terrestrial, the baby struggles. The chemistry doesn't trigger what it should, the gasp doesn't come. What the British physiologist Joseph Barcroft called *anticipation,* the respiratory system's readiness to engage with breathing's complexities, has a glitch. One of the remedies during Barcroft's era, the late nineteenth and early twentieth centuries, for a glitch that caused the ultimate emergency—a baby unable to breathe—was a Ferris-wheel-like maneuver called Schultze's swing. Basically, the alarmed doctor grasped the baby under both arms, held it between his legs, had the nurse scoop any fluids out of its mouth, and then swung the limp newborn up over his head and back between his knees, trying to fill its lungs with air almost like you would a small balloon with a large mouth. Today, physicians use different techniques, but the goal is the same, to get that child breathing as quickly as possible.

"After the squeegee effect of coming out the vagina, with some babies it's wraaa-aa-aa in a couple of seconds," Dandurand says, describing a normal delivery. "But others, they're stressed, kind of limp. They're not breathing. You can't go two or three minutes then. A minute is a friggin' long time in that situation. You have to stimulate the shit out of them. You warm them very quickly. You dry them. You rub their feet. You rub them around. That doesn't work, the next step is to give them oxygen. Oxygen breaks the cycle. That's not working, you start bagging them. You get oxygen into the lungs, expand the lungs instead of having them do it. But usually, things are fine; [if not] there's some indication there's going to be a problem."

Using an illustration in a medical school text as a guide, Dandurand explains the relationship between the lungs and the heart to me. He describes how sudden increases in oxygen and air pressure instigate all sorts of changes. As he talks, it strikes me that a child's first few hours in the world of air are a three-ringed circus—physiological, biological, and metaphysical—with oxygen as a sort of ringmaster. With blue and red colored pencils, Dandurand sketches the circulation pattern: Blood goes from the right ventricle to the lungs, picks up oxygen and gets rid of carbon dioxide, then returns to the left atrium, passes to the left ventricle, and is pumped into the body. He points out something he calls the *foramen ovale*. To me, it looks like a flap of skin over the oval door of a teepee. In the heart, it separates the upper chambers, the atria.

"In the fetus a fair amount of blood goes across this foramen ovale," Dandurand says.

"Is *foramen* the name of the person who discovered it?" I ask.

"No," he says. "*Foramen ovale* is Latin. It's another term for 'hole.' The reason the hole closes right after birth is that in fetal circulation the pressure on the right side of the heart is as high as the pressure on the left side. So blood gets dumped across. But right after birth, in the newborn's circulation, the pressure is suddenly higher on the left side and pumps the foramen ovale shut. The flap of skin holds it shut at first." He taps the pencil tip on the closed flap of skin. "Eventually, it seals the hole shut."

The closing of the foramen ovale ends the easy flow of blood between the heart's upper chambers. This sealing of the hole in a newborn's heart helps transform the organ from a somewhat passive single pump into a two-sided pump that must quickly take over blood circulation duties from the mother, supplying blood to the baby's lungs and body. The right ventricle starts pumping oxygen-hungry blood through the pulmonary artery to the lungs, Dandurand explains, steering his pencil along the route. The left atrium receives the oxygenated blood back from the lungs and passes the blood to the left ventricle, where it gets pumped into the body. Triggered by blood's rising oxygen content ("You never get pure oxygenated blood in the fetus," Dandurand says. "You've always got this mixture of blue and red blood in a low-oxygen system"), and by changes in pressure, the heart is switched into full operation. Its fetal life ends and its working life begins.

A key player here is partial pressure, an attribute of gases. In air, for instance, each of the constituent gases has a partial pressure proportional to its percentage of the total volume. That means that the partial pressure of nitrogen is different from that of oxygen, the partial pressure of oxygen is different from carbon dioxide, and so on. Differences in partial pressures are what allow gases in a blend to pass individually through membranes, leaving their companion gases behind. Partial pressure was first understood in the late eighteenth century by an English mathematician named John Dalton. He collected air samples in the streets of Manchester and on the surrounding hills and then experimented with them. Partial pressure has existed since the genesis of the earth's atmosphere more than four billion years ago. It is a fundamental force behind the creation of life, a veritable sculptor of animal and plant physiology, and particularly important to the evolution of mammals, whose respiratory systems tend to be complex. In a human child the partial pressure of oxygen in its arteries is three or four times what it was in the fetus after only a few hours in air. The partial pressure of carbon dioxide, meanwhile, has dropped slightly from the fetal state. Here is an example of the gases in air carving their signatures on the biological inner surfaces of life. Air carves systems that can survive in its environment.

A female friend of mine, when I was rattling on about some of the marvels of a baby's first breath and how we'd all been there, interrupted and said that a baby can breathe and drink at the same time, something adults can't do. She was right. A baby's larynx, which is extremely sensitive to foreign matter and chemicals and can pass along an order to the brain to shut respiration down *now, this instant, this is an emergency,* is high in the throat after birth. Later, the larynx drops. Physically, the displacement of the larynx lower in the throat makes it easier for a child to start to talk. But for the first months of life, an elevated larynx plays a more crucial role; it allows the baby to suck a breast and breathe through its nose at the same time.

If all the changes and transitions I've mentioned aren't enough to have going on almost simultaneously in a new child, every second air must perform its most basic function of all: firing the child's metabolism for growth and warmth. The challenge is to have oxidation, the burning of food inside cells, taking place throughout the body all the time, often at different rates to satisfy varying needs—for instance, providing more energy when a baby is thrashing its arms and kicking its feet, and slacking off when it's in REM sleep—and at the same time keeping the temperature relatively steady at around 98.6 degrees Fahrenheit. Different systems, though interdependent, monitor each fundamental task. The controls for the oxygen supply and for carbon dioxide removal, as well as for temperature, are in a primal switch. The switch is a speck of flesh in the brain stem, the part of the brain called medulla.

Pulmonate snails, monitor lizards, blue whales, coyotes, baboons, you, me—we all have the primal switch in our medullas. It's sensing concentrations of carbon dioxide and oxygen in your bloodstream this instant. It's making all kinds of adjustments. Of course, you're totally oblivious to it. Just like you're probably not thinking much about air at the moment, either.

That a speck of flesh in the most primitive part of the brain holds the command headquarters for the breath was suspected back in the third century. A physician named Galen was tending the medical needs of gladiators at Pergamum, a major gladiator training center, for the Roman emperor Marcus Aurelius, when he noticed that some neck wounds terminated breathing immediately. Others, lower on the neck, allowed a gladiator to gasp for a while. Galen, who had a good understanding of human anatomy from having performed autopsies, deduced that a message center of some sort existed high in the neck, and that it communicated with the diaphragm, which in turn worked the lungs. Galen thought the messages probably traveled through the blood, or along the nerves, although his knowledge of either blood or nerves was limited.

Until 1811, Galen's concept of an upper-neck triggering mechanism that controlled the lungs remained the last word on breathing controls. That is, until a myopic and short-fingered French doctor came along. Cesar Legallois was born to a farmer who had insisted that his boy get an education. And what an education it was: Legallois studied medicine on the run as a kind of outlaw physician-in-training during the French Revolution, a time when medical teaching faculties were abolished. In a delicate experiment the doctor removed all of a baby rabbit's brain, except for a piece called the medulla oblongata, and the rabbit kept gasping, just as some gladiators with neck wounds had at Pergamum under the watchful eye of Galen. When Legallois removed all the rabbit's medulla, the gasping stopped. He repeated the experiment, slicing away the medulla oblongata of another baby rabbit, three millimeters at a time. On the fourth slice he found what he was looking for. All breathing stopped. He had located the primal switch.

In place at birth the primal switch is filled with the anticipation described by Barcroft: "By that, I mean the organization of some mechanism apparently in advance of the time when that mechanism is used and against some contingency which has not yet arisen." A baby doesn't need to be swung or bagged or slapped on the butt. It just has to experience the shock of air, and the switch clicks on.

One final mystery of respiration—how the primal switch communicates with the lungs from the first of a half-billion-plus breaths to the last—was solved in the 1860s in a series of classic experiments conducted by a physiology professor in Vienna named Karl Hering, and his student, Joseph Breuer. Hering and Breuer showed that the breath is self-regulating. That is, we virtually have nothing to do with it. That's one reason we think so little about something so fundamental. We don't have to. Evolution has taken care of that need marvelously.

Imagine if it hadn't...if you can. Start thinking support system, such as you would need on Mars, with every single breath created technologically. For humans, there's not much wiggle room with respiration. The window given us here on Earth to breathe is surprisingly narrow and astonishingly unique.

Hering and Breuer also proved that the chemical trigger of respiration was carbon dioxide, not oxygen. As one of their contemporaries, the German physiologist F. Miescher-Rusch, put it, "carbonic acid spreads out its protecting wings over the oxygen needs of the body." A crucial piece in the mystery of respiration, Hering and Breuer's findings showed that a gas in the bloodstream passing through a speck of flesh in the top of the neck controlled the human breath. In the 1920s, a secondary, peripheral site that fine-tuned the breath by detecting carbon dioxide levels elsewhere was also discovered. Clusters of chemoreceptors in the carotid bodies, which nestle alongside the carotid arteries, major blood vessels leading to the brain, do the fine-tuning. A breakthrough in microscopic photography allowed the carotid bodies to be photographed. The images showed sensors registering carbon dioxide and oxygen levels in blood flowing up toward the brain, and passing the information along to the medulla, so it could integrate it into a system of such sophistication that its owner was oblivious to its efficiencies, interior music, and scientific marvels, all of which played a symphony with air.

The Next Half Billion or So Breaths

coming up for air
rising to a very new
somewhere
coming up for air
on the last breath around

—PATTY LARKIN, *COMING UP FOR AIR*

We come into this world of air from the world of our distant ancestors, water and air breathers, amphibians, gills and lungs, destined to breathe

600 million or so breaths. Breaths come quickly at first, about thirty-three a minute when we're first born. During the first twelve months, we breathe around twenty-five times a minute while asleep, and as fast as sixty times a minute when awake. We hit the million-breath mark at around three weeks, the five-million mark sometime during the fourth month. After eighteen months, our lungs have doubled in weight; by the time we're six to eight years old, they have doubled again. All that time, the alveoli, the little sacs in which air exchange occurs, increase in number, and our tidal volume expands. Tidal volume, the air you breathe in and out in a normal breath, is close to its maximum in a girl when she reaches puberty. A boy continues to add tidal volume into his late teens. For both sexes, age twenty-three marks the zenith of respiratory system growth. After that, the lungs, the muscles, the diffusion efficiencies all begin to slowly deteriorate. But, as a cardiovascular technician told me after I was breathing hard during some lung exercises, the declines in function were steeper "before everybody became an exercise nut."

Out of the womb and the world of water, we have a respiratory system more efficient than the one found in fish but put to shame by birds. Birds have developed highly efficient respiration not only to help with the rigors of flight, but for the demands of song as well. Most birds have a syrinx, a small appendage off the windpipe in the throat. The syrinx is like a harmonica. It emits the trills, warbles, cries, tweets, and haunting melodies of birds. However, simply perched on a branch and not singing a note, a bird is a turbocharged respiratory system in a looped configuration worth our admiration. Its lungs don't inflate and deflate like ours. Instead, air passes directly through them, and the lungs take up oxygen and give off carbon dioxide simultaneously in a unidirectional flow, like a traffic circle. The volume of air is regulated by numerous air sacs that ventilate the bird as needed through nostrils in its beak. Overall, the circular flow system maximizes diffusion rates in the feathered body. Such a system makes it possible for geese flying above Mount Everest to have enough oxygen-fired energy both to flap their wings and to honk loudly. You couldn't do that, even if flying an ultralight. Birds carry more oxygen per unit of blood than do mammals like us and are better at shunting oxygen-rich blood to exactly where it's needed in times of stress or high demand.

Though not as efficient as a chick's, the respiratory system of a child is very busy during the early stages of life. The air sacs, the alveoli, increase in number from around 50 million to 300 million in eighteen months. A growing child utilizes the increased capacity for oxygen and carbon-dioxide exchange to toddle, then walk, run, dive, and ascend high hills with barely a second thought. As she grows, her lungs become more elastic. They produce more surfactant, a structural lubricant that maintains the lungs' surfaces and keeps them from collapsing. The bronchial tree defines its many branches in the womb and then stays the same shape, like a baby bonsai, as we grow, simply getting bigger.

In the nineteenth century the English anatomist Richard Owen, the avowed enemy of Charles Darwin and natural selection, suggested that the shape of the lungs was immutable and God-sent, a perfect structure for the breathing of air. Owen was wrong about the perfect structure. There are more efficient designs for gas exchange, as birds prove. But human lung design is clever and highly evolved. That each of our lungs, the left being slightly smaller to make room for the heart, is shaped like an upside-down tree is more evidence for form following function. And the tree shape appears again and again in nature; it's a biological archetype. As a real tree in a field is surrounded by air and efficiently takes carbon dioxide from it, so is a lung suffused with air, from which it takes oxygen.

If you were to somehow shrink very small and wind your way into the human bronchial tree of an adult, you would enter through the nostrils, pass the sinuses and the larynx, and then go down the trachea, or the windpipe. The treelike design would become apparent with the first bronchial bifurcations. If you were to count the bifurcations, the places where the route branches, you'd get to around sixteen before alveoli appeared, blood flowing through their membranous tissue and ready for oxygen molecules to get over there and bond to hemoglobin for the ride

back to the left atrium, the left ventricle, followed by a gentle slosh to billions of cells via the circulatory system. The bifurcating branches of the
bronchi continue well past sixteen. If you were to continue on your respiratory journey, you'd enter smaller and smaller channels until you had
counted twenty-five or so total bifurcations. Then you'd reach a cul-de-sac.

As we grow, the resistance to air flow in this many-branched
bronchial tree decreases. There's a simple explanation. As the branches
grow, they become larger in diameter, so it is easier for air to pass
through, both entering and leaving. Breathing becomes less work, but
not much less. Throughout our lives, from birth to death, we only invest
about 1 percent of our energy in respiratory work. The energy use stays
the same, whether for an infant breathing an average of thirty-three
times a minute and exchanging about a pint of air a minute, or for an
adult breathing an average of fourteen times a minute and exchanging
about eight quarts of air a minute. The best single indicator of how big
your lungs are is your height, just as the British physiologist John
Hutchinson declared in the 1840s, after measuring the lung capacity of a
broad cross section of people in London, including soldiers, policemen,
dwarves, wrestlers, dancers, and giants. And your overall respiratory
potential, also suggested by Hutchinson, is related to your lung size.

To protect the lungs, evolution has provided us with a series of
defenses against unwanted things' entering our upper and lower airways.
Millions of waving cilia in the nostrils snag dust and particles out of air.
Reflexive sneezing and coughing are barriers to unwanted entry, driving
particles back out the throat on the cilia. Foreign substances that get deep
into the lungs are attacked by macrophage cells, which are produced by
bone marrow, travel in the blood, and keep the alveoli region clean and
free of obstructions. Macrophages can ingest and destroy toxins, bacteria, and viruses. They rely on white blood cells for help in steering the
chemical junk they intercept to lymph nodes and then to the bloodstream, where the kidneys flush the offenders out in urine.

One breathing defense that has nothing to do with unwanted stuff
and everything to do with survival back in the aquatic environment is the

diving reflex. An evolutionary safeguard well tuned in children but unreliable in adults, in whom the appropriate sensors may have lapsed, the diving reflex is not something to test on one of your kids. Physiologically, though, this is how things might proceed in a diving reflex experiment.

Take one kid. Plunge her into cold water. Keep her under for fifteen minutes. Pull her out. She'll be OK, probably. What happens—if her diving reflex does its job—is that sensors in the nose and larynx take umbrage at sudden immersion and instantly shut down the breath. Respiration virtually stops. Your daughter could be a seal on a dive or a loon under the placid surface of a lake. Both are dependent on the diving reflex. In the cold water your daughter can't breathe even if she wanted to. A loon, of course, or a seal diving beneath ice flows doesn't want to breathe. It only wants to push its unique metabolic envelope. That envelope is ounce-for-ounce larger than your daughter's, however. Diving birds and diving mammals have more blood in their systems for their size and their blood also has more hemoglobin. These two factors bump up the oxygen levels, which the diving reflex locks in, when diving creatures take a plunge. Divers have another adaptive mechanism; they divert blood to the heart and brain, bathing the fundamental organs with oxygen, and shut down circulation to the muscles, relying on anaerobic metabolism, energy production without oxygen, to fuel them. Humans do that on occasion. If you've ever been on a long run and had a sudden cramp in your calf, it's probably a result of anaerobic metabolism. Your muscles weren't getting enough oxygen and, out of a kind of metabolic desperation, resorted to converting food molecules into the common energy currency of all cells, adenosine triphosphate (ATP). But the conversion was done inefficiently. One consequence was a buildup of lactic acid, and the burn you felt in tissues. At any rate, when you pull your long-immersed child out of the cold water, she won't be able to stand up right away or shake herself dry. She'll be oxygen depleted and partially poisoned by the excess carbon dioxide that her lungs were unable to dump. She'll probably be comatose and have profound lactic acid buildup. Her circulation may have collapsed and demand artificial resuscitation. Still, she's alive and, given good treatment, should recover. Much like the automatic responses in the loon and the seal, her reflexes shunted oxygenated blood to her vital organs, keeping them going for much longer than they would have in an adult. When I ask the Dartmouth Medical School physiologist Jay Leiter at what age the diving reflex blinks off, he says, "Probably under age ten. It's part of the variation in the human population."

Something you might talk to your daughter about, once she's recovered brain function and is alert, is the frogman of the spider world, *Argyroneta aquatica*. Like all spiders, *A. aquatica* has lungs, lungs a bibliophile might envy since they suggest little books with leaflike pages that receive air that perfuses the abdomen. *Argyroneta aquatica* hunts, courts its mate, breeds, and lays eggs that hatch into embryos—all underwater, despite a need for oxygen that only breathing can provide. The spider defies its environment by trapping air bubbles against its abdomen before diving. It weaves a kind of silk gondola, a diving bell that it tethers to underwater plants, and then ferries bubbles of air from the surface to this rather remarkable home, making it inhabitable. Another neat creature with adaptable breathing skills your daughter might like to hear about is the lungfish. There are three kinds: Australian lungfish, African lungfish, and South American lungfish. They first appeared on earth during the Devonian period, 400 million years ago, and have kept their ability to breathe in both water and air through either gills or lungs. Lungfish can survive long droughts by digging cocoons, climbing in, and breathing air with their metabolisms on very low, living without food or water for months, or even years. The red eft likewise has gills at birth, crawls out onto land with the respiratory help of lungs, and, a few years later, returns to the water and gets oxygen through its gills. Frogs, amphibians we tend to be more familiar with, breathe through their tails as tadpoles, then poke their mouths through the water for a breath as they develop lungs. As adults, along with lungs, they respire through skin on their backs and legs, which explains why you see frogs with just parts of their backs and forelegs out of the water in ponds.

How humans evolved into lung-breathing creatures remains a bit unclear (except maybe to creationists), but evolution doesn't invent a new process just because a creature develops a bigger brain. Or do something that fails anatomically, like putting lungs in our legs. Our ancestors may have crawled out of the primordial muck with gills fading and lungs growing, but we upright *Homo sapiens* now only deal with air exchange with lungs located in our chests. In evolution, form follows function. If your lungs were in your thighs, they could inflate and deflate, but there wouldn't be much room left for muscles so you could walk. To keep the distance from exterior air to the lungs short, a functional consideration, thigh lungs might require butt breathing—a rather comic idea even in a scatological period of the Holocene. Anatomical unity would not be well served with thigh lungs and butt breath. Elegance, precision, function, all would be compromised, not to mention aesthetics and the thinness so prized in models at fashion shoots.

Leiter told me that the evolutionary line linking us to the early animals was probably not straight, as I had thought, but rather a zigzag, with creatures going in and out of the sea for hundreds of millions of years before settling full time on the land with lungs only. "Most investigators think animals have invaded the land and had a terrestrial existence at least four times," he says. "Possibly more. The crustaceans and insects invaded, mollusks invaded, fish invaded. I think the fish invasion is what ended up in the mammalian radiation." Some fish and all mammals have hearts and lungs, he reminds me. They're vertebrates. The fish probably invaded swamps and rivers with brackish warm water, where oxygen was not very soluble. "These early fish sometimes had a specialized structure in their mouths," Leiter explains. "It had a rich blood supply. They'd take a big gulp of air and then the oxygen would circulate in by diffusion through the tissue of this structure into the blood. Catfish are like this. Gar. They tend to be fish we don't think are so good to eat."

An amphibian invasion probably came next. Frogs and salamanders. "Many of them were aquatic in their newborn phase. They had gills. Then, as they went through metamorphoses, they invaded the land and became less dependent on water, and the gills actually regressed. The problem with gills is that they collapse in air. The surface tension collapses them and reduces their surface area, so they're not very good for gas exchange."

Reptiles, probably the next invasion and the one from which birds evolved, followed. Unlike amphibians, reptiles aspirated air and did not return to the sea to lay their eggs. Speculating about the zigzag from sea to land and back again—the apparent path taken by evolution—Leiter said he thought the first invaders of the land, although they found it easier to breathe, returned to the sea to eat. Or to eat sea life that migrated offshore. Or possibly they plunged back into the salt water to elude predators that appeared on land. "Of course, there are predators in the sea, too," he acknowledges. "You're damned if you do, damned if you don't. There are going to be predators one way or the other."

Finally came mammals. The first invasion probably began about 370 million years ago, and the creatures had already developed lungs in the sea. "Structurally, all mammals have very similar lungs," Leiter says. "They aspirate them. Most investigators think that the first mammals we'd recognize looked something like modern-day rats. They were small. They stayed out of the way. Most people think they started in trees." He chuckles quietly. "One of the ironies is that if our world ends, there will be cockroaches and rats. Both are excellent survivors."

Whatever the exact path, our mammalian forebears were probably furtive tree rats with strong survival instincts and physical swiftness abetted by oxygen-rich circulatory systems that efficiently powered muscles and brains. And they were small, none weighing over five pounds until after the dinosaurs disappeared, 65 million years ago. These small mammals had lungs in chest cavities that handled their breathing needs pretty well. And totally involuntarily.

Today, you can hold your breath until you almost pass out if you want and you do your lungs no harm. You can go into an oxygen bar and pay for a hit of pure oxygen and handle it OK, although you don't want to do that too often because of something called oxygen toxicity. You can

inhale nitrous air, or laughing gas, and talk funny. You can save a life with
CPR, forcing your breath down someone else's bronchial tree until respiration revives. You can even blend your breath with malefactors, plotting
a conspiracy, for *to conspire* means "to breathe together."

Ever since athletes have existed, regardless of age or nationality or
gender, it seems some have been disappointed with their limited lung
capacity. No competitor can train to run like a horse or a dog, both of
which have spleens that store extra oxygen needed for that burst of speed
seen at times in these fast-moving creatures. But an athlete, with rigorous
training, can practically double her aerobic scope. She can boost the volume of air in her lungs and the oxygen reaching her muscles with exercise. If that doesn't do the trick and get the big win, or the desired
endorsement, she can try enhancing the oxygen-carrying capacity of her
blood. One popular red-cell booster of the 1990s was erythropoietin
(EPO). Questions of EPO use tainted both the 1998 Tour de France and
the 2001 Tour of Italy. Because tests for the drug were poor at detection if
given more than seventy-two hours after EPO use, the drug may have
been taken more often than reported. Large doses of the drug can bump
up the blood's oxygen capacity by 50 percent or more. They can also kill
you. During the height of EPO's popularity, dozens of bicyclists suffered
strokes and heart attacks. Blood doping clotted their blood. More than
two dozen very healthy athletes died from it.

Lance Armstrong, the American biker who won his fifth consecutive
Tour de France in 2003, was hounded by accusations that he used EPO.
But Armstrong never failed a drug test. Asked how he could pedal longer
and faster than all the other guys, and why his blood and heart and aerobic scope were better, he replied, "I never miss a day of training. Never."
Armstrong worked out two to six hours a day, stayed away from swimming and upper body work because it bulked up muscles and added
weight. His regime emphasized one thing, aerobic scope. And he made
sure that he steadily pushed himself to just below his threshold, which
continually boosted it. This regime helped him oxidize fats and carbohydrates longer than his competitors could during a race. For instance, on

steep mountain ascents in the Alps during the Tour de France, Armstrong didn't feel the burn, caused by lactic acid buildup from anaerobic processes in stressed muscles, until long after most of his competitors were gritting their teeth and gutting it out. "When you see Lance on the climbs in the Tour," says his longtime coach, Chris Carmichael, "it doesn't seem he's hurting. You know what? He really isn't. Lance is almost entirely aerobic. When he attacks, then he goes anaerobic and everybody else has already been there."

To learn how far I was below Lance Armstrong on the aerobic scale, and hopefully how much higher than the average couch potato, I decided to take a series of breathing tests called pulmonary function tests (PFTs). With Leiter's help, I put myself in the skilled hands of Roy Ward, a cardiopulmonary technician at Dartmouth Hitchcock Medical Center. To measure my lung volume, mechanics, and diffusion, Ward had me blow into a tube distressingly similar to those used by cops suspecting that you're inebriated and driving.

Ward exhorted me with cheerleading-like refrains: "Push, push, push, push, push," was one. "Deep and fast, deep and fast, keep going, keep going, deep and fast, couple more, deep and fast, deep and fast," was another. He ran me through the spirometry test and the maximum voluntarily ventilation test, and he checked my diffusion capacity. Finally, he directed me to sit in an enclosed glass-and-plastic box for—what else?—the body box test. It gauged the size of my lungs. With each pulmonary function test, Ward made me feel a little better about my relationship with air. He kept glancing at the normal ranges of predicted results on a computer. My numbers were above them.

"You're perfectly normal," he said, after we'd finished the spirometry test.

I'd thought I was above normal, but let it ride.

"Three point two six is your predicted FEV1," he said. "That's the amount of air you can get out in the first second. You did 4.26 liters, so you're getting out a whole liter more. And your forced vital capacity, which is your total volume after you've taken that breath and blown it all the way out—that's 5.25 liters. And 4.55 is your predicted. That's very good."

That made me feel better.

My blood diffusion figures were normal, he went on. I had inhaled a special mix of gases, a little methane and a little carbon monoxide. The methane was inert, so it mixed with the residual air in my lungs and the carbon monoxide diffused in the blood, Ward explained. On the printout that he gave to me, my diffusion rate was 146 percent of what was normal for my age and height. My total lung volume was 7.7 liters, well above the norm.

"So I'm in good shape?"

Roy Ward nodded.

"How do I compare with Lance Armstrong?"

Roy said he didn't think Lance had much to worry about from me, even if I did decide to dope my blood with EPO. "For a man your age, though, you have the ability to move a lot of air if your muscles are in good shape." He scanned the data again. "You could move a lot of air, probably enough for two or three liters of oxygen for a brief time."

A good athlete can double that. The best athletes Roy Ward had ever tested for moving air had been rowers. But even their aerobic scope paled compared to this one Norwegian skier, he recalled suddenly. That guy had been able to get nine liters of oxygen into his system every minute, Roy said in a tone of astonishment.

The PFTs were not such great diagnostic tools for lung disease, Jay Leiter told me later. "They give us a place to start. But we have to think about what's wrong, what causes the abnormalities."

At this point in my respiratory evolution, thankfully nothing was abnormal. But what if I'd been a coal miner? Or had asthma? Or smoked? Or had spent most of my life breathing the smoggy air of Los Angeles or the petrochemical-polluted soup of the Black Triangle in the Czech Republic, a region once notorious for black snow?

No matter how clean or dirty the air we breathe, someday we all will come to our final breath. For most of us, it will be strained, the muscles tired, the lung tissue stiff, the fluids thick. People familiar with death talk about the death rattle, the sound the windpipe makes as it struggles to get life's sustaining gases in and out. I listened to the death rattle of a friend's elderly father, a Jew who had escaped from Nazi Poland and the Holocaust and was now in his nineties and dying at home in Montreal. Each breath had a rasping, shuddering quality to it. Every few minutes, breathing came to an abrupt and sudden stop. He lay there not breathing, like a newborn infant yet to take its first breath. Then came a sharp gasp. It was what physiologists call a gasping event. A respiration jolt, it can save a person, keep death at bay awhile longer, or introduce a child into life. With my friend's elderly dad, the gasping event was probably triggered by erratic breathing controls in the medulla because of the morphine in his blood. But his *prana,* his life force, lingered on. It refused to leave his body as long as he could gasp and fill his old lungs with some oxygen.

2

A Whole Lot of Air

As a boy I wondered where the sky ended. Was there a wall out there somewhere, with a golden door in it for those headed to heaven? Were there bunnies on planets we hadn't discovered yet? A little knowledge of air would have answered some of the flurry of questions in my boyish brain. As I learned later, this "thin seam of dark blue light," as the German astronaut Ulf Merbold described our atmosphere from outer space, exists nowhere else—at least not in this universe. Our nearest planetary neighbors, Venus and Mars, each had an atmosphere of sorts, but comparing the air of Mars to ours was like comparing a bouncing, laughing child who can run and do somersaults in a field of dandelions to a slab of icy rock on a vast, windswept expanse of lifelessness.

Without an oxygen mask at high altitudes, things can go wrong quickly, as they did for the first balloonists to reach the stratosphere in 1862. Despite the dire states of the pilot, Henry Coxwell, and his technician, James Glaisher, in this illustration of the famous flight, the two men lived.

MY PARENTS were not the kind of people who thought about the atmosphere on Mars (it's a frigid, dry, oxygen-poor place), or anywhere else, or who tried to get their minds around space. We did have a tenant upstairs whose son went to Dartmouth College, and I recall her using the word *atom* and giving me books about science and math that were far beyond my comprehension. She may even have mentioned *atmospheric perspective,* on occasion, referring to light. My father's rare references to air usually involved the back stairs. After climbing them, he'd sigh and say he was out of breath. My mother's awareness of air, at least in my memory, is summed up by an exclamation she delivered in a startled voice after a beloved parakeet flew out the kitchen door. Flow-through lungs on high, it navigated between the branches of a tree in our yard and rose on excited wings. The parakeet disappeared over a neighbor's roof, never to be seen again. Sounding heartbroken, my mom said, "Sam just flew away through the air."

I believe my parents limited attention to air was the norm, then and now. Persuasive commentators claim that the average person prefers to think about things his or her own size. About monkeys rather than microbes, for instance, about the weather in town rather than across the state, about lilacs in the yard rather than the ozone layer in the stratosphere. At the same time, we're urged to come to grips with the modern world and its information explosion. We're told we should be able to understand the very small—an electron, a gene, a silicon wafer—and the very large—the human genome, a superconductor, the immune system—simultaneously. To keep up with the times, the mind should be nimble enough to go in different directions, sort of like an explosion.

I've found I can't do that. My mind—except in dreams—seems programmed to plod along, handling one thing at a time. It heads toward home plate, looking for a slow roller; it can't spring back and be out by the fence to catch a high fly without some synapses clicking into gear. That takes more than a few nanoseconds.

This issue about my brain's lack of modernity worries me a little, but not much. I'd love to be able to envision radio waves from Prague sending a live performance by a Romany band in a smoky café off the Kennelly-Heaviside belt, an electrically charged strata in the ionosphere that acts like a parabolic mirror for radio waves, bouncing them back to earth. And I'd love to simultaneously be aware of carbon dioxide molecules in my medulla kicking up my respiration rate in anticipation of dancing. But I can't do that.

I know that understanding air, which is both big and amorphous, and small and right in front of you, demands a few mental oscillations. I concede the importance of this and even the value of thinking about two things at once, if you can do that. But really, is that kind of thinking crucial for knowing that air is entering my lungs this instant and also encircling the planet? I don't think so. Shifting your thinking from angstroms to miles or from alveoli to the ozone layer doesn't have to occur instantaneously. It just has to occur.

Now, take a deep breath. Hold it and feel how inflated your lungs are. Think about the air inside you pushing against ribs and expanding your chest cavity. A lot is inside you right now: water molecules as old as Earth, dust from rugs you haven't cleaned, oxygen molecules making their way to the razor-blade-thin membrane of a single air-exchange sac. One by one, millions simultaneously, the molecules are passing through your lungs, the largest organ in your body. If you took it out and flattened both lobes, your lungs would cover a tennis court. Beneath the court flows a layer of blood as thin as the membrane of the lungs themselves. In the blood, oxygen is attaching itself to hemoglobin with a Velcro-like grip for the ride to your cells, where the oxygen will fire your mitochondria.

Now breathe out. Think of your breath—some oxygen molecules, more carbon dioxide ones that look the same on Venus and on Mars, water vapor you can see if you breathe against a mirror. The gases and vapor slip into the air in front of you. Lift your hands into that air, fingers spread. Imagine holding your breath, your exhalation, in your hands. It's pretty exciting, having basic life-building materials of the universe right there in front of you: carbon and oxygen, hydrogen and nitrogen, along with a smattering of argon and some really strange stuff, like exotic radiation from live volcanoes on the Jovian moon Io and molecular traces of some of the thousands of commercial compounds appearing this year. Give this breath of yours a color you like: blue or green or red, aquamarine, chartreuse, or mango. Now replicate the breath in your hands ten times with your imagination. Suddenly, your room has a colored splash of air partially visible in it. Replicate the splash a thousand times. Your house or apartment is now filled with air. Replicate the splash ten thousands times, and you can visualize a piece of colored air floating over the neighborhood like a real big balloon identifying your house for a party.

The atmosphere is a composite of such breaths—little ones from bacteria and plants, bigger ones from trees and volcanoes, pint-sized ones from you and me. Different-sized breaths, different gas and matter compositions. All invisible yet there. Individually, each breath consists of virtually unimaginably small atoms, which collectively are vast in scale. But, as astronaut Ulf Merbold said the first time he saw the atmosphere surrounding the earth from space, it is vulnerable looking from up there. "I was terrified by its fragile appearance," he said.

Now, that's a mind-expanding contradiction to get your mind around: Earth is surrounded by a fragile atmosphere.

Playing with your breath and visualizing it floating over the neighborhood may seem a bit silly, given the problems afflicting air, from smog to ozone loss to global warming to nuclear radiation to anthrax spores sent in envelopes. At the same time, being a little silly with air at least makes you aware of it. Holding air in your hands, you have the complex, invisible stuff right there, right now. You know then, as Lewis Thomas said about air, that it's "totally impersonal, yet as much a part of life as wine and bread." I like this idea of thinking of air as that basic, which it is. I'm not an atmospheric scientist. I have a limited grasp of chemistry. But wine and bread I can understand.

So take another breath and think: atoms entering your lungs by the hundreds of millions right this second, vast spheres of air surrounding the planet up to the height of five hundred miles, and the origins of both the atoms and the spheres stretching back in time to the Big Bang, more than fourteen and a half billion years ago. These dimensions of air can be looked at one at a time. They are the discovery sequence by which humankind has learned about air. First came the infinitesimally small, then the finitely vast, and, most recently, the incredibly distant past.

Air As Atoms

My head is too full of triangles, chymical processes and electrical experiments, etc., to think much of marriage.

—JOHN DALTON

John Dalton was a big-thinking Quaker bachelor who had a thing about air. It began in 1781, when he was fifteen. At a Dissenting academy in Kendal, England, where he took a job as an assistant, Dalton met John Gough, the blind natural philosopher memorialized by William Wordsworth in *Excursion*. Dalton was impressed with Gough, who spoke Latin, Greek, French, and English and understood mathematics, astronomy, chemistry, medicine, and meteorology. Dalton liked Gough's habit of jotting down daily observations about the weather. As a teen, Dalton started doing the same. He would make observations about the weather every day of his life, including the last one. And Dalton lived into his late seventies. At the start of his weather fixation, he also built his own measuring instruments—barometers, rain gauges, thermometers. Dalton designed tables of wind, humidity, barometric pressure, and rainfall. He described atmospheric phenomena, including the northern lights, snow, and thunder. *Meteorological Observations and Essays,* Dalton's first book, published in 1793, included his tables, but more importantly, clarified a discovery he had made: the law of partial pressures. He defined partial pressure as the nature of the atoms in one gas to repel its own atoms but not to repel the atoms of other gases. In other words, the atoms of gases distributed themselves evenly among one another. This wasn't a law yet, only a hypothesis put forth by a provincial professor in Manchester, far from London, still the center for science in England. Dalton's new theory of partial pressures attracted little attention.

Dalton had moved to Manchester in 1792 to teach mathematics and natural philosophy at New College, a successor to the Dissenting academy where Joseph Priestley, the preacher scientist who had first isolated oxygen in 1774, had once taught. Cut from similar cloth—humble origins, little money, few prospects—Dalton possessed a drive and a freethinking spirit similar to Priestley's. He also liked to experiment and spent much of his meager income on equipment. Unlike Priestley, Dalton thought big and theoretically and wasn't shy about it. He also was intolerant of the demands teaching made on his time. In 1800 he resigned from the New College staff and opened his own Mathematical Academy, hoping it would give him greater freedom and more flexibility, as well as a better income. During the school's start-up years, Dalton managed not only to administer and instruct, but to give shape to one of science's big ideas, atomic theory. He did it by thinking deeply about air.

In those years, it was not unusual to see John Dalton walking the smoky streets of Manchester, England's nascent textile center, or hiking through the surrounding valleys and up their adjoining hills, gathering samples of air in various kinds of weather. "You can sometimes collect snow still lying high in the hills and keep it for a day or two in a basket for experiments," he wrote to a friend. Dalton sometimes took a bottle of air into the shop of a textile dyer, where the temperatures might be a hundred degrees Fahrenheit. Instantly sweating, he watched the rapid condensation of water vapor inside his bottle.

At the time, little was known about water vapor in air. In the last fifty years, scientists had just begun to understand that air was not a fundamental element, as Aristotle had said, but a collection of gases. By the early 1800s, air was known to contain nitrogen, hydrogen, oxygen, and carbon dioxide. A vapor rising from a liquid, such as steam from water, was still considered a fluid, but a different kind. Dalton proved that water vapor, which rose off the land, was always in air. He showed that temperature determined how much water vapor air could hold and that there were no chemical reactions taking place between air and water vapor. Moreover, as he observed in Manchester dye shops, water vapor condensed and separated itself from air as the temperature rose.

Dalton's fascination with the constituents of air soon led him to a puzzle: Why were gases mixed together in air instead of layered? From the new chemistry of the French genius Antoine Lavoisier, Dalton knew that carbon dioxide was heavier than oxygen, that oxygen was heavier than nitrogen, and that hydrogen was the lightest of all. So why didn't carbon dioxide, the heaviest of the four, separate from the others? Pulled down by gravity, why didn't it form a layer close to the ground? Why didn't oxygen form a layer above it, with nitrogen next, and hydrogen on top? In other words, why wasn't air stratified, like rock or Neapolitan ice cream?

Dalton was fairly convinced that the explanation was his theory of partial pressures, with each gas repelling the other so that the gases stayed evenly distributed. Regardless of where he collected samples of air on his long walks—in the city, on a hilltop, in a valley, the proportions of the gases stayed the same. To understand why, Dalton conducted simple experiments. Basically, he poured one flask of gas into another. For instance, he took bottles of the lighter hydrogen and the heavier carbon dioxide, connected them with a vertical tube, and observed "that a lighter elastic fluid cannot rest upon a heavier." Dalton reasoned that the thorough diffusion of gas particles was caused by the particles of each trying to move away from their similars and mixing with their dissimilars.

Dalton also suspected that atoms were the elementary building blocks of the gases, and of all matter. Many other natural philosophers and scientists before him had also made this connection. Yet none had been able to prove it, despite the fact that the concept of atoms had been around for more than two thousand years. Back in 400 B.C., Democritus of Abdera had made this observation: "Nothing exists except atoms and empty space; everything else is opinion."

Democritus's theories and wit had come down to the celibate bachelor in the Midlands thanks to a poet, Lucretius, who so admired Democritus that he consolidated some of his thought in verse in *De rerum natura,* and to Constantinople, the Byzantine stronghold whose libraries managed to survive the Dark Ages while other famous libraries in Alexandria, Rome, and elsewhere were razed and their books sent aloft as smoke. Atomism, buried in Lucretius's verse, survived deep in the stacks of Constantinople libraries, only to be resurrected by Renaissance scholars, who translated the texts and distributed them out into the world by the printing press in the 1490s, the decade during which Columbus sailed to America.

Christianity opposed the salvaged atomism of the long-dead Greek Democritus, but the highly charged concept refused to die. Pierre Gassendi, who attempted to create a complete system of the world during the mid–seventeenth century, said the size and shape of atoms explained the properties of substances; for instance, small, round atoms caused heat, and pyramidal ones with sharp points made things cold. Atomism also hid conspicuously in the contention of Robert Boyle, the seventeenth century's great English scientist of air, that all matter was made of invisible particles, each a "perfectly unmingled body." By the time John Dalton gathered, separated, and pondered different elements in air, atoms were widely accepted as nature's basic building blocks but hardly understood. Dalton imagined the atoms in air being moved around and mingled by his law of partial pressures. He pictured the atoms as tiny, solid spheres, each surrounded by its own tiny atmosphere, a zone of heat. Extrapolating from his vision and turning to the new chemistry and math, both of which he excelled at, Dalton calculated the weights of the gases. Then, using symbols, he devised the first table of atomic weights. He had his atomic theory.

Like Newton's calculus before it and Einstein's theory of relativity later, atomic theory was a grand synthesis. It was an intellectual edifice whose construction began with Dalton's hikes around Manchester collecting samples of air. The theory generated much confusion and disagreement. It was wrong in some respects and burdened with arbitrary assumptions. But it was basically correct and gradually became widely accepted.

The impact of atomic theory on the understanding of air was monumental. The theory completely toppled Aristotelian logic. Aristotle had said atoms didn't exist. Atomic theory explained that atoms made up everything. The theory said atoms could be neither created nor destroyed, that they simply united, broke apart, then reunited. Atoms enjoyed an eternal dance. "We might as well attempt to introduce a new planet into the solar system or annihilate one already in existence, as to create or destroy a particle of hydrogen," Dalton wrote in his *New System of Chemical Philosophy,* published in 1807.

Dalton's work affirmed the growing power of the small, and of reductionism, to define science and to explain natural processes. Since Antonie van Leeuwenhoek had invented the microscope around 1600, a gradual inversion of classical thinking had been taking place. Instead of giving most of their attention to the celestial sphere, to peering into the vastness of space and toward the stars, scientists came down to the earth. They focused on natural forms, including the human body. Reductionists liked to take things apart: hearts, frogs, flowers, philosophical texts. They discovered that the small mirrored the cosmos, that the minute had depth and structure and mystery. The English anatomist William Harvey plotted human circulation after working on cadavers at Padua, the famous anatomy school in Italy. The irritable hypochondriac Robert Hooke, first curator of the Royal Society of London, coined the word *cell* from his examination of cork under a microscope. Peering through another, stronger microscope, an Italian professor of medicine named Marcello Malpighi observed alveoli, the air sacs in the lung, and said they "form an almost infinite number of orbicular bladders just as we see formed by wax plates in the walls of the honeycomb cells of beehives."

John Dalton went deeper yet into the small. He went past the visible and into realms demanding an intuitive feel for multiple dimensions. Dalton went so small that it was almost, but not quite, impossible to imagine the terrain.

Still, Dalton didn't get the atom quite right. A Newtonian, he thought gravity, not electricity, held atoms together. Another century passed between Dalton's discovery and a clear understanding of what an atom really looked like. Even in the early twentieth century, a physicist as brilliant as Max Planck, who conceived quantum mechanics, scoffed at the notion of the atom. Planck, a German, said the invisible particle was an invention of English scientists. It took an English scientist, though one born and raised in New Zealand, to prove Planck wrong. Working at the Cavendish Laboratory in Cambridge, Ernest Rutherford showed that an atom consisted of a heavy, positively charged nucleus balanced by the opposite charge of a cloud of light electrons swirling about it. A decade later, Niels Bohr, a Dane, refined atomic structure. Bohr proved that it was the outermost shell of electrons whirling around the distant nucleus that determined an atom's potential to bond with other atoms. The outer shell gave an atom its reactive properties.

By 1920, physicists had unlocked the atom and were on the verge of unleashing its incredible energy. The number of working scientists in the world had increased exponentially since Dalton's era, from about one thousand to more than one hundred thousand. All that while, few first-rate minds had been attracted by the mysteries of the atmosphere. Reductionism had made the cosmos passé. Ironically, the world knew more about the interior of the atom than it did about the atmosphere surrounding the earth. In 1920, the atmosphere remained terra incognita. No one had yet reached the stratosphere. In fact, the stratosphere had only been identified as a distinct layer of air, with unique properties, shortly after 1900. Unlike the pursuit of the atom, a process of reduction and theorizing, which fit the momentum of science during the nineteenth century, an exploration of the atmosphere demanded a cornucopia-like expansion, an ascent into vastness and toward the void.

Compared to looking through a microscope, traveling into the sky was out of step with the times.

The Atmosphere
As a Great Big Mystery

Etienne Montgolfier...achieved international renown by constructing a balloon that flew for eight minutes above the royal palace at
Versailles, bearing as passengers a sheep called Montauciel (Climb-
to-the-Sky), a duck and a rooster.

—ALAIN DE BOTTON, *THE ART OF TRAVEL*

Speculation about the atmosphere and the longing to explore it are as old as humankind. In ancient India the sky was said to have gushed from the navel of the man with a thousand heads and a thousand eyes (and presumably a thousand navels, although mythology does not clarify this). In the Amazon, air was said to rise from the moisture of the lush and verdant forests. In Egypt, around 1600 B.C., the sky was thought to be an iron lid, an idea not as far-fetched as it might sound today. Early Egyptian implements were crafted from meteorites, suggesting that bits of sky had flecked off and fallen to the earth. A mythical union of the earth and the sky created the atmospheres of many cultures, atmospheres ruled by sky gods, mythical forebears of the Christian "heavenly father" in the Lord's Prayer. The sky-loving Greeks created an entire pantheon of gods to explain and empower the atmosphere. Their first sky god, Ouranos, was said to have emerged from Gaia, the earth mother. Ouranos wore a crown of stars. In sexual union he spread himself over Gaia. This mythical version of earth's creating its own atmosphere is very close to the latest scientific concept in which emanations from a cooling and hardening planet evolved over billions of years, leveling off around 300 million years ago with the atmosphere we have today. The atmosphere is probably still evolving, albeit very slowly. Scientists now view the earth and the sky as one continuous entity, a living biosphere. The atmosphere—another layer of gases, liquids, and airborne solids—is considered an extension of the earth's crust.

In the history of atmosphere exploration, the 1780s were a milestone. Two brothers, Étienne and Joseph Montgolfier, paper magnates from Vidalon, sent a sheep, a rooster, and a duck up in a basket attached to a balloon that was inflated by hot air. The air was heated from a fire of straw and wool scraps. Human passengers soon replaced animals on balloon flights, and suddenly, scientists, philosophers, and folks on the street all wanted to know what was up there. Questions ranged all over the map. Did air keep getting thinner and colder the higher you went? Could a mercury thermometer tell the temperature a thousand feet above Mont Blanc? How high could a bird fly? At what altitude might the devout glimpse the gates of heaven? And what about Pascal's void—could an aeronaut, as the first balloonists were called, touch it?

The void. It had been an alien concept throughout the Renaissance because Aristotle had said it did not exist and the church had agreed. "God abhors a vacuum" was the dictum. To challenge it had been to challenge the church, not a very good idea if you wanted to stay alive. Yet, in the early seventeenth century, a student of Galileo, Evangelista Torricelli, who lived with the old, blind, and sick genius during the final months of his life, had proven with a primitive barometer that a vacuum existed. Following a suggestion from Galileo (the old man was under house arrest in Florence for having offended Pope Urban VIII), Torricelli filled a three-foot tube with mercury. He covered the open end of the tube with his finger and flipped the tube, placing the open end in a bowl of mercury. He removed his finger. The mercury in the tube dropped as mercury inside it flowed into the bowl. Torricelli kept an eye on the height of the mercury in the tube. It varied a little each day. Torricelli deduced that the change was caused by slight variations in air pressure. As for the space in the top of the tube, the one the mercury had vacated and no air could reach, well, it was a vacuum. And a playful one at that. Both the church and Aristotle might agree on the *horror vacui*, or "nature abhors a vacuum," principle. But the youthful Torricelli concluded, "Nature would not, as a flirtatious girl, have a different *horror vacui* on different days."

Blaise Pascal agreed. In 1648, a year after Torricelli's death, the twenty-five-year-old mathematician repeated the experiment with a rudimentary barometer, but more flamboyantly and with a greater impact. First, Pascal walked up and down the stairs in the tower of Saint Jacques de la Boucherie in Paris, lugging a mercury barometer and making certain that it responded to even slight changes in height. Then he talked his brother-in-law, Monsieur Perier, into repeating the process on Puy-de-Dome, a 4,800-foot mountain near Clermont, Pascal's hometown. Keeping accurate measurements and relying on the help of several clerics, a doctor, and some officials, the brother-in-law did as asked. Leaving one four-foot-tall barometer as a control at the base of the mountain with a certain Father Chasin, a Jesuit "as pious as he is capable, and who reasons very well upon these matters," Perier started to climb with a second barometer filled with mercury. As he got higher, the height of the mercury column dropped. With enthusiasm, Perier wrote to Pascal:

We were so carried away with wonder and delight, and our surprise was so great, that we wished, for our own satisfaction, to repeat the experiment. So I carried it out with the greatest care five times more at different points on the summit of the mountain, once in the shelter of the little chapel that stands there, once in the open, once shielded from the wind, once in the wind, once in fine weather, once in the rain and fog which visited us occasionally. Each time I most carefully rid the tube of air; and in all these experiments we invariably found the same height of quicksilver. This was twenty-three inches and two lines, which yields the same discrepancy of three inches, one line and a half, in comparison with the twenty-six inches, three lines and a half which had been found at the Minims [the location at the base of the mountain]. This satisfied us fully.

Pascal's quirky "great experiment" proved that air had substance and that its weight decreased with altitude. Making the assumption that air would continue to weigh less and less the higher one went, Pascal decided that eventually air must not weigh anything at all. If air became weightless, it must become nothing. And nothingness was a vacuum. It was the void.

At issue, of course, were not the properties of air but the authority of the church. At issue was atheism. If a vacuum existed somewhere overhead, what did that say about the whole grand scheme of spheres of air that fit inside one another, as described by Aristotle, and of a final sphere, called the quintessence, that was said to be God's home?

Not that Blaise Pascal, who underwent a religious conversion shortly after his great experiment, cared much about God's home. Pascal's theology was summed up in a concept of his called *deus absconditus*. If God had created the earth, the concept said, he'd turned his back on what he'd made, leaving it for man to figure things out. And frankly, Pascal decided, man wasn't doing such a hot job of it. Pascal did, however, believe in miracles. He said so after his favorite niece recovered from a bad illness because of the Holy Thorn, a remnant of the crown of thorns that had been placed mockingly on the head of Jesus by Roman soldiers. Most clerics weren't like Pascal, however; they thought God's hand appeared everywhere, not just to do miracles. And they got real upset over the notion of a vacuum, because it challenged their concept of God and the cosmos. A cosmos of air and a material called ether and the quintessence, where God hung out.

Pascal immediately came under attack by the Jesuits. He was not cowed into submission. He would not abandon science, though he didn't champion its truths, either. Pascal had not set out to cause a controversy over the nature of air. He'd set out to validate a fellow scientist's proof and had done that well. He became somewhat of an ascetic, joined the strict Jansenists, and distanced himself from science and the rational mind, both of which were in ascendance in the 1650s. Pascal rejected science as any kind of savior. He asked whether a Cloud of Knowing offered people any easier way out of the predicament of life than the murk of myth, alchemy, and folklore that science was trying to displace with its experiments, theories, and laws. An alluring light emanated from science, but those drawn to it were often godless. Full of pride in their new instruments and ideas, on the verge of seeing themselves as new gods, the gods of reason, they preached quietly so as not to upset the church but had their own ideas about what made the wind blow and the earth turn and the leaves fall. And it was not God. For Pascal, science merely beckoned humans toward another predicament. If they veered away from God, they trod a path rampant with dangers and gaudy with diversions intended to preoccupy their desires and to capture their attention. The only way out was to beg for a reprieve. Pascal's big question was whether a person could wake up enough to beg. He woke himself up with prayer and meditation. He worked to transform himself into a scientist in love with God instead of with math. Pascal never abandoned science completely, however, and thought quantitatively about the human condition, and about science's unlikelihood of making it better, all his short life. Like Torricelli, he died just shy of forty.

A third debunker of the *horror vacui* principle was the showman Otto von Guericke. A contemporary of Pascal's and Torricelli's, von Guericke staged a masterful experiment in 1654. He secured metal half spheres together, then used an air pump he had invented to siphon air from the sphere, creating a vacuum inside. Two teams of horses, their traces hooked to either half of the sphere, failed to pull it apart because of the atmospheric pressure on its exterior. In the audience, enjoying the spectacle, was Emperor Ferdinand III. Once the horses worked up a lather, von Guericke opened a stopcock on the sphere, allowing air inside, and the horses walked off with the two half spheres bouncing in the dust.

Regardless of the proofs of von Guericke, Pascal, Torricelli, and others that a vacuum could exist, the church clung resolutely to its position that such a thing was impossible, conceding the point only in the nineteenth century. Instrumental in science's victory in the fundamental argument was an adventurous young French chemist named Joseph Guy-Lussac.

In 1803 Guy-Lussac began taking balloon flights outside of Paris. An assistant of Claude Berthollet, a chemist whose lab in Arcueil anchored the famous Society of Arcueil, a salon for the scientifically young and restless, Guy-Lussac soloed to twenty-three thousand feet in 1804, setting an altitude record that would stand for fifty years. Aloft beneath his hydrogen-filled balloon, Guy-Lussac collected air samples, recorded the temperature, and wrote down the barometric pressure at various altitudes. He confirmed that rarefied air was thin and cold, and that atmospheric pressure dropped steadily with altitude, as expected. Surprisingly, though, he found that the percentage of oxygen at all altitudes was the same as near the ground, around 21 percent. That was hard to understand; if the oxygen concentration stayed the same at different altitudes, why the shortness of breath and the heart palpitations when you went as high as Guy-Lussac dared?

Wisely, he didn't go higher than twenty-three thousand feet to see what happened. He vented his balloon and descended long before becoming disoriented. He passed up the opportunity to become the first man to touch the void, then theorized to begin around forty-five miles above sea level. This misconception had been popularly accepted since 1714, when the royal astronomer, Edmond Halley, of Halley's comet fame, calculated the figure based on his best-guess estimate of where air, known to get thinner and lighter as Pascal had demonstrated in 1648, would thin out to nothing. After the forty-five-mile mark, Halley had agreed air feathered off into Pascal's void.

Balloon flights during the first half of the nineteenth century left Halley's atmospheric ceiling in place and Gay-Lussac's reports as the final word about air at altitude. More money was spent on military balloons for reconnaissance flights behind enemy lines than on scientific exploration. Shelley, the romantic poet, said he imagined a long balloon flight over Africa, the balloon's shadow gliding "silently over that unhappy country" and in the process emancipating every slave and annihilating slavery forever. More practical aeronauts restricted their flights to regions they knew and came down with air samples containing pollen from plants, grains of sand from beaches, soot from smokestacks, and water vapor now understood to form clouds and make rain. An American aeronaut, James Esprey, took numerous balloon trips in the 1840s and wrote the *Philosophy of Storms,* which explained how air pressure caused the counterclockwise rotation of wind in storms. Esprey discovered that certain factors, such as the earth's rotation, mountains, large bodies of water, and friction from land surfaces, created complex air flows and shaped high- and low-pressure systems. If Esprey had his back to the wind, a high-pressure system was always rotating clockwise on his right, and a low-pressure system rotating counterclockwise on his left. Pressure gradients between the two systems caused air to accelerate between them and toward the center of the low-pressure system in a counterclockwise direction. Learning all this, Esprey must have had some wild balloon rides.

Compared to the advances being made in other fields, however, the knowledge gained about the air from balloons was pretty meager. Biology had been revolutionized by Charles Darwin's *Origin of the Species;* geology leaped forward as a major science, thanks to the Scottish gentleman farmer James Hutton's theory of the earth, which argued persuasively that the planet had been evolving for eons, long before humans ever showed up; and organic chemistry put carbon at the center of all organic matter after the groundbreaking work of Baron Justus von Liebig, a German chemist. Meanwhile, atmospheric exploration languished.

In 1862, the British Association for the Advancement of Science decided to do something about it. Soon, the country's foremost aeronaut, forty-three-year-old Henry Coxwell, was fabricating a special fifty-five-foot-diameter balloon to be inflated with coal gas, a by-product readily available in the nation leading the world in coal mining. James Glaisher, fifty-three, an influential member of the association's balloon committee and an amateur meteorologist, agreed to accompany Coxwell into space. Glaisher would man the instruments to record the temperature, humidity, altitude, and magnetic fields, and toss the pigeons out to see how they flew, if Coxwell was otherwise disposed with the balloon's controls.

Into Thin Air

What happened to my pencil?

—JAMES GLAISHER, *SAGA OF THE OXYGEN-STARVED BALLOONISTS*

Bad weather delayed the launch. For three days, the coal-gas-inflated balloon sat tethered in Wolverhampton, a village outside Birmingham. On September 5, 1862, against their better judgment, but urged on by impatient members of the balloon committee, Coxwell and Glaisher climbed into the balloon's wicker basket. They called for the ropes to be freed. Immediately they were swept off horizontally in wind and rain, Glaisher clutching several instruments to his chest in case of an accident. Out of control, the balloon lifted off, only to disappear in a rain cloud. It emerged briefly, then disappeared again into a yellow thunderhead.

Inside the cloud, winds pummeled the balloon. They tossed Coxwell and Glaisher against the sides of the basket, which twirled beneath the rigging like a mouse being played with by a huge yellow and gray cat. Surviving the second battering, the balloon and its soaked aeronauts settled into a somewhat normal ascent at ten thousand feet.

"The instruments are fine—thank God!" shouted Glaisher, his mutton chops puffed out like wet feathers.

Coxwell scrutinized his balloon. Thankfully, it looked fine; he was relieved.

Sky, blue and clear, stretched invitingly overhead. Fluffy white clouds spread below toward the horizon. Ballast was dumped. Glaisher took instrument readings. Up they went.

At fifteen thousand feet, the wet-bulb thermometer froze. Glaisher had a backup, a dry-bulb thermometer. He began recording air temperatures with it.

The first pigeon, one of four brought in a cage lashed to the rigging, was released by Glaisher. The bird headed straight down, apparently out of control. At twenty thousand feet, a second pigeon followed. It too dropped like a stone, but then extended its wings and spiraled out of sight. At twenty-five thousand feet, the last two pigeons, cowering in the rear of the cage, were ushered into space by Coxwell. One dropped, gravity taking it straight down. The other flapped its wings wildly in the cold and disappeared around the balloon, never to be seen again.

Not that either aeronaut was all that observant by this time. The temperature had dropped below zero degrees Fahrenheit. Glaisher's eyes bothered him; he could barely read his instruments, never mind write down data. Coxwell, though sluggish, managed to hoist up another bag of sand, slit it open, and dump it out. Was he thinking all that clearly?

At twenty-eight thousand feet, his brain feeling like mercury, Glaisher looked up at the balloon and vaguely realized they were in big trouble. Not only was he drifting toward sleep and terribly cold, but the cord connected to the release valve for venting the coal gas had snagged. It was out of reach. All the thrashing around below during the launch had put the valve handle in the rigging overhead. They couldn't vent the balloon now and descend, even if they'd wanted to.

Handling the cold and breathlessness better than his older partner could, Coxwell managed to climb up on the rim of the basket and hook a knee around the iron hoop encircling the shrouds, and would have grabbed the vent cord if he'd remembered to pull on gloves. But his hands were frozen fast to the metal loop because of their moisture and the freezing temperatures. Below him and gazing up, Glaisher wasn't much help. He could hardly see Coxwell and could no longer talk.

"I dimly saw Mr. Coxwell," Glaisher recalled later, "and endeavored to speak, but could not. In an instant intense darkness overcame me, so that the optic nerve lost power suddenly, but I was still conscious. . . . [I thought] death would come unless we speedily descended; other thoughts were entering my mind when suddenly I became unconscious as on going to sleep."

So Coxwell, fingers turning black from the cold, consciousness drifting from the lack of oxygen, was on his own. Although extremely sleepy,
he felt no pain, even vaguely euphoric—odd, given the circumstances. He
managed to catch the end of the vent cord, which was swinging in a
frigid breeze, in his mouth. He wedged it behind his teeth and pulled so
hard that the cord gashed his cheek and snapped a tooth. But then suddenly he was overcome by nausea and vomited, losing the end of the
cord as everything in his stomach launched into the thin air. Thankfully, a
little lucidity returned and Coxwell managed to catch the cord in his
teeth again. Then he thought he was going to pass out. Clamping down
hard on the cord with his molars, he rolled over backward in an act of
complete desperation. His hands tore free, the taut cord popped the
valve in the balloon, coal gas gushed into the air, and England's number
one aeronaut somersaulted into the basket, landing by the comatose
kingpin of the balloon committee.

High-altitude sickness was not new. But the two Brits had pushed it to
new extremes. They knew about a recent famous high-altitude flight over
the Adriatic. Three Italian aeronauts had tossed their instruments over-
board when a storm lifted their basket so fast they couldn't hear each
other, even when shouting. Neither Glaisher nor Coxwell had probably
ever read the Chinese classic *Ch'ien Han Shu*. It describes men, asses, and
cattle all getting feverish and sick as they cross the high passes between
"the Great Headache Mountain, the Little Headache Mountain, the Red
Land and the Fever Slope" in around 35 B.C. Nor had they seen the
account of José de Acosta, the Spanish priest who hiked across the Andes
in the sixteenth century and wrote that he was "suddenly surprized with
so mortal and strange a pang, that I was ready to fall."

I know how Father de Acosta felt. At 17,500 feet, an altitude I reached
once in Bolivia to ski on a glacier, I thought I was going to die. I took the
biggest, longest, deepest breaths I could muster. They made little difference in how I felt. I sprawled on some loose hay by the Swiss-looking hut
that housed the ski tow for those adventurous souls not slammed as hard
by asphyxia as me. I drifted asleep and never did ski Chalcaltaya that day.

In a popular adventure book of my youth, *Annapurna,* the mountaineer Maurice Herzog wrote that he had felt full of "lassitude," his
breathing "short and uneven," at twenty thousand feet. At twenty-four
thousand feet, "every 30 seconds I had to rest. I felt as though I were
suffocating, my breathing was out of control, and my heart pounding
away." Herzog was a healthy, acclimatized mountaineer in his thirties;
James Glaisher was a middle-aged gentleman who had done some hiking
in Wales to prepare for a balloon flight that took him much higher than
the top of Annapurna, all the way to the bottom of the stratosphere.

Age makes a big difference to how the body adapts to environmental
stress. Regardless of age, though, within a certain range of responses, what
happens physiologically at high altitude is the same for everyone. The oxygen level in the blood drops. This stimulates rapid breathing, which makes
the carbon dioxide level fall. But as you go higher, the air pressure falls and
is inadequate to make gases diffuse efficiently. The first of your body
defenses to take note of what's happening are the chemoreceptors in the
carotid bodies located alongside arteries leading to the brain. They alert
the respiratory command center in the brain. Arteries dilate, the heart
beats faster, and the lungs work harder. At eighteen thousand feet, for
instance, you breathe twice as fast as normal to dump the carbon dioxide
glut and to take in more oxygen. As you go higher, a complex series of
defensive responses protects core systems in the body while shutting down
peripheral ones. It's a relentless retreat if the oxygen deficit continues. The
crisis has its humorous side, however; the lack of oxygen alerts the carotid
bodies to an emergency while simultaneously the same lack depresses
brain function so you hardly care—and you exercise poor judgment. The
breath command center in the medulla, of course, also depends on oxygen
to operate and on carbon dioxide levels to make its biochemical decisions.
It keeps everything else functioning as long as it can. But it too will stop at
heights at which partial pressures fail to do crucial tasks, like load and
unload hemoglobin of gases circulating in the blood.

Luckily for Glaisher, the human body can handle diminishing oxygen
levels for a little while before experiencing irreparable damage to key

organs. As the balloon descended, more oxygen became available and the barometric pressure rose, so his hemoglobin carried more oxygen to all his organs and the navigator gradually revived. First he heard only the loud gush of coal gas venting from the balloon, and Coxwell's entreaties that he wake up. Wake up? He couldn't see, couldn't talk. At twenty-five thousand feet, he recovered his senses enough to focus. There was blood on Coxwell's cheek, his hands looked damaged. Slowly able to appreciate that they were still alive, Glaisher dutifully shrugged off his lethargy and returned to the instruments.

Once the balloon touched down outside Wolverhampton, no member of the impatient balloon committee was there to congratulate the atmospherically battered aeronauts. They walked seven miles into town, about the same distance they had gone into the sky.

The men claimed to have reached 35,000 feet. The exact altitude was tough to prove. Both of them had been out of it, their senses awry. The primitive altimeter may have failed in the cold. Skepticism about the record-breaking figure only increased thirteen years later, when the French aeronaut Gaston Tissandier reached a verified 28,820 feet, but two men with him died. They had taken along tubes of oxygen, just in case, only to be paralyzed so suddenly at altitude that they couldn't pick up the tubes that were lying right by their feet.

Today, historians give Coxwell and Glaisher the benefit of the doubt about the thirty-five-thousand-foot figure, claiming that Glaisher's instruments did in fact validate it. Record or no record, the men brought down little new information. Gay-Lussac's fifty-year-old reports on temperature and pressure and Edmond Halley's century-plus-old theory of a forty-five-mile-high ceiling retained their authority. If Coxwell and Glaisher proved anything, in addition to their audacity and good fortune, it was that life above twenty-five thousand feet doesn't last long without support systems. Their flight, soon celebrated in magazines that portrayed them as heroic explorers who'd survived an ordeal in an attempt to probe the upper atmosphere, together with the Tissandier's fatal flight in 1875, marked a hiatus of sorts. Any daredevil aeronaut contemplating bagging air in the nether regions now knew he was putting his life, and the lives of those with him, on the line, even if he brought along oxygen since "one does not suffer in any way," Tissandier advised cautiously, warning other aeronauts that lack of oxygen would blunt any awareness of their dire situations. "One feels an inner joy, as if filled with a radiant flood of light. One becomes indifferent and thinks no more of the perilous situation or the danger."

Keenly aware now of the perils of altitude, those intent on exploring it began building kites and unmanned balloons. The first good sounding balloon, one with instruments like Glaisher had placed onboard, took years to refine. Barometers, temperature gauges, and wind-measuring devices were heavy. Researchers needed to devise some means for recording the instruments' readings. And, most importantly, the instruments had to be retrieved. In 1892, a sounding balloon finally parachuted its instruments back to the ground without damaging them. It had only risen a few hundred feet, but the balloon burst as planned from hydrogen expanding inside its skin, and the parachute deployed successfully. After that, the balloons improved quickly, becoming the mainstay of atmospheric research. Sounding balloons had drawbacks, however. Someone with good legs and a stout heart had to run, eyes up, tracking the balloon and hoping it didn't disappear into clouds and not emerge. Or, in open country, a rider had to keep his steed on the correct course, leaping obstacles and avoiding holes, while also keeping one eye on a dot in the sky. In the late 1890s, the French meteorologist Leon de Bort skipped the expense and hassles of mounting instruments and parachutes to balloons and organizing their pursuit. Instead, he sent hundreds of tougher balloons aloft with nothing aboard except a nice note in six languages. The note asked anyone who found one of the balloons to please return the note, which was stamped and prepaid, to de Bort's observatory in Trappes, outside Paris. And to jot down where and when the balloon had landed. The meteorologist received his note cards from as far away as Sweden; they helped him draw the first map identifying some of the air currents over northern Europe.

De Bort also sent aloft conventional sounding balloons from Trappes. But he designed them so they only exploded at exceedingly high altitudes, 20,000 feet and above. The data de Bort gathered from the balloons proved perplexing. It suggested that the atmosphere was not unified and homogeneous, as had been believed since the record-setting balloon trip of Joseph Gay-Lussac almost a century before. What alerted de Bort to the flaw in popular thinking were odd temperature readings. Meteorologists everywhere then accepted as fact that for every 360 feet you went up in elevation, the temperature dropped one degree. Manned and sounding balloons alike had confirmed this relationship. But they had not gone as high as de Bort was sending his balloons. These balloons reported that above 25,000 feet the temperature leveled off and then, instead of continuing to fall, started to rise. De Bort's data threatened to undo a new atmospheric ceiling, one that had only recently lowered Edmond Halley's 45-mile figure to approximately 26.5 miles, or, more precisely, to 140,659 feet. That was the maximum altitude where air could exist, based on the uniform-drop-of-temperature theory and something new, called absolute zero.

Back in the 1780s, one of the first aeronauts had been a Frenchman named Jacques-Alexandre Charles. A freelance physicist, Charles was the first to go aloft in a hydrogen balloon. Today, he's pretty much forgotten as a figure in the early history of flight but remembered for a law that he discovered and that now bears his name. Charles's law describes an observation related to the uniformity of gases: At constant pressure, the volume of any gas is directly proportional to the temperature. In the 1850s, a British physicist named William Thomson, Lord Kelvin, realized that Charles's law set a limit to how cold a gas could get. If you turned the heat up, a gas such as hydrogen would expand. But if you turned the heat down, the volume of the gas would shrink. Theoretically, based on Charles's law, the gas would shrink to a point at which it took up no volume at all. And since a gas's volume was related to the temperature, when the volume theoretically became nothing, the temperature was as cold as it could get. This temperature, at which a gas would occupy no space at all, was called *absolute zero*. Lord Kelvin calculated it to be negative 273.16 degrees Celsius, or negative 459.69 degrees Fahrenheit. Transposing the math to a balloon rising through an atmosphere where the temperature dropped uniformly, and absolute zero arrived at 140,659 feet. At this height, a gas would occupy no space, so air could no longer exist. It was the end of the atmosphere.

That is, if the assumptions of both absolute zero and the uniform drop of temperature with altitude were correct. Leon de Bort's data told him the latter was suspect. A dedicated empiricist, de Bort launched more than two hundred sounding balloons, each carrying a needle that scratched temperature lines across a smoke-saturated cylinder, to confirm his growing conviction that the atmosphere was not as scientists believed. In 1902, he went public with his findings. He announced that air did not end where presumed. Nor was it the same all the way up to wherever it did end.

De Bort soon teamed up with an American wunderkind named Lawrence Rotch, and the two carried out the first international study of air. In 1901, Rotch had become the first person to take weather readings in the middle of the ocean; he flew instrument-carrying kites from the deck of a ship while crossing the Atlantic. He was director of the Blue Hill Meteorological Observatory, which he had opened while an undergraduate at the Massachusetts Institute of Technology, just south of Boston. Encouraged by de Bort, Rotch repeated his French colleague's temperature measurements of the upper atmosphere. A natural grandstander, Rotch flew his sounding balloons from a science exhibit at the 1904 World's Fair in St. Louis. Two years later, he and de Bort were onboard a tramp steamer in the tropics, flying kites from its deck and launching sounding balloons. They were verifying that the atmosphere overhead varied in the same ways that it did above St. Louis and Paris.

More balloon flights and data convinced de Bort that the atmosphere consisted not of a unified medium but of layers of air, each with its own properties and characteristics. He named the lower layer, the one closest to the earth, the *troposphere* (from the Greek *tropo,* for "turning," and *sphere,* for "circle"). The upper layer, he named the stratosphere, or "layered sphere." He deduced from the data collected to date that the stratosphere was calm and crystal clear, absent of water vapor and of the dust, smoke, pollen, and beach sand that earlier aeronauts had brought down from the troposphere.

At the outset of the twentieth century, the stratosphere was thought to be *it,* the end of air. It wasn't. Higher layers would be identified, investigated, and analyzed. Their layered and synergistic dance above an industrialized world, Ouranos spread over Gaia, revealed a new rhythm in air. The rhythm said that what went on below and what went on above were one. That the atmosphere was created over time, just as the earth's rock had been. And that air, like rock, had a beautiful, complex, and vulnerable unity of design.

First Trip to the Stratosphere

The cosmic rays are rattling down on my gondola!
—AUGUSTE PICCARD

One element of air's design was cosmic rays. Mysterious, ubiquitous, powerful waves of radioactivity that bombarded the earth from outer space, cosmic rays tantalized physicists during the opening decades of the twentieth century. The physicists suspected the rays might possess unique powers, reveal how to split the atom, even become a future energy source to replace coal and its new competitor, petroleum, once fossil fuels were exhausted by an energy-hungry world. The cosmic rays bombarding the earth through the atmosphere were measured by electroscopes, precursors of Geiger counters. Readings said that cosmic rays had less energy in the lower atmosphere than in the newly discovered stratosphere. Yet even weakened by their ride through the thicker gases and water vapor of the lower atmosphere, cosmic rays easily penetrated several feet of lead. Even X-rays, discovered by Wilhelm Roentgen in 1895, couldn't do that.

Where did cosmic rays come from—the sun? Deep space? Could they be analyzed in the stratosphere, where their power seemed greatest? Sounding balloons could not carry aloft the bulky instruments needed to measure the rays. A manned balloon could have hauled the instruments to forty thousand feet, but such a feat would have been extremely dangerous. In the late 1920s, a balloonist, Captain Hawthorne Gray of the U.S. Army, had managed to reach the stratosphere in an open cockpit, only to die there. Dressed like an Arctic explorer—fur-lined suit, hat, goggles—and still wearing his oxygen mask, Gray had crashed into a tree in Sparta, Tennessee, unconscious and out of control. If a physicist intended to analyze cosmic rays up close, he needed a safe, reliable, pressurized gondola, or he would probably share Captain Gray's fate.

Enter a memorable name in the history of atmospheric exploration: Auguste Piccard. In 1930, Piccard, a professor at the University of Brussels, persuaded the Belgium government to fund a hunt for "cosmic bullets."

Piccard's first launch, from Augsburg, Germany, was a fiasco. His pressurized aluminum gondola rose ten feet rather than ten miles. Piccard became an object of scorn and ridicule. Other balloonists mocked his space capsule, which Piccard said was inspired by a similar spacecraft described in Jules Verne's 20,000 *Leagues Under the Sea*. Undeterred, the professor with the wild hair and wire-rimmed glasses and his twenty-seven-year-old partner, Charles Kipfer, lifted off again in May 1931, but a bit prematurely. Crew members untied the ropes that tethered the aluminum gondola and its hydrogen-filled balloon without notifying the occupants. Piccard, busy checking gear inside the gondola, only knew they were aloft when Kipfer told him there were some factory chimneys passing by the portholes. Scrambling, Piccard managed to plug a big hole in the floor, which was supposed to have been sealed before the launch. Using petroleum jelly and cotton threads, Piccard made a pliable composite, but it took a while. The hole wasn't closed until the altimeter read twenty thousand feet. By then the capsule's interior contained far less oxygen than anticipated, because it was filled with high-altitude rather than ground-level air. To rectify the situation, Piccard dumped liquid oxygen from a canister on the floor. The oxygen vaporized, and the cabin pressure then approximated that of the atmosphere at only twelve thousand feet, a significant, and as it turned out, probably life-saving change.

Once the modern aeronauts passed fifty thousand feet, where they were supposed to start measuring cosmic rays, they discovered that the venting line, which controlled the inflation of the balloon above the gondola, was snarled. A rope unintentionally hauled aloft had become intertwined with the vent line during the sudden launch. Just like their English predecessors, Henry Coxwell and James Glaisher, who, sixty-nine years earlier, had found themselves in a similar predicament, Piccard and Kipfer couldn't have descended if they had wanted to.

The next time you fly in a commercial jet, recall Piccard and Kipfer's predicament and what might have happened to them if more things had gone wrong. Even at thirty-five thousand feet, the average cruising altitude for an airliner, if one of the cabin windows happens to pop, you will have about thirty seconds to do what the flight attendant explained before takeoff but what you probably paid little attention to. That is, get the oxygen mask, which drops down automatically from overhead, over your mouth and nose promptly. If you flub around, in just fifteen seconds you'll probably be a little disoriented. But not unhappy. If you didn't have your seat belt on, however, and just happened to be next to the popped window, circumstances would be dire. You'd be jammed in the window like a cork. Body parts outside in the stratosphere would freeze rather quickly, since you're moving at around five hundred miles per hour, and the temperature is approximately minus 50 degrees Fahrenheit. If you've ever wondered why the windows in a commercial jet are so small, one reason is that they would slow the loss of cabin pressure should one of them pop. The pilot immediately puts the plane into a steep dive. If you'd been flying on the high-speed, supersonic Concorde, however, other factors would have come into play. Before the ravages of time and the high price of fuel put the famous planes out of commission, they flew where, as the French aeronaut Joseph Gay-Lussac had discovered in 1804, the air is thinner. In the Concorde's case, much thinner. The planes cruise between fifty thousand and sixty thousand feet. Above forty-five thousand, you're unlikely to survive, even with pure oxygen entering your nasal passages, because of the sharp fall in air pressure. So, a popped window in the Concorde would probably be fatal, "a fact of which many passengers may be blissfully unaware," noted Frances Ashcroft in *Life at the Extremes*. At sixty thousand feet, the gases in solution instantly vaporize. It's Dalton's law of partial pressures again. At that altitude, the gases inside your blood are acting as if you're at sea level and you're not. No longer under pressure, the dissolved gases form bubbles in your blood. Think of the depressurization that happens when you open a bottle of soda pop. With one twist of the cap, the high pressure inside the bottle meets the lower pressure outside. The dissolved carbon dioxide is no longer dissolved, and the sweet liquid instantly becomes bubbly and fizzes. Similarly, your blood gases would diffuse so quickly that the blood would virtually boil.

The windows in the specially built gondola of Auguste Piccard and Charles Kipfer stayed put. The temperature at fifty thousand feet was rising, however, because Piccard, attempting to create a crude heating system, had painted one-half of the gondola black and the other half white. The gondola was supposed to rotate, black side toward the sun to warm the interior, white side toward the sun for a cooling effect. But another mechanical malfunction scotched the rotational cooling system. Its motor wouldn't work. Gradually, the gondola became a low-grade oven.

Natural processes, and a lot of luck, saved Piccard and Kipfer. When the sun set, the gondola stopped rising as the hydrogen in the balloon cooled and contracted. Then the capsule began to descend as the temperature continued to fall. The cosmic-ray-hunting sphere descended at a reasonable speed but with absolutely no control, the Austrian Alps taking shape below. At ten thousand feet, Piccard pulled open a porthole for fresh air. He got hold of the vent line and piloted the gondola down on a glacier high above the alpine village of Ober Gurgl in the Tyrol. The deflated balloon plopped over like a shapeless sock. Inside the black-and-white orb, two very relieved passengers fell fast asleep.

The next morning's *New York Times* carried an article about Piccard's flight that read like an obituary: "The Piccard stratosphere balloon is floating aimlessly over the glaciers of the Tyrolean Alps, out of control and occupied only by the dead." A rescue party climbing up from Ober Gurgl, where an innkeeper had spotted the balloon descending in the twilight the night before, soon encountered a jaunty Piccard and Kipfer. Roped together for safety, they were making their way down.

A year later, Piccard tried to reach the stratosphere and take measurements of the energetic cosmic rays that had pinged the roof of his gondola on the ill-fated voyage. Launched from outside Zurich, with no unstoppered holes in the floor and all systems go, the airship ascended above fifty-three thousand feet and returned without incident. Looking out a porthole into a calm and windless realm, Piccard described air that was "bluish purple, a deep violet . . . ten times darker than on Earth."

Cosmic rays failed to live up to their advanced billing as life-altering energy sources that humans could saddle for special purposes. These energetic rays did help physicists probe the atom before high-speed accelerators were designed especially for that purpose. Today we know that cosmic rays are primarily the nuclei of hydrogen and helium atoms racing at nearly the speed of light through our solar system and the rest of the Milky Way. Deflected by the earth's magnetic field, most of the rays spread into a primary layer at the edge of our atmosphere. The others plunge through the layers of the atmosphere, losing much of their energy before reaching the ground or the oceans.

It's ironic that the atmosphere, which first revealed its true shape to Leon de Bort and Lawrence Rotch's balloons and kites, then to Auguste Piccard's much-maligned gondola, and later, in the twentieth century, to planes and satellites, resembles a series of spheres, just as Aristotelians and clerics thought during the Renaissance, when the earth was considered the center of the cosmos. The spheres don't sing, nor are they filled with luminescent ether or the quintessence, which the church claimed was God's home. But the spheres do fit concentrically together. The boundaries are tough to discern and not that easy to describe, either. What's most important for people, the spheres interact in complex ways that modulate the radiant source of all life on the planet—the sun—and provide life with air.

To most of us, the nature of air in the farthest spheres is strange. Temperatures range from 50 degrees Fahrenheit to negative 150 degrees. Jet streams cruise through at hundreds of miles per hour. Belts of electric particles spread through the air like nets.

Even when you're at only thirty-five thousand feet, well short of the stratosphere, about 90 percent of the weight of the atmosphere is below you. You're still in the troposphere, the home of the earth's clouds, birds, dust, and most air pollutants. Most of the earth's water vapor is in the troposphere, making this layer the source of much of our weather.

The stratosphere begins between five to ten miles up, depending on your distance from the equator. The bulge in the shape is caused by the inertia of the earth's rotation, which is greater at the equator and least at the poles. Air gets warmer as you rise through the stratosphere, because of ozone. Oxygen molecules bombarded by shortwave ultraviolet radiation knock electrons free from their orbits. The free electrons bond with normal oxygen molecules and form the unique three-atom ozone molecule. It, in turn, intercepts additional ultraviolet light. The resulting ozone layer protects organic life from radiant light that penetrates cells and damages DNA.

The stratosphere ends about thirty miles, up. Above that, air thins out rapidly. The temperature turns cold, then extremely hot. Oxygen molecules appear as single atoms rather than compounds. Above fifty miles, oxygen and nitrogen atoms form belts in the ionosphere, an electrically charged strata. When an ultraviolet ray, which is a stream of photons, strips away an electron from oxygen and nitrogen atoms, what's left is a positively charged atom. When the stripped electron bonds to an atom, what's created is negatively charged. The result is ionized air. It exists in charged belts. The Kennelly-Heaviside belt, named after its discoverers, keeps radio waves from going deep into space; the belt of charged air reflects the waves back toward the earth. In this way, radio waves reach points on earth thousands of miles from their origins.

Distant space gets real strange, and very lonely. Two hundred miles out, an atom may travel a mile or more before bumping into its nearest neighbor. Lighter gases rule. Helium, one-fourth the weight of oxygen, dominates the vast space of the atmosphere between six hundred and fifteen hundred miles from the earth. Here you enter the region of the Van Allen radiation belts, magnetically controlled and donut-shaped clusters stuck there since Earth's origin. The Van Allen belts also protect life. If their radioactive particles ever rained down on us (they are held in suspended equilibrium by the earth's magnetic fields), it would be curtains, radioactive curtains, Chernobyl ad infinitum, over the entire planet.

You go two thousand miles out in space, and you enter the realm of the lightest element of all, hydrogen. The wispy reaches of our air are feathered with hydrogen, the element that makes up 63 percent of the atoms in our bodies, that bonds with oxygen to form water. Hydrogen is out there tickling Pascal's void, cohabiting with cosmic rays wrapped around the biosphere like arms millions of years old reaching out of deep space.

3

The Origins of Air

So where do all the atoms stacked up in our atmosphere come from? What role did air play in shaping the evolution of our airy bodies, bodies reliant on air every second of our lives? A couple thousand years ago, in the *Metamorphoses,* the exiled Roman poet Ovid said that air, like everything else, came "out of blind confusion," then found its place between fire and earth. More recently, the biologist Lewis Thomas wrote that the earth's air formed a membrane because "it takes a membrane to make sense out of disorder in biology ... [and] when the earth came alive it began constructing its own membrane, for the general purpose of editing the sun." Others claim that the history of the atmosphere is really "one of the great true detective stories," or a fabulous mystery film with an incredibly long playing time.

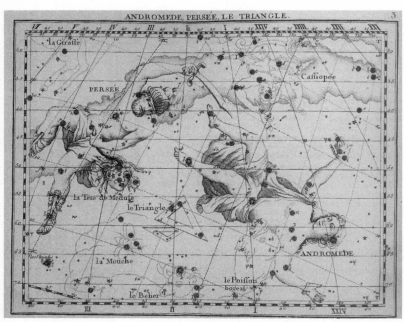

This star map drawn by John Flamsteed in the 1770s captures the imagination. It probably would have pleased the ancient Greek astronomers. It fails, however, to tell us much about the scientific origins of the sky, which were not quite as romantic or human involved. A truer map would depict bacterium constellations, for without bacteria there would be no airy sky as we know it, or a human race either, for that matter.

WHETHER A MYSTERY, a poem, or a tale of a membrane, the story began 14 billion years ago with an explosion: the Big Bang, which created the universe. About 4.5 billion years ago, in our corner of this universe, a vast, swirling cloud of dust and gases called a *cocoon nebula* encircled a newborn star. The cocoon was rich with so-called planet-making materials that had been set free by the explosion, including minerals, carbon compounds, and water vapor. The debris shaped itself in a vast disc and slammed into itself and began aggregating into asteroids, moons, and planets. Scientists say it probably took between 50 million and 100 million years for the cocoon nebula to virtually disappear and for our solar system to appear. Then, as one of the planets, Earth, cooled, the gases released from its surface remained held in place by gravity and formed our planet's atmosphere. Over the next 4 billion years, the atmosphere evolved. Around 350 million years ago, it reached roughly the proportions of gases it contains today. Despite small changes, and a few catastrophic ones, the proportions have stayed relatively stable until very recently.

In effect, our planet grew like a composite of chewing gum—millions and millions of different-sized pieces sticking together as gravity hauled them out of orbits and brought them crashing into something bigger and stickier than themselves. If there was an embryonic piece of gum, it was probably an asteroid the size of Connecticut. The asteroid pulled in any other, small whirling bodies it could absorb. As more and more debris was hauled in by the pull of the growing asteroid, heat and gravity shaped it into a sphere. Meteorites pounded it almost like solid rain. The bigger the planet got, the more debris it attracted. The biggest pieces, so-called planetesimals, hit with such force that they left craters and caused eruptions from the molten interior, which in turn helped create an atmosphere.

Earth's distance from the sun influenced the amount of water and the

kind of elements trapped during the formative period. The distance
acted as a thermostatic control on the outgassing, the natural blowing off
of gases from the hot and molten interior through vents and volcanoes.
During this gurgling phase, parts of Earth's surface probably resembled a
bubbling pie crust in an oven. The hot liquid interior allowed heavy metals to migrate toward the core, further strengthening gravity, which
insured that few molecules in the air would ever escape the planet's pull.
The outgassing put hydrogen, water vapor, carbon dioxide, and other
gases around the planet as its first atmosphere. The blend created Earth's
first greenhouse effect. It helped sustain the average temperature around
50 degrees Fahrenheit, leaving water in its liquid form.

The importance of water as a fluid cannot be overstated. Without
water, there would have been no atmosphere, no basis for life. A dry
Earth would probably have been much like nearby Venus and Mars,
which had also taken shape out of the nebula, hauling in their own
gummy aggregate of rocks and becoming solar-system spheres to be
reckoned with. Venus, which is just slightly smaller than Earth and closer
to the sun, is now cloudy all the time. The typical temperature is 800
degrees Fahrenheit, and the air as heavy as a room full of bowling
balls—not a great place for a vacation, even if Venus is our closest planetary neighbor. Mars is a little more inviting. Its average temperature is
minus 63 degrees Fahrenheit, tolerable in the right gear if no wind is
blowing. Cloud cover is variable. The air contains about 5 percent of
Earth's oxygen level. The atmosphere on Mars, like that on Venus, is
dominated by carbon dioxide that vented out of the planet soon after it
was formed. Even though the clouds on Mars aren't a blanket of sulfuric
acid, as they are on Venus, there are no towering cumulonimbus or wispy
cirrus feathers of water vapor, either, as there are on Earth. Dust, water
vapor, and carbon dioxide form the clouds on Mars. Neither Mars nor
Venus evolved atmospherically because of a variety of conditions that
distinguish them from Earth. Their atmospheres leveled off relatively
soon after their creation. They became carbon-dioxide-enshrouded places

with no life, because the ingredients for it mostly blew away. In other words, the stories of Earth's nearest neighbors, though interesting, were short and relatively boring in comparison to the ongoing drama taking place here.

Here's Lewis Thomas describing the appearance of an atmosphere, of a sky, over the earth:

> Taken all in all, the sky is a miraculous achievement. It works, and for what it is designed to accomplish it is as infallible as anything in nature. I doubt whether any of us could think of a way to improve on it, beyond maybe shifting a local cloud from here to there on occasion. The word "chance" does not serve to account well for structures of such magnificence. There may have been elements of luck in the emergence of chloroplast, but once these things were on the scene, the evolution of the sky became absolutely ordained. Chance suggests alternatives, other possibilities, different solutions. This may be true for gills and swimbladders and forebrains, matters of detail, but not for the sky. There was simply no other way to go.

Sounding a similar refrain, the Biblical scholar Jeffrey Burton Russell says that "the imperishability of the human body is related to the salvation of the whole cosmos." He uses the image of the new body "sprouting and growing" out of a dead or decaying one. The resurrected body after death, Russell writes, "is in some essential manner the earthly body," with the soul still in it. Teasing his reasoning out a little further, you could say that the earth kept its soul as it formed, but the other planets lost theirs. And the soul required a living, breathing shell, which God insured by placing this rock in such a way that carbon, water, hydrogen, nitrogen, and a few other basics stayed put.

Whatever the spiritual nuances of the earth's beginnings, the evolution of air and life was symbiotic. Nonliving molecules created the first living bacteria probably deep in the oceans. The bacteria evolved because of the air above the ocean surfaces and its interplay with the sun.

Biologists call the first bacteria *archae,* or "the ancient ones." The oldest of three types are called *bubblers.* They lived in the ocean, where the water shielded their DNA, the molecules responsible for their genetic code, from being damaged by ultraviolet light. The DNA floated around inside these minute creatures in looped chains. Bubblers were anaerobic; that is, they created energy by fermentation, not by burning oxygen. These primitive microorganisms gave off acid wastes that rose to the surface as bubbles, hence the nickname. Their descendants are the blue-greens, the first living things to expel molecular oxygen, and the breathers, bacteria that got so evolved they had propeller-like tails to move around and help themselves get up on land.

During the earth's atmospheric infancy, as bubblers inhabited the oceans, our planet was probably surrounded mostly by carbon dioxide. The result was an intense greenhouse effect and rain: centuries, millennia, millions of years, of rain—falling through carbon dioxide and ammonia, through lightning flashes and erupting volcanoes, onto a colorless, blue-gray world where there was no green, no blue, only a landscape that probably resembled a bubbling pie crust and that gradually cooled to rock across which winds blew and rain splashed. And where there was no free oxygen, none.

The dramatic event that transformed earth's story from a prolonged drift of elemental life, which sparked a dead ball of warm gases and rain inhabited with fermenting bubblers clustered around deep ocean vents, into a billion-year thriller that begat plants and animals, was the appearance of oxygen. And this crucial element got its start because of the archae, the ancient ones.

The Archae Shape the Earth's Early Air

The first archae must be hardy pioneers to survive the violent upheavals, extreme temperatures, and harsh chemistry of early Earth.

—SIDNEY LIEBES, ELISABET SAHTOURIS, AND BRIAN SWIMME, *A WALK THROUGH TIME*

The story of the archae on a cooling, wet, meteorite-bombarded earth is a remarkable chapter of the earth's past. It's mysterious, poetic, a bit mind-boggling.

The big question, of course, is How did the bubblers, these primordial organisms, get here to begin with? This remains the missing link in evolutionary biology. Tracing the genetic string from a bacterium to us humans is relatively easy compared to figuring out how the raw elements of life, hauled out of the cocoon nebula circling the sun, were transformed into the archae. Even a single-cell bacterium is a biological wonder, the complexity of which science is still probing. How did the first such creatures come into being? Were the chemicals in the atmosphere and the oceans combined into organic life by lightning that sparked synergistic interactions and generated cells, as new theories claim? Did the first bacteria roil into life in near-boiling temperatures by the lips of deep ocean vents, a theory scientists seriously entertained decades ago? Was life extant before the Big Bang, as the metaphysically inclined argue—that is, were the seeds of life brought from afar in the cosmos only to lie dormant during the earth's embryonic period, ripening at the first opportune moment?

In the ongoing debate over life's beginnings on earth, I tend to side, although not 100 percent, with the British atmospheric chemist James Lovelock and his Gaia hypothesis, and the American biologist Lynn Margulis and her theory of symbiogenesis. Put very simply, Lovelock's theory asserts that the living and nonliving elements here are part of the same fabric and that the life they have shaped in turn regulates the optimal conditions for life to continue. In other words, the entire biosphere is alive and has its own primal switches, controls, and homeostasis. At least that's the way I interpret the Gaia hypothesis. Margulis theorized that life came about through a complex union of molecules in the primordial soup. The closeness, motion, and chemical interactions of elements and then compounds nourished a biological synthesis, probably abetted by electricity from the early atmosphere and various so-called mergers, which set molecules into a geophysiological union when conditions were right.

Though persuasive, the Gaia hypothesis and symbiogenesis theory are not totally convincing to me. Maybe I'm not rational enough, or I'm too drawn to myth, which makes its own claims on the origins of life. I doubt science will ever totally trump myth. The missing link of evolution will, like the philosopher's stone, the fountain of youth, and the Holy Grail, forever remain missing.

Regardless of how life came into being, it seems unlikely that there was a precise instant when it snapped to. No *presto!* No sudden leap out of the hat of time, like a bunny bouncing around on a distant planet, as I fantasized as a boy and was appalled to learn not to be true. There was no touch from the finger of God, no apocryphal bolt of lightning zapping the perfect combination of chemicals into the first archae. Still, the image of an instantaneous beginning of life, a time certain, is hard to shake. But like most great discoveries of science, which take place over time, shuffling along until all the needed circumstances merge and—Eureka! There it is!—life probably slipped into being gradually. That's a conclusion that now irritates us because we have a proclivity for momentous beginnings. We want to say *there*—that's when it all began. And we can't.

At any rate, once the archae appeared on the earth, they stayed. They consumed, reproduced, and repaired themselves. They were creative in crisis. No food shortage, no air pollution, no ultraviolet radiation, no huge meteorite hitting the earth ever completely wiped them out. They evolved because of the changing atmosphere above the ocean's surface and its interplay with the sun.

Science writer Jennifer Ackerman argues in *Chance in the House of Fate* that people today see bacteria as "a penal form, the Alcatraz of nature, as if recognizing the import of the small would diminish our own personal importance." Yet bacteria ruled the earth for over two billion years. There was absolutely no other living competition. The scale and complexity of colonies they created left slimy mats of algae by river mouths and the first living crust on the shore, but they also built environmental monuments such as the white cliffs of Dover and the Great Barrier Reef because bacteria have the ability to induce minerals to aggregate into large, diverse platforms brimming with life. Looking at a bacterial colony through a microscope, "scientists are amazed to see them packed as tightly as our own urban centers, but with a decidedly futuristic look," write the authors of *A Walk Through Time*. "Towers of spheres and cone- or mushroom-shaped skyscrapers soar one hundred to two hundred micrometers upward from a base of dense sticky sugars, other big molecules, and water, all collectively produced by the bacterial inhabitants.... The cities are laced with intricate channels connecting the buildings to circulate water, nutrients, enzymes, oxygen, and recyclable wastes. Their diverse inhabitants live in different microneighborhoods and glide, motor, or swim along roadways and canals." Checking out one of the incredible bacterial cityscapes, the researcher Bill Keevil exclaimed, "It looks like Manhattan when you fly over it."

Ackerman adds this about the glory of bacteria: "It is they who, in the last half-century, taught us nearly all we know about the nature of our genes. We may as well look down for kinship."

Few of us do. Admittedly, it's hard to relate to things so small, and so tough, even if they do build their own beautiful and intricate cities. It's the scale dilemma again, the need to shift from the compelling allure of a huge reef to the intricate chemistry of its creators, for a better understanding. The conditions the early bacteria thrived in—ash flying out of volcanoes, pools of steamy water toxic with sulfuric acid—are not what we associate with fountains of life. Yet they were. And the archae exploited them. When conditions changed, so did these microorganisms. Bacteria developed new DNA. They adapted and changed their lifestyles. Their driving force always seems to have been efficiency. Whenever conditions allowed, they became more efficient. Bacteria, relentless opportunists, were always seeking an edge. And when they found one, evolution took a step either forward or sideways or somewhere—it didn't stay still.

For instance, as the earth aged toward the one-billion-year mark, the opaque atmosphere caused by volcanoes, asteroid splashdowns, and constant rain began to clear a little. Earth was cooling and crusting over, the cocoon nebula thinning out. As a result, more ultraviolet light, damaging light, began to reach the bubblers in the oceans. But instead of receding deeper away from the light, these bacteria exploited its energy, evolving into bluegreens, the earth's first photosynthesizers. Bluegreens used photons of light now reaching them in the water to unlock carbon from carbon dioxide, which was highly soluble and therefore plentiful. With light, carbon, and hydrogen extracted from the water, bluegreens made their own food. They emitted oxygen as waste. The cells of bluegreens, whose color came from the green chemical compound chlorophyll, would later become incorporated into all the earth's plants and trees.

The third type of bacteria, the breather, survived by respiration; it took in oxygen and gave off carbon dioxide. So the breathers and the bluegreens had a symbiotic relationship, the one dependent on the other. The cells of breathers later became insects, mammals, humans. Because of their efficiencies compared to the anaerobic bubblers, both bluegreens and breathers, once they appeared, increased rapidly. Adaptive and wily, they turned the evolving atmosphere of the planet to their advantage. The bluegreens relied on light to power a new chemical reaction, and the breathers took the reaction's waste product to fire their mitochondria, just as we do.

The oxygen emitted as waste by bluegreens was an element with the power to change the future of a young planet. But to accumulate in the air, first it had to deal with iron. Earth's cooling crust contained large amounts of iron that had been thrust up from the molten interior. The iron had a strong affinity for oxygen; it bonded instantly with any free oxygen rising from the seas. Because of the large amount of iron in the earth's crust, for hundreds of millions of years virtually none of the oxygen made it into the atmosphere. The phenomenon probably slowed down the development of life for a billion years or so, but it also helped shape a surface on which life could anchor itself. Today, we see the result of more than a billion years of the oxidation of iron partially fringing the coast of Nova Scotia, in western Australia, and in sedimentary layers of the Grand Canyon. Geologists call this atmospheric phenomenon *red beds*. Red beds were the product of the chemical affinity between the breath of bacteria and the mineral exhalations of a gassy planet on the earth's crust. One nice by-product of their chemical liaison was color. A reddish tint probably suffused much of the air around the earth about 2.2 billion years ago, when red beds formed landscapes.

The tinting occurred as the level of oxygen reached about 1 percent of what it is today. Oxygen had exhausted iron's availability. The waste product of the ever-growing marine colonies of bluegreens then did some remarkable things. But first, oxygen turned against the oldest bacteria, the anaerobic bubblers, and began to kill them.

Air containing 0.2 percent oxygen does not sound as though it would have been lethal. But as atmospheric scientists have made abundantly clear, the history of air, though very long, has certain crucial moments. And it is guided by a maxim: the scale of the atmosphere is very large, but the scale of the chemical processes that drive it is very small. The milestone level of oxygen in the air proved catastrophic for the bubblers. It tore their molecules apart because bubblers were not designed for oxidation. They wanted nothing to do with the new level of oxygen. It killed probably 90 percent of the bubblers. Survivors slipped deeper into the oceans. They hid under mud, dug into sediment. Even bluegreens emitting oxygen found its toxicity threatening. But bluegreens, as they had done with the photons of light allowed in by a clearing sky, handled the emergency chemically. They learned to make enzymes, catalysts from special proteins that reacted with oxygen before it damaged a bluegreen cell's DNA.

Meanwhile, overhead, oxygen began burning up meteorites before they hit the earth. The higher level of oxygen in the atmosphere increased the friction against the rocks from space during entry and intensified their combustion. The antimeteorite safeguard protected the vulnerable skin of the planet. It also must have lit up the night with what we now call shooting stars, fleeting scratches of light caused by galactic rocks burning to a crisp and sprinkling down through the earth's membrane rather than scorching it and hammering the ground or the sea, causing damage and other change in proportion to their size. Oxygen's protective role kept the earth's young landscape from resembling "the pounded powder of the moon," noted Lewis Thomas. The meteorites, fossils from the great explosion that whirled into our solar system with their rare elements and gases, still bombard the atmosphere's upper reaches. "Even though our receptors are not sensitive enough to hear it," Thomas added, reflecting on the presence of the meteorites, "there is comfort in knowing that the sound is there overhead, like the random noise of rain on the roof at night."

More important than its meteor-busting quality was oxygen's ability, even at low concentrations, to act as earth's sunscreen by forming ozone in the stratosphere. When ultraviolet light struck an oxygen molecule that had migrated there, the light energy drove out an atom, breaking the molecule apart. The oxygen atoms bonded with other oxygen molecules, forming ozone, a three-atom molecule which in turn could absorb ultraviolet light of a different wavelength. For the earth, ozone was double protection. Both ozone formation and ozone destruction blocked ultraviolet radiation from reaching living cells below. An ozone layer eventually surrounded the earth almost like a veil, a very efficient veil for absorbing ultraviolet radiation.

The appearance of red beds, the killing of the bubblers, the burning up of more meteorites overhead, and the appearance of ultraviolet-light-absorbing ozone all converged at the time that the earth became more hospitable to organic life. They contributed to this improving environment, each in a different but interconnected way. New evolutionary paths began to open. One path led to nucleated cells. It is from the nucleated cells, the *eukaryote* (pronounced "you-CARRY-ote"), that all animal and plant cells would evolve. Two billion years ago, almost halfway between the earth's beginnings as an embryo the size of a small New England state and now, the oxygen level was rising. Over the next billion years, the atmosphere lost its red tint and became blue, the clouds white. Algae colonies crawled onto shores, followed by fungi and lichen, which gripped rocks and dissolved their minerals. Oxygen continued a long, slow buildup to the 10 percent mark of its level today, or to 2 percent of the atmosphere. During the time it took for that milestone to occur—it happened roughly about 700 million years ago—how many bluegreens came and went to emit all that oxygen is anybody's guess. Maybe a computer could, or already has, figured it out. The entire planet was undergoing protracted change. The earth's land mass, shifting on the molten interior, unified into one vast continent we call Pangaea, then broke apart into continental drift. Mountain ranges lifted and became the spines down which glaciers crawled. New species increased and flourished, all underwater, powered by the possibilities afforded by advanced cell design and complex organs thriving on oxygen to burn food and create energy. In the oceans there was also a hunger for carbon dioxide, which was continuously stripped from the air and locked in the seas as carbonated rock.

All this while, Venus and Mars had not changed much. They were whirling around the sun exactly like Earth, but they still lacked the right conditions to combine the four elements most basic to organic life: carbon, nitrogen, oxygen, and hydrogen. Neither planet was surrounded by a thin ozone layer; nor did either have any bacteria or water.

Probably the most important thing to remember about the origin of the unique atmosphere surrounding Earth is that for the planet to become life friendly, the carbon in its early atmosphere had to be removed and locked up somewhere. Air had to shed carbon. The carbon removed from air became a storehouse of energy in Earth, a storehouse to be burned only by returning carbon to the sky from the traps in which it was sequestered by processes as mysterious as they were beautiful. The processes, primarily photosynthesis and carbonate rock formation in the oceans, were what made the planet a living system striving for balance, and created a place quite unlike its stoned-dead neighbors, Venus and Mars.

A good story should tell a reader what happened to important players that in earlier chapters seem to have disappeared. So here is what we now know happened to the anaerobic bubblers driven into hiding by the gas that brings you life. They didn't disappear, not at all. Large numbers of them made it to the ocean floors. Over the last two billion years, the bubblers spread, many migrating to beneath the earth's polar ice caps. As bubblers do, they kept reproducing and giving off gases, one of which was methane. A consequence, only recently discovered and not that well understood, is that we now have a reservoir of energy called crystal fuel because of the crystalline shape of its molecules. Referred to as gas hydrates by geologists, the reservoir may be the earth's biggest untapped fuel source. Being methane, of course, it's a hydrocarbon, just like coal, natural gas, and oil. A result of several billion years of accumulation, each bit of crystal fuel has a carbon atom at its center. Each carbon atom is surrounded by hydrogen atoms and is very unstable. Pulled up from beneath the ocean floor by energy-exploration machinery and exposed to air, crystal fuel globes the size of softballs sizzle like bacon in a frying pan. Little is left but some drops of water. Elusive and strange, gas hydrate is the product of the ancient, deep-sea anaerobic bacteria that oxygen banished from life's busier playing field. With its unusual beginnings and volatile properties, crystal fuel sounds both compelling and scary. If we do find a way to release all the methane thought to be stored as gas hydrates, the hydrocarbon would theoretically cover the earth in a layer 130 feet thick. The contribution to global warming of such a basic greenhouse gas would make present-day emissions from fossil fuels seem laughable, if there was anyone left to laugh.

*

How do we know about all this: the cosmic fecundity of Big Bang debris; the earth's hoarding of the right elementary particles for life; the biographies of bubblers, breathers, and bluegreens; the milestone increases of oxygen; the lockup of carbon dioxide into rocks, plants, and weird stuff like crystal fuel? The simple answer is from rocks. Layers of sediments in western Australia and South Africa, despite all the tossing and shifting of the land and seas, have been identified as 3.5 billion years old. They hold telltale remnants of bacteria action and evolution in their fossils, which scientists isolate, analyze, and compare. The red beds of Nova Scotia and elsewhere testify to oxygen's bonding with iron. Especially fossil-rich places, such as the Burgess Shale in Canada, the mountains jutting out of the Karoo Desert in South Africa, and the 1,200-foot thick Scaglia rossa limestone walls in the Bottaccione Gorge in northern Italy, have kept detailed records of the march of air and life. They can be read by scientists and technology almost like books. As can the small number of meteorites that still thread the atmosphere and hit the earth, bringing with them peculiar treasures, bits of the raw materials of our universe. These fossils from space provide scientists with materials to compare with those found on the earth.

The means to truly investigate and interpret the rocks and fossils has been radioactive dating, a technique discovered by a French physicist, Henry Becquerel, and a German physicist, Wilhelm Roentgen, in the 1890s. Radioactive dating allows anthropologists, paleoclimatologists, geologists, biologists, atmospheric scientists, hydrologists, and others to research the distant past and to cross-reference their findings with those in other fields. One result is a more three-dimensional portrait of the past, a story with parallel chapters and multiple themes, than we have ever had until recent decades. The invention of the electron microscope, in the 1940s, opened up entirely new dimensions of the very small. Traces of air locked in rock for hundreds of millions of years, the cellular structure of microscopic fossils billions of years old, the chemistry and physics of transformation—all became more clearly visible with the development of electron microscopy. Ancient air was found to have left signatures on the surface of rocks just as fossils had become pressed between the stratified sheets that built up over time. With the right technology and techniques, scientists could read the gas mixes, water vapor levels, and dust content of ancient air as anthropologists could ponder tusk and bone, skin and hair. Atmospheric conditions once thought lost in time were right under foot in stone. Air, too, was shown to have left a remarkable history of its past.

The history of air has also been well served by the drilling of ice cores beneath Greenland, Antarctica, and Siberia. Ice cores, as well as cores hauled up from the bottoms of the oceans, provide scientists with time-sensitive evidence of air makeup and change for more recent time spans. Bubbles of air found in layers of compressed ice stacked up like rings in an old tree help answer questions about the earth's air during the last 500 million years, rather than billions of years ago. Extracted and analyzed, the bubbles provide clues about catastrophic conditions that wiped out the dinosaurs, about volcanic eruptions that altered the air for tens of millions of years, and about the circumstances that brought the atmosphere back into balance. When the data is correlated with specie die-offs found in fossils in terrestrial rock, ice cores have helped chart the rise and fall of groups of animals and plants. They have provided a timeline for air pollutants, beginning with emissions from primitive smelters that made coins 10,000 years ago.

Still, much of the information is new, and the conclusions drawn from it are hypotheses. Atmospheric scientists such as Paul Crutzen and Thomas Graedel caution against drawing hasty conclusions that might be wrong. They use *probably* a lot, and *maybe* almost as often, in their writings. For instance, they make this observation: "Earth appears to have been especially favored from birth as a potential reservoir for life." Not that it was favored. They say "clues...necessarily circumstantial ...at least paint a scenario."

The details of the scenario get finer the closer the records are to modern time. Evidence is more plentiful, for one thing. And the cross-fertilization of fields seems to result in an exponential increase in understanding.

Not that creationists agree with any of this oxygen-is-the-key-to-life business. The Bible describes the earth's origins as the Book of Genesis, not as the theory of symbiogenesis. "In the beginning God created the heaven and the earth." On the second day, God split the waters with land. And on the third, gathered the waters under heaven and separated land as a dry place. "And God called the dry land Earth; and the gathering together of the waters called he Seas: and God saw that it was good." Adherents to this script can hardly be expected to agree that bacteria made love billions of years ago, and from their promiscuity came you and me.

For the record, bacteria didn't really *make* love in any conventional or emotional sense. But bubblers and bluegreens did have sex. Physically, they got it on. Our ancestral cellular precursors rubbed up against each other, breaching membranes and dumping in some DNA. Then they went about their business in the community. They're still at it today.

Evolutionary theory was still being fought hard in many corners of the world in the early twenty-first century. America's president, George W. Bush, a man with one foot on the Bible and the other trying to stomp out terrorists, claimed the jury was still out on evolution. What jury was President Bush talking about? A jury of scriptural industrialists dismissive of evolution? Hard-liners of Bush's ilk routinely went to court to challenge the right of schools to teach Darwin's theory of natural selection to kids. Reluctance to march forward in the light of evidence that the world is a physical rather than an ethereal production, a place where morality arises from the consciences of men and women, is nothing new. As the famous naturalist and geologist Louis Agassiz said back in the nineteenth century, "Every great scientific truth goes through three stages. First people say it conflicts with the Bible. Next, they say it has been discovered before. Lastly, they say they always believed it." The Harvard paleontologist and natural science writer Stephen Jay Gould added this to Agassiz's observation: "No scientific truth can pose any threat to religion rightly conceived as a search for moral order and spiritual meaning."

Today, most scientists and a significant portion of the earth's dominant species, *Homo sapiens,* believe the Big Bang begat the earth, the earth begat the atmosphere, and the atmosphere begat life. If our atmosphere had leveled off when those of Mars and Venus did, there might still be life on the earth. But it would be bubbler life.

Air Shapes Life

Pure oxygen was the champagne we drank and clicked our glasses, one to one.

—ANNE SEXTON, *THE DEATH OF THE FATHERS*

In an old issue of *The New Yorker,* I think it was, there was a cartoon topped with a dateline that read "500 million years ago." In the cartoon, a primitive-looking creature was crawling up a beach on stubby fins and saying over a shoulder to one of its own, hesitating warily in the shallows, "Because this is where the action is, baby."

The cartoonist was right. Air was where the action was. Five hundred million years ago, the earth completed the Cambrian period. Life had blossomed, and evolution had gone sort of crazy. Odd-looking little creatures had begun crawling about and digging burrows in the ocean floors. During a span of 10 million years, a blink of the eye in geological time, half of all animal groups, called phyla, came into being during what scientists now call the Cambrian explosion. It wasn't an explosion at all, of course, but rather a blossoming, a flourishing of life and biodiversity. But all the biological action was underwater. Not terrestrial. So the cartoonist's rendition of a creature flopping along the beach, with a chest cavity implying lungs, was a little in advance of the appearance of anything resembling it.

What really crawled out of the sea around 500 million years ago was a slimy carpet of algae. It stayed damp because it was the bluegreens moving onto land and surviving by photosynthesis. But it was red algae in places as well. The first life on land probably resembled nothing better than a soggy, tartan-plaid carpet you might come across on the shore. But the oxygen-enriched atmosphere nourishing it, along with sunlight, enticed more creatures to emerge from the sea. They were simple creatures. The more complex would take another 50 million years or so to appear, evolutionary biologists tell us, and included centipedes and beetles with hard, protective shells. It would take an additional 100 million years before creatures with stubby fins, destined to become legs, and with rudimentary lungs, lurched out of the slippery shallows, hauled in on a little air, took a look around, and then dashed back into the sea. The beach might have been where the action was, but life had to get used to it as a new, airy home.

But life came up on the beach because air is a metabolic gift. Oxygen and carbon dioxide both diffuse about ten thousand times faster in air than in blood or water. Probably before the Cambrian period, but certainly during it, oxygen contributed to the faster growth of complex cells and of muscles. Creatures began crawling and digging in the oceans. Worms were stretching themselves thin, like wires, to maximize gas exchange through their skins. Jellyfish kept their inner workings to a minimum, eliminating any need for circulation. But after the Cambrian period, oxygen enticed life to come out where the living was easier, out into air.

Following algae and fungi onto the land came lichens and then the first moving creatures, insects. Beetles in armorlike shells lumbered across algae and lichen-gloved rocks and beneath mushrooms dusting their backs with spores. Feasting on oxygen, the newcomers helped shape the earth's first communities, ecosystems in which the different species were interdependent. In the adjacent seas, the first vertebrates continued developing, but were not quite ready to take the leap to land yet.

Chordates, the first vertebrates, were a new animal phylum, or primary group, one that emerged during the Cambrian explosion, a span of
10 million years during the period when species blossomed as they never
had before, or have ever since. Think tadpoles, and you're thinking chordates. The earliest chordates looked more like worms, however. Flimsy
spinal rods helped them undulate about. Nerve cords evolved into spines,
nerve bundles became brain nubs and eyes. Our evolutionary forebears
grew tails and gill slits, which were not true fish gills—they have capillary
beds for gas exchange in either folds or plates—but the precursors to real
gills. Chordates grew bigger tails, teeth, and jaws for clutching prey.
Probably around 360 million years ago, when the vertebrate invasion that
followed the insect invasions brought the first chordates onto land, their
gills disappeared, fading into functional obscurity and distinguishing the
land branch of the group from the sea branch.

Today, we see such gill-like structures briefly in the human embryo.
During the first weeks of a fetus's life, vestiges of gills emerge, along
with tails, and then quickly recede. Of the many compelling aspects of
evolution, one is that little that came before seems to have been forgotten. Gills and tails in humans, hands and ears in whales—they seem to
flow by in the womb, memories of the ancient past, of odd biological
forebears. Biology seems reluctant to throw anything totally away, even
as it remodels families and species into unrecognizable heirs. Today, what
walks and talks often looks very little like what once crawled and swam.
"Ontogeny recapitulates phylogeny," Jay Leiter told me on several occa-
sions, seemingly fond of the alliterative quality of the phrase as well as its
succinct message: The history of every individual echoes that of its
species. Leiter was quoting Ernst Haeckel's famous dictum. Haeckel, a
German physician and evolutionist, had coined the famous phrase in the
late nineteenth century, and also added the words *phylum* and *ecology* to
the modern biologist's vocabulary.

Sea worms, plankton, slugs, humans—all come from chordates that
first appeared in the seas during the Cambrian period, 540 to 500 million
years ago, while on the bacterial-bulldozed margins between sea and

shore, algae moved, adapted, and lived by photosynthesis. One theory about the oxygen level of air claims that it spiked around 450 million years ago, achieving roughly the same proportions as today before dropping and then rising again 100 million years later. If correct, this theory may help explain the appearance of the first insects, many of them beetles. Insects evolved quite remarkable respiration systems, the likes of which no animals had previously demonstrated, that exploited oxygen for metabolic needs. Systems of tubes often led to organs beneath their exoskeletons, so that the insects did not need a circulation system with blood to carry gases from lungs to tissues. They got the job done with diffusion and air tubes fit with dust filters. Insect design was a masterful stroke. It insured that arthropods, the phylum to which the insects belong, became the earth's largest, numbering well over a million species today. One peculiar structural feature that helped some insects move a lot of air was a partially empty shell. Certain beetles today still have cavernous, ominous-looking exteriors. The inner space is for air, a supply of which thus stays micron-close to tissues that might suddenly need oxygen for a burst of action, and to dump carbon dioxide. Insects developed abdominal muscles that could contract, increasing cell ventilation and pushing more gases around. Flying insects, though, emerged as a special category because of their ability to burn oxygen. Hundreds of million years ago, with neither a strenuous exercise regime nor blood-doping with erythropoietin (EPO), these flying insects evolved into highly efficient little creatures. Once airborne, many burned more than a hundred times the oxygen they used at rest.

Joining the new ecosystems on land, as the stratospheric ozone overhead was thickening and protecting their genes, and meteorite hits were fewer because of increasing atmospheric friction, amphibians slid out of the slimy, algae-thick shallows on stubby fins, gill arches regressing, or got washed up on shore during storms or tides and found themselves stranded. The first amphibians probably had fishlike scales, which would have prevented the use of their skins as a respiration membrane, although that would change. Some fossils suggest these early amphibians had ribs to work their lungs. Frogs, salamanders, and other amphibians never did develop strong spines. Their spines were stronger than those found in fish, which didn't need the rigidity because the buoyancy of water lent them support, but weaker than those in reptiles. Reptile backbones tended to be anchored to their hips and arched, like a suspension bridge, to carry heavier loads and to deal full-time with gravity, which wanted to force the animals flat on the ground. The amphibian invaders united the water world and the air world. The amphibian invasion probably took place around 360 million years ago, the start of a geological span of time especially meaningful to the way we live today.

The Carboniferous period stretched from 360 to 290 million years ago. Thick veins of coal were laid down by dying forests "heaped like corpses in drawers," as Annie Dillard put it so vividly in *Pilgrim at Tinker Creek*. These were carbonized corpses of trees and other plants, all of which had gorged on carbon dioxide, boosting the earth's oxygen concentration and depleting its load of carbon. The Carboniferous period was the earth's energy-making blast. It was the end of the birth of air as we know it, and the locking up of carbon into a vast storehouse of energy in sinks under the ground and beneath the oceans. Probably around 350 million years ago, oxygen reached its present level of just under 21 percent and, despite some minor fluctuations, has stayed there since.

On land, trees provided some of the first vertebrates with safe harbor. Plants covered the ground and flourished in part because of their symbiotic relationship with insects. Ants, flies, beetles, and so on, dug in their roots, spread their seeds, lived everywhere in their branches and leaves, almost as though plants were their green condominiums. Toward the end of the Carboniferous period, the first birds evolved from reptiles that also liked trees. Some reptiles lived in trees; others climbed to the high branches to sun themselves or to avoid predators on the ground. The generally accepted hypothesis detailing the evolution of birds says that small reptiles probably began to leap from one tree to the next, using air and leg muscles to escape predators in the trees or else to avoid having to climb down and back up again. Reptiles had the brains for such thinking. Their brains were larger and more complex than those of the amphibians that seasonally moved back and forth from land to water and that the reptiles eyed and sometimes caught and ate. The leaps of reptiles led to parachuting maneuvers. Limbs wide and bodies flattened, they caught air for distance and to soften landings. Then came gliding, followed by flailing—desperate attempts to add a little bit of lift to reach a limb or an outcrop that the mind had misjudged—the destination was just a bit too far. Wings gradually appeared. Today, we can look to reptiles as the predecessors of our feathered friends with their ventilated song boxes and their flow-through lungs and their hollow bones filled with air to make flight easier.

Birds, like reptiles, laid hard-shelled eggs, another masterful adaptation to gases and the pressure of air. Bird embryos require almost one thousand times as much oxygen at full term, when they are about to hatch, than when first laid. The morphology, the form and structure, of the egg shell is such that gases can pass through it at different rates as the chick grows and, of course, the shell doesn't. This little marvel of nature pays homage to John Dalton's law of partial pressures and to the life-shaping power of membranes. In an egg shell the membrane is beneath the outside shell, which is perforated with air-filled pores. Inside, just before the egg hatches, an air pocket expands. A baby bird pokes its beak into the pocket to first ventilate its lungs, to introduce them to its new world before bursting its shell.

After the milestone of oxygen's becoming 21 percent of the atmosphere, reptiles came to command an oxygen-lush, plant-thick planet. And they branched into a wild menu of creatures: turtles and snakes, lizards and dinosaurs, crocodiles and birds. During the Triassic and Jurassic periods, 240 to 140 million years ago, some dinosaurs became huge. A few had lungs the size of refrigerators and appetites for sun-loving plants that would have cleared the side of a Tennessee highway of kudzu. Descendants of the gliders and flailers, the pterodactyls (from *pterdactylus*, meaning "feather fingers") and the *Archaeopteryx* (ancient wings) coasted overhead through clear blue skies. Mammals scampered furtively through the ferns and across the savannas, up into trees and out on the ends of limbs; they were in all likelihood the size and shape of rats. Flowers, their pollen spread by airborne insects, added splashes of color to a landscape we would recognize today. But air was about to betray the life it had ushered in. Betray it again.

Air Extinguishes Life

For if God is in one sense the igniter, a fireball that spins over the ground of continents, God is also in another sense the destroyer, lightning, blind power, impartial as the atmosphere.

—ANNIE DILLARD, *PILGRIM AT TINKER CREEK*

Sixty-five million years ago, the atmosphere was violated by a bolide, a large meteorite that penetrated the planet's protective membrane without burning up. The bolide, around six miles in diameter, came angling through the atmosphere at a relatively slow rate of speed. On the ground, a dinosaur or mammal saw something huge and glowing, with a fiery tale, like a comet, sweeping across its line of sight for about ten seconds. Then the ground shivered and the world changed forever.

The bolide hit Earth like a bullet roughly the size of Mount Everest. It released energy equivalent to several billion atomic bombs, each like the one that flattened Hiroshima, exploding all at once. Kingly dinosaurs such as *Tyrannosaurus rex,* soaring birds and diving fish, lichen and algae and plankton, flowers and mosses and everything else close to the Yucatan region in Mexico, where the bolide came down, were incinerated because the temperature around the fiery rock briefly touched the 5,000-degree Fahrenheit mark. Forests ignited like Kleenex. Seawater flash-boiled into steam. After the impact, seismic waves radiated in spokes from ground zero. Smoke from fires, steam, and debris sent aloft turned the air opaque, blocking sunlight as dust and ash drifted, soon damaging lungs and clogging air tubes of creatures thousands of miles away. Halfway around the planet, photosynthesizers could not get enough sunlight. During the bolide's plunge through the earth's thin membrane, the intense combustion had created vast amounts of nitrogen oxides, or NOx, the same pollutant emitted from the tailpipe of your car and a component of acid rain. So acid rain caused by the NOx began to fall. Depending on which way the winds blew, ponds, rivers, lakes, and forests were damaged by the rain. As synergistic interactions had abetted the evolution of life for millions of years, once again the same types of synergy, except dark and destructive, eliminated it.

Yet, some life survived in protected niches. Over time, the planet slowly repopulated itself with an astonishing diversity of creatures. Some were new, others had forebears that had made it through the catastrophe. All the creatures were biological descendants of the bacteria that had held on, no matter what the air looked like or how cold or hot it became. But as the bubblers had learned the hard way, when the composition of the atmosphere changed, whether it took a billion years or happened in a few minutes when a rock came to visit after tumbling through space since the Big Bang, life was transformed. Some species died, and some hung on.

Therein lies a central irony about air: It makes life possible but doesn't give a damn about life's survival. Air just doesn't seem to care—the same accusation that Blaise Pascal made about God.

If you're wondering what the odds are of a bolide shredding the atmosphere like a bull through a spider's web and wiping out much of life in the twenty-first century, you can relax. A study by a team of astronomers at Princeton University put the odds at five thousand to one against a bolide larger than one-half mile in diameter ending modern times in smoke, a shiver, and carnage—at least during the present century.

The bolide hit of 65 million years ago caused only one of five great extinctions the earth has endured, scientists now tell us. Death's sidekick in all five extinctions was air. Once a vibrant membrane of air encircled the planet and nourished life toward greater biological complexity, it almost seemed to relish wrecking everything. Air seemed absolutely Old Testament–like in its wrath.

Today evidence is accumulating that another extinction may be in progress. Exactly what size, we don't know yet. But in the Holocene epoch of the late Cenozoic era, there are markers remindful of the ancient past, markers that science has become familiar with. Species have been dying off for several thousand years, but the rate has been accelerating. And the single most crucial marker of an extinction event is species die-off. If present trends continue, by 2100 it is estimated that we may surpass the 50 percent species die-off milestone associated with a pending extinction. In *The Diversity of Life,* Harvard entomologist E. O. Wilson wrote: "Humanity has initiated the sixth great extinction spasm, rushing to eternity a large fraction of our fellow species in a single generation. Any number of rare local species are disappearing just beyond the edge of our attention. They enter oblivion like the dead of Gray's *Elegy,* leaving at most a name, a fading echo in a far corner of the world, their genius unused."

The first of the extinctions we know about occurred 440 million years ago during the Ordovician period. The others occurred 365 million years ago during the Devonian period, 250 million years ago during the Permian, 210 million years ago during the Triassic, and 65 million years ago during the Cretaceous. The worst wipeout, dubbed "the great dying," was the one 250 million years ago. It killed virtually everything that swam, crawled, flew, or had roots. Between 80 and 96 percent of all species vanished. All of the wipeouts pushed what Wilson called "the far end of the curve of violence." That was putting it mildly. The recovery times also were prolonged. Wrote Wilson: "In general, five million years were enough only for a strong start. A complete recovery from each of the five major extinctions required tens of millions of years.... These figures should give pause to anyone who believes that what *Homo sapiens* destroys, Nature will redeem."

The story of the bolide that caused the carnage 65 million years ago, closing out the reign of the dinosaurs and opening a window for the ascent of mammals, was only clarified recently, in the 1980s, thanks to a father-and-son team, Luis and Walter Alvarez. What seeded the now widely accepted theory was a trip by Walter Alvarez, the son, to the spectacular 1,200-foot-high cliffs of Bottaccione Gorge in northern Italy in the mid-1970s. The cliff face is stratified, rose-shaded limestone. It's called Scaglia rossa and for centuries has been much sought after by the designers of villas. The limestone had built up in layers over a period of 100 million years because this section of Italy had once been the bottom of an ocean. The natural accretion left a mother lode of information for geologists such as Walter Alvarez, paleontologists, and others investigating the earth's past. For them Bottaccione Gorge was a quarter-mile-thick stone book of rose-colored pages embedded with the story of the planet since the Cretaceous period.

Walter Alvarez chipped away a few chunks from a unique layer, a half-inch scree of clay thought to have been laid down at the end of the Cretaceous period, when the dinosaurs disappeared. He had high hopes that, in the chunks, he'd find the key to the extinction. Once Walter Alvarez returned home, however, his search for the key didn't pan out. His father, physicist Luis Alvarez, suggested Walter take a fresh look at the chunks. To look at them not as clay but as a clock, as a clock of interstellar dust.

The senior Alvarez was no quack. But he wasn't your normal physicist, either. He'd X-rayed the Egyptian pyramids in search of hidden tombs. He'd invented the bubble chamber, which helped prove the existence of subatomic particles and won him a Nobel Prize. Luis Alvarez was the type of physicist whose curiosity was unrestrained, whose interests were boundless. He made the odd suggestion to his son, probably because he was sympathetic to the younger man's sense of frustration at coming up empty-handed after having a strong hunch. But he also had a broader, and deeper, grasp of existence than his son did. And he knew a few things Walter did not. For instance, he knew about iridium.

Iridium is a rare element that scientists had traced to a single source. It spewed out of the atomic inferno of supernovae. A considerable deposit of iridium had accompanied the formation of the earth, but probably sank to the core of the planet when it was molten. But since then, a steady pattering of iridium had been coming down as interstellar dust as meteors suffused with the element had ended their celestial careers and been shredded into particles by the earth's membrane. Over the last 100 million years, Luis Alvarez knew, iridium-tinged dust had collected in the layers of limestone at Bottaccione Gorge. It was this dust that he thought his son might consider as a kind of clock.

Bringing diligence and imaginative thinking to a challenging puzzle that scientists had pondered for centuries, the Alvarezes eventually showed that the iridium clock ran consistently in the layers of limestone in the gorge, both above and below the thin layer of clay, but went bonkers in the thin, one-half-inch of clay itself. The clay contained about thirty times the iridium found immediately above and below it. Why did iridium spike in this beautiful and strange geological clock in Italy? Because, Luis Alvarez hypothesized, about 65 million years ago an iridium-embedded asteroid or comet from space smacked the earth and sent an unusually large amount of the rare element into the atmosphere. Some of this iridium had ended up in the odd little layer of clay stuck in the 1,200-foot cliff of limestone in northern Italy, upsetting the smooth functioning of the limestone clock with a clue to the cause of an extinction event.

Luis Alvarez likened the event to Krakatau, the tropical island in Indonesia that had blown apart in August 1883. Except magnified a couple million times.

Krakatau is the biggest and best-documented eruption modern science has for a base line. The island's eruption provides investigators with a taste of extinction. There is also a fine description of it penned in a hurry by the excited first office of the *W. H. Besse,* an American bark that just happened to be sailing about fifty miles away from the island when it climaxed. The first officer scribbled in the log, noting that the barometer plunged, and a squall hit the ship, rapidly increasing to hurricane force:

It was darker than any night I ever saw. . . . A heavy shower of ashes came with the squall, the air being so thick it was difficult to breathe, also noticed a strong smell of sulfur, all hands expecting to be suffocated; the terrible noises from the volcano, the sky filled with forked lightning, running in all directions and making the darkness more intense than ever, the howling of the wind through the rigging formed one of the wildest and most awful scenes imaginable, one that will never be forgotten by any one on-board, all expecting that the last days of the earth had come; the water was running by us in the direction of the volcano at the rate of 12 miles per hour, at 4 p.m., wind moderating, the explosions had nearly ceased, the shower of ashes was not so heavy; so was enabled to see our way around the decks; the ship was covered with tons of fine ashes resembling pumice stone, it stuck to the sails, rigging and masts like glue.

Vaporized rock blew across the Sundra Strait at close to three hundred miles per hour. Tsunamis left the epicenter in ripples. For weeks after the eruption, luminous red sunsets were seen around the world and the hearts of doomsayers were warmed because here was evidence that Satan had finally come. The earth's weather was cooler for the next two years. Luis Alvarez understood, however, that compared to a bolide hitting the earth, Krakatau was a minor event.

Combining Luis's ideas about the iridium clock of the Bottaccione Gorge and an exponentially bumped-up Krakatau, the Alvarezes shaped their theory. It claimed that the extinction 65 million years ago was caused by a bolide that gouged a crater somewhere, lofted debris thick with iridium into the air, and shrouded the earth in a sun-absorbing and suffocating cloud that soon pelted the landscape with acid rain and set off a death spiral. Plants died, as did the sea's largest food stock, plankton. Fish and other animals found little to eat and starved to death.

Big, rough, but persuasive, the theory was attacked. But critics had a tough time killing it once additional evidence from shocked quartz (globules of glass that form when hot volcanic eruptions hit a cooler atmosphere) directed the search for the telltale crater toward the Yucatan. In the 1950s, Mexican geologists made a discovery that suggested a huge crater had been formed off the Yucatan. The Yucatan discovery was dusted off, and too many pieces—the iridium, the shocked quartz, the crater—fit together for the theory to fall. Presently, the Alvarezes' theory stands as the explanation for the disappearance of the dinosaurs and for the fifth great wipe of life on earth.

After the bolide, *T. rex,* along with other large land reptiles and marine creatures, was history. Smaller creatures eked through, as did mammals, none of which weighed more than five pounds. Insects and plants, although hammered as well, survived in large numbers because of their symbiotic partnership and their innate toughness. Much of the devastation that swept creatures and ecosystems from continents relied on air for its dirty work. On air's fumes, its greenhouse effect, its opacity from dust and smoke, its transport currents in the upper layers, its acid rain, its temperature shifts, and, ultimately and probably most importantly, its long-term climate change. The variables did not function in isolation but synergistically, seemingly indifferent to life scrambling to sustain a foothold below.

Yet life again slipped through.

Possibly the most amazing thing about the big extinctions is that life did hang on, did find pockets of respirable air, usable water, regions where the temperatures neither froze life to death nor fried it in the heat. The survival of life in niches, in pockets, despite the catastrophes, moved Stephen Jay Gould to reverie in *I Have Landed:* "If anything in the natural world merits a designation as 'awesome,' I nominate the continuity of the tree of life for 3.5 billion years. The earth experienced severe ice ages, but never froze completely, not for a single day. Life fluctuated through episodes of global extinction, but never crossed the zero line, not for one millisecond. DNA has been working all this time, without an hour of vacation or even a moment of pause to remember the extinct brethren of a billion dead branches shed from an ever growing tree of life."

Surely, the self-serving platitudes often heard that praise human destiny, human goodness, human intelligence, human morality, or anything else especially "human" as playing a more meaningful role in the ability of *Homo sapiens* to survive all the cuts, so to speak, is either egotism, fantasy, or denial. As Kenneth Kardong, a physiology professor and an author of a widely used college textbook on vertebrates, writes about human privilege in evolution, "in reality, the human species is one of thousands of recent evolutionary products. We're not the final cause of evolution, we're no pinnacle of achievement, we're just one of a long march of creatures whose common mammal ancestors are now thought to have been very few."

How few? During one of the conversations I had with Jay Leiter about the evolution of mammals, when the subject suddenly shifted to common ancestors, Jay said with a note of astonishment, "You know, for humans, we're probably talking about eight women."

As for the extinction events, we're not done with them yet. The latest is different from the others, however. It's being caused not by a bolide or volcanoes, but by a single species: us. We initiated this turn of events several thousand years ago by hunting down "the large, the slow, and the tasty," as Wilson described the large animals, the saber-toothed tigers, the woolly mammoths, the large flightless birds called moas, the giant wolves, and all the other animals unprepared for the expertise of their predators. It was a "hunters' blitzkrieg," noted Wilson. A protracted wipeout not lacking black humor. A Mexican truck driver, asked why he shot one of only two remaining woodpeckers, which were the largest in the world and becoming extinct, said, "It was a great piece of meat." The Maori who slaughtered the moas into extinction had their expropriation of life memorialized in a popular ditty sung often in New Zealand:

> *No moa, no moa,*
> *In old Ao-tea-roa.*
> *Can't get 'em.*
> *They've et 'em;*
> *They're gone and there ain't no moa!*

The role of air in the relentless extinction of species has accelerated the last two hundred years, since the start of the industrial revolution in coal-burning, smoky England. But air's contribution began long before that, with the smelters making coin in pre-Biblical times, and has vastly increased in our time, with worldwide reliance on carbon extracted in various forms from the sinks in which the planet's evolution to a life-friendly place had relegated it. The loss of tropical rain forests, the increase in backyard burning around the world, and odd novelties, such as the brown cloud that loomed over Indonesia in 2002 almost like a reminder of Krakatau, but in reality a result of human burning of biomass and garbage, have all jived up the tempo of atmospheric change. If we characterized the change as a dance, we might say that what had been a leisurely waltz toward a threshold of extinction for a couple thousand years has recently turned into a mad jitterbug.

There probably is little to worry about tomorrow, next week, next year, even five years from now. You probably won't have to buy air like you now buy water. Humans are too smart to be wiped out. We're not the dinosaurs. We'll avoid the big bullet. Or will we?

Not at the rate we're going, according to a chorus growing louder by the year. The extinction our species seems intent on masterminding is probably psychologically driven, I've decided. It's straight fear. Fear of the dinosaurs.

And don't laugh.

Imprinted somewhere in our molecular memory like those gills glimpsed briefly in the human fetus, there's a fear switch. And it's on. It's been switched on for over 300 million years, ever since all this oxygen gave our precursors big brains, the energy to think, a leafy place to think in, *and* cursed them with a nemesis—all those damn reptiles that were trying to chase them down to eat them. The fear switch that we inherited molecularly from our precursors that lived 300 million years ago rules our existence today.

Evolutionary biologists plot a similar story line leading back to why, on the molecular, biological level, we're hell-bent for destruction, even if our big brains are now highly capable of laying out the end game toward which we're heading. Got to make life safe. Be protected. You can't stop doing what you're doing, because your deep fear instincts, stirred up to even greater heights by marketing ploys and war-driven politics, tell you to protect yourself. You can't shut your instincts down. There's no going back to the den, to the nest in a tree, to the hunter-gatherer community, where individuals contributed according to their skills to chase down those woolly mammoths and moas and then to butcher and cook them, licking their fingers happily afterward.

Fear, psychiatrists tell us, can be a complex astringent. It erects protective spheres of different types: spheres of materials, of prestige, of money, of arrogance, of love. The spheres fit together concentrically as invisibly and as neatly as the sky defined by Aristotle and the Christian church during the Renaissance, and as confirmed by modern science in the twentieth century. It's the micro reflecting the macro again. Our past is the shape of air, our present is the shape of air, and our future is the shape of air. With one very big difference between us and it. Air has no fear. Air is indifferent. Air is *deus absconditus*.

As brilliant a commentator on the human condition as Charles Darwin was not fond of the idea of extinction, especially extinction that happened rather quickly, over a time frame such as I have been referring to here, ten or twelve thousand years. "For Darwin," wrote Carl Zimmer in *Evolution: The Triumph of an Idea,* "extinction was simply the exit that losers took out of the evolutionary arena." And it happened slowly, over epochs, via natural selection and survival of the fittest.

Today, however, we know that Darwin was wrong about losers, that natural catastrophes weeded the evolutionary arena with a grand disdain just as surely as natural selection did. When we ponder the colossal death lists of the five big extinctions, and the list being writ now, it's hard not to draw the conclusion that any embrace of life as something inviolate and precious is farce. One calamity after the other after the other, a sequential collapse of the Rube Goldberg structure called life, a stampede of causative factors stomping over the suddenly precarious hold most of life has on life—that's the evolutionary past, as surely as natural selection and survival of the fittest. The tenuousness of life seems nowhere better epitomized than by the thread linking every cell in our bodies to air and its elixir, oxygen. We have no backup. Four minutes without oxygen, and we're dead.

That furtive tree rats took control after the giants that had hunted and dined on them for almost 200 million years were eliminated by a big rock and a life-strangling atmosphere strikes me as absurdly ironic, evolution knocking its funny bone: Let's see what happens next. As Dillard wrote: "Evolution loves death more than it loves you or me. This is easy to write, easy to read, and hard to believe."

What happened next, of course, involved air. It had to become clear for sunlight to get through; it had to go through thousand-year cycles, it had to coax life again out of hiding in niches, just as air had coaxed life out of the sea onto the land. Even after Krakatau, it was air that brought the first inhabitant to the lifeless place: a spider. The little beast probably crossed the Sundra Strait by catching air in a balloon, a series of silken threads extruded from its abdomen so that it could take a ride, become a pioneer. The spider was followed by other air travelers: aphids, fungus spores, tiny seeds, and an assortment of arthropods, including more spiders, together with earwigs, moths, flies, beetles, and crickets. Inside these creatures and also setting up new homes on the reviving landscape were countless parasites, including bacteria, worms, and lice. Other colonizers included little creatures dropped in the feces of passing birds. A similar rehabilitation, though much slower and over large areas of the earth, followed each of the five extinctions.

A final note here: In the path of evolution, as in most other things, there have been contrarians. One reversal that humans are quite fond of today is the whale. About 50 million years ago, the sea-to-land momentum that had started with the tartan-plaid carpet of algae creeping onto shore around 500 million years ago was reversed by a coyote-sized creature with lungs. Ignoring the Do Not Enter sign on the path of evolution, this creature headed back into the sea.

Over 20 million to 30 million years, the creature evolved into the whale. First it appears to have become a wader, then a crocodile-like shoreline predator. Named *Ambulocetus* or "walking whale," by the paleontologist who discovered it, this fabulous critter had "giant feet that looked as if they could fit in clown shoes and a bulky skull," and it lurked in the shallows, waiting to pounce on anything smaller that wandered within range. Eventually, the contrarian mammal adopted an otterlike swimming motion, an arching of the back and a shoving with the feet, and it had the means to swim offshore. The peculiar swimming motion remained, but the legs withered and became fins. The nostrils, once embedded in a long snout, shifted to the top of the head as a blowhole. It was a lot more efficient for a whale to undulate with small fins and a large tail, its blowhole skimming the surface of the ocean to breathe, than to lift its head above the water every time it needed to inhale and exhale. Warm air exhaled through the blowhole condensed, making the exhalation look like a fountain, something that delights whale watchers to this day.

Contemporary investigations put the DNA of whales very close to that of a particular land mammal, the hippopotamus. The two are very close genetically.

When I learned this, I had to laugh. Years ago, a friend had warned me quite earnestly, "Never get between a hippo and water." Now I knew why. A hippo's evolutionary destiny can be said to be the sea. Somewhere in its senses, the lumbering beast knows that. Threatened, a hippo heads toward the nearest water in a beeline. No puny, two-legged mammal is about to stand in its way.

Now I've found myself speculating that maybe, 50 million years ago, the coyote-like ancestor of the whale, called *Pakicetus* or "whale of Pakistan," for where the paleontologist Philip Gingerich, from the University of Michigan, first discovered its fossil in 1979, heard something that warned it of future predators. These were the blitzkrieg of hunters that E. O. Wilson mentioned as the killers of the large mammals once numerous around the globe. Forewarned somehow, *Pakicetus* veered back toward the sea, if not in a beeline, in a relentless quest for adaptability. The land may have been where the action was during the Cambrian period, but this creature was taking its morphology, its ability to function on air as a mammal, back into the water. But it wasn't going to be a fish. It was going to be better at using oxygen; it was going to bring the advantages of lungs and two respiration pumps into the world of gills and diffusion.

On its new evolutionary course, *Pakicetus* grew ear bones, "two shells shaped like a pair of grapes...anchored to the skull by bones in the shape of an S." The ears amazed Gingerich when he first came across the fossils in a region of Pakistan once dotted with islands and in the middle of a sea. The ears in the fossil told Gingerich he'd found a relative of the whale. Maybe, as I said, the ears gave *Pakicetus* a sixth sense, so that it could avoid the sixth extinction of large wild mammals on land. *Pakicetus* heard what other land-bound creatures did not and got out of the way of *Homo erectus*.

If this sounds a little absurd, listen to what the naturalist writer Diane Ackerman said in her book, *A Natural History of the Senses*, about ears: "In Arabic, absurdity is not being able to hear. A 'surd' is a mathematical impossibility, the core of the word 'absurdity,' which we get from the Latin *surdus*, 'deaf or mute,' which is a translation from the Arabic *jadr asamm*, a 'deaf root,' which in turn is a translation from the Greek *alogos*, 'speechless or irrational.' The assumption hidden in this etymological nest of spiders is that the world will still make sense to someone who is blind or armless or minus a nose. But if you lose your sense of hearing, a crucial thread dissolves and you lose track of life's logic."

In *Homo sapiens*, is the inability to "hear" what we are doing to the air and the sea and the other species on the earth anything other than absurd? "Waves of sound roll like tides to our ears," Ackerman wrote, "where they make the eardrum vibrate; this in turn moves three colorfully named bones (the hammer, the anvil, and the stirrup), the tiniest bones in the body. Although the cavity they sit in is only about a third of an inch wide and a sixth of an inch deep, the air trapped there by blocked

Eustachian tubes is what gives scuba divers and airplane passengers such grief when the air pressure changes. The three bones press fluid in the inner ear against membranes, which brush tiny hairs that trigger nearby nerve cells, which telegraph messages to the brain: We *hear.*"

Collectively, though, we seem not to hear the sixth extinction loud enough to respond yet. We are like the baby in the crib unaware of its face in the sheets and its primal switch screaming at it to turn its face, to wake up to breathe. The movement of air molecules radiating from wings of extinct moths, the tensing fibers in the legs of endangered grasshoppers, the carbon-dioxide-rich roar of fewer and fewer lions—these diminishments of sound do not reach our ears despite being harbingers of an impact not terribly different in its dark promise than a six-mile-wide bolide the size of Mount Everest arching out of the sky, striking the Yucatan, and wreaking global cataclysm.

Part 2

Exhale

Now, whenever air comes into being out of water, light out of
heavy, it goes to the upper place. It is forthwith light: becoming is
at an end, and in that place it has being.

—ARISTOTLE, *ON THE HEAVENS*

4

Once Air Was
Simply Marvelous

Once air was simply marvelous, filled with demons who lurked in
shadows, with sylphs who lived in thunderheads and caressed the
back of your neck as warm summer breezes, with angels speaking
Enochian, with particles from distant planets trying to invade your
body—all watched over by celestial ether, the big air made splendid
by the light of God. In this world Aristotle still ruled, although he'd
died eighteen hundred years ago. His grand system, based on four
elements, one of which was air, dominated scientific thought and
was a bedrock for Christianity. Beginning in the sixteenth century,
into this world came a procession of men who challenged many of
Aristotle's precepts. Natural philosophers, preachers, and explorers,
they dismantled a world filled with marvels and mysteries and
replaced it with one of mechanics and science. The new world dominated intellectual thought until the arrival of quantum physics. This
newest discipline revealed that even air is mostly empty, a void. That
everything we think of as solid is mostly emptiness. And that the universe is a very strange place in which "it might be said that the vastness of space, the nuclear conflagration of stars, the explosions of
galaxies are simply mechanisms for producing that first glimmer of
awareness in your baby's eyes."

Gulliver, the famous traveler created by the English satirist Jonathan Swift, spots a flying island in "Voyage to Laputa," published in 1726. Seeing "a vast opaque body between me and the sun," an excited Gulliver waves his hat. The island is about two miles away, and held aloft by a magnetized lodestone. Says Gulliver, "The reader can hardly conceive my astonishment, to behold an island in the air, inhabited by men."

IT'S HARD to overstate Aristotle's impact on the way the earlier world, the world of marvels and mysteries, saw itself. Aristotle didn't just come up with theories. He shaped the universe. Wrongly, as it turned out. But right or wrong, Christianity embraced his vision like a mother wrapping both arms around a brilliant child.

In two works, *On the Heavens* and *Meteorology,* Aristotle defined his system of four fundamental elements: earth, water, fire, and air. Beginning in the thirteenth century, the Christian church buttressed its vision of the Great Creator's universe with his thought. In this scheme, daily doses of heavenly influence, almost like drugs handed down by a Great Pharmacist, poured from the sky into the bodies and affairs of all living creatures. The microcosm of the body mirrored the macrocosm of the cosmos. All things possessed some final cause, some God-willed reason for being here. And the entire universe hummed with morality. In that world, to challenge Aristotle was to question the authority of the church. It could get you labeled as a heretic.

Structurally, the Christian, Earth-centered cosmos was a lot like a set of Russian dolls. At its core was Earth. Surrounding Earth was air. Surrounding air was a sphere of "pure air." Then came a succession of spheres holding the known planets—Mercury, Venus, Mars, Jupiter, and Saturn—with the moon's sphere nestled inside that of Mercury and the sphere of the sun between those of Venus and Mars. Finally came the stars and the end of the physical world, the *primum mobile,* "the outer skin of the cosmos," as Jeffrey Burton Russell described it in *A History of Heaven.*

But the theologians weren't done yet. Beyond the outer skin spread something they called the *quintessence.* The quintessence was where God hung out. A blend of Aristotle's physics and the mystical theology of Dionysius, a sixth-century Syrian monk who preached that "we can know neither what God is nor what he is not," the quintessence was a fifth element, the *quinta essentia,* which the church cobbled on to Aristotle's basic four. Both matter and not matter, in the back of the beyond, a space imbued with the light of glory, the distance, strangeness, and mystery of the quintessence filled numerous voids. Depending on need, the quintessence could be heaven, the celestial paradise, or the empyrean.

The precise, almost mechanical, way this universe fit together gave it strength and grandeur. It was simple enough for a peasant to grasp but grand enough to awe the pope. Everywhere there was something, from the four elements of the inner spheres to the quintessence of the most outer. Thinking about all these spheres holding each other in place, with God's light surrounding them, you can easily grasp the system's appeal. It was imaginative. It was fun. It was comprehensible. The spheres were even said to sing. God himself might be inscrutable, but this gigantic egg-like universe he'd created wasn't.

Aristotle had not arrived at his grand system alone. Modern scientific thought can be traced back to ancient Mesopotamia and to Egypt of the pharaohs. Both sandy, river-sided, hot cultures were sky addicted. In 1600 B.C., the Egyptians thought the sky was made of iron plate, possibly because meteorites containing iron, which they called *baa-en-pet* (iron of heaven), fell out of it. Living along the fertile shores of the Tigris and Euphrates rivers in what is now Iraq, the Babylonians loved to watch the stars at night. Their philosophers said deities ruled the heavens and through heavenly bodies shaped people's lives. The shaping began at birth, often with a clash of cymbals accompanying the first cry of the newborn, which notified the family astrologer it was time to plot the position of the stars, sun, moon, and five planets to determine the child's fate.

In the ancient world, air was considered a medium, a means of communication and transport, almost like a river, between the heavens and the earth. Through air came powers not only to rule human destinies but to effect important minerals, an idea later embraced and expanded by the alchemists.

In the sixth century B.C. a group of Greek philosophers seeking knowledge for its own sake began to shape the physical world into systems. The philosophers were loosely grouped in the Ionian school, and they absorbed many of the older theories and myths of the Babylonians and Egyptians. The Greeks thought big, conceived grand systems, and did not experiment. Speculation ruled. The only restraint on far-out ideas and sweeping generalizations was reasonableness.

One of the early questions the Ionian school wrestled with was where the world came from. Their answer was the indefinite and the infinite; they called it the *aperion*. Out of the *aperion* floated worlds inside bubbles. The worlds were thought to be composites of opposites, hot and cold, dark and light, good and evil, moist and dry, male and female, and so on. Then, around 550 B.C., an Ionian philosopher named Anaximenes rejected the *aperion*. Anaximenes said the source of the primitive chaos shaping the world was air.

Air, said Anaximenes, was primary. Not water, as the Babylonians had thought. And definitely not the *aperion,* as his fellow Ionian philosophers believed. Air was not the lowly cousin of fire, earth, and water, as some believed, said Anaximenes, but their superior. He reasoned that air was primal because it rarefied into fire and condensed into water, as well as into earth and stone, according to his observations. He also proclaimed that all things, including air, were in constant flux. Things were always changing, this philosopher of the physical world deduced brilliantly, changing into one another. Anaximenes' theoretical speculations lent shape to the concept that all matter existed in one of three states—gas, solid, and liquid—and could change into the other two under different conditions.

Another member of the Ionian school, Heraclitus, built off Anaximenes' theories but rejected air as the source of all things. It was fire, Heraclitus said. Fire begat the world and took a "downward path" to become air, water, and earth. Heraclitus expanded the concept of opposites, a favorite of the Greeks, and applied it to science, insisting that the stars and the planets were composed of opposites, as was man.

In 494 B.C., the center of Greek philosophical thought shifted to Athens once the Persians captured Miletos, the Ionian thinkers' strong-

hold on the western shore of what is now Turkey, and forced the philosophers to leave. In Athens, the philosopher Empedocles, who was actually
a Sicilian and a doctor but who thought like a Greek philosopher, soon
became enamored with air. Empedocles insisted air was fundamental,
not a transitional state between fire and earth, as Heraclitus had said.
Empedocles also agreed with one of his contemporaries, Anaxagoras,
that small particles made up everything visible, although the particles,
which Anaxagoras thought of as "seeds," were invisible. Empedocles
thought the particles came in four shapes, one for each of the four fundamental elements, putting into theory for the first time the idea of four
elements made of small, never-changing particles or atoms. He also
spiced up the physical world with some emotion. He argued that Love
and Strife explained why the world was in constant motion and flux. To
Empedocles, love and strive were as real as air and water. A final accom-
plishment of this groundbreaking philosopher was an experiment, a radical break from the status quo Greek approach to theory, which was
relentlessly intellectual and disdained the hands-on.

The Greeks of Empedocles' time used something called a water
clock, a hollow wooden cone with small holes in the top and bottom, to
give them a rough estimate of the passage of time. They dropped the
clock in water, and since the time it took to sink was approximately the
same each time, the device provided a reliable measure. Empedocles
used his water clock to demonstrate that air, a fundamental element
made of particles, took up space, that air was material. What he did was
place the wooden clock in water while keeping one finger over the hole
in the top. When Empedocles pushed the clock down, water could not
fill it completely as it would have with both holes open. When he lifted
his finger, air rushed out the top hole, and water slowly displaced it, sinking the timepiece. Though simple and less than rigorous, the experiment
was eminently reasonable and is one of very few ever recorded from the
centuries of creative ferment in Greece that marked the origins of modern science. Here was the use of a physical proof to substantiate an intellectual argument.

The two Greeks who most influenced the future view of air emerged almost in tandem. Teacher and pupil, genius and genius, philosopher of a final cause behind all things and philosopher of a grandiose system to explain all things, Plato and Aristotle lived, thought, and taught in Athens during the third century B.C.

Plato, the older of the two, adopted Ionian theories. Plato was a reductionist. He accepted that four elements consisting of geometric atoms (those of air were shaped like octahedrons, he thought) formed physical matter and that the matter existed in a state of constant change. But Plato wanted to know the reason behind constant change. What drove change? What made a tree grow, gold become gold, clouds disperse in the blue sky? He came up with an answer that changed the way the world thought about things. Plato shifted the dominant paradigm. Instead of mechanical causes driving things forward, making them happen, Plato theorized that there was something ahead of matter and events, something inscrutable, something pulling things into the future almost like a magnet. In other words, constant change had a final purpose, he deduced, a destination to reach. Plato's concept of final cause was a metaphysical idea, a mind game. It was optimistic. Architecture and art, politics and social systems all wanted to achieve harmony and form, to become beautiful. Final cause would become a boon to the Christian church a thousand years after the philosopher's death, with the church proclaiming that God was the final cause. God shaped the present on its route to the glory of heaven.

A pagan, Plato didn't know about God. When he looked up, he probably thought about Zeus and Hera, about Apollo and Dionysus. And he thought about space as the source of all matter, sort of a retreat to the *aperion*. This distant space beyond air was called ether, or the "nurse of generation," by Plato. From it sprang air, fire, earth, and water.

Many of Plato's scientific theories appeared in the *Timaeus*, a series of his dialogues that Aristotle, the master's most famous student, was no doubt familiar with. Aristotle put much of what he learned from Plato, sizable bites of Ionian theory, and a prodigious amount of his own novel and expansive thought into the biggest, grandest, most detailed vision of the world yet. He explained everything. And he had a knack for visualizing. Philosophers, priests, warriors, and eventually even the masses could understand him. Interestingly, one popular idea that Aristotle rejected was the atom. Tiny particles, whether they were the seeds of Anaxagoras, the octahedrons of Plato, or the unchanging atoms of the laughing philosopher Democritus of Abdera, were not the building blocks of Aristotle's cosmos. He did embrace the notion of contraries. But those he trimmed down, from a broad array of physical and emotional pairings, to two sets of opposites, hot and cold, dry and moist, whose qualities could easily be detected. As for primal matter, Aristotle stuck with Plato on this core issue. It came from space, from the ether, from the nurse of generation. Aristotle did add something new and fantastic to space, however: the quintessence. Beyond fire and ether and circular in motion, the quintessence was the home of the stars, planets, and other heavenly bodies. When Christianity got hold of the quintessence a thousand years later, Zeus, Hera, Poseidon, and all the other Greek gods were removed. The quintessence was turned into the Great Maker's home.

Aristotle also said exhalations rose out of the earth. You saw them as vapor or smoke. Like many of his theories, this one tickled the imagination. It suggested a living, breathing planet, a fitting notion in a world rich with the myth of Gaia, the earth mother who had created the sky, Ouranus, from herself. When the exhalations were trapped beneath the earth's surface, Aristotle said, they changed forms. For instance, vapors, which were usually moist and opaque, became metals such as copper, gold, and iron. Smoke, dry and thick, became stones. Congealment, the fusion of the trapped exhalations under pressure, turned them into solids. Aristotle described the process in the fourth book of the *Meteorologica,* a compendium of ideas, bewildering in their breadth but ultimately unified by a tenuous fiber of the philosopher's thought and honored today as chemistry's first text.

It's hard not to admire Aristotle's audacity. Like a big-brained spider he wove together a web that ensnared the world. He did his weaving not by exact, reductionist experimentation, the kind so valued today, but from musing, discussion, reading, speculation. And, of course, from mental leaps into the void, a physical reality that he denied existed because everywhere there had to be something, from the core of the earth to the outer fringe of the quintessence.

If your teacher was Plato, and one of your students Alexander the Great, you might be excused from spreading your intellectual wings and flying like a god up into the quintessence or the ether or anywhere you damn well pleased. Like all Greek philosophers, the soaring Aristotle worked without a net. It was all observation and deduction and dialogue all the time. No experiments, no tests.

I think this, therefore it is. Disagree? Challenge me.

That Aristotle handled the challenges so well can be attributed to many reasons. He was brilliant, number one. He was also practical. In contrast to Plato, he preferred language that was more concrete than abstract. His theories stimulated visual thought. Aristotle was apparently a man in touch with his senses and knew that most other people were, too. He distanced himself from some unresolvable intellectual entanglements, such as Empedocles' Love and Strife contraries as a murky explanation of motion.

During Aristotle's lifetime, Alexander the Great united the warring Greek city-states, bringing peace to people who had known mostly conflict for centuries. Stability may have been good for politics and commerce, but it proved deadening for philosophy.

Inventiveness and originality of thought went into hiding, almost like the bubblers chased into the deep by the rise in oxygen billions of years earlier. A golden age from which democracy and modern science took their first clues ended. The rise of Alexander the Great and the death of Aristotle brought down the curtain on three centuries of sustained and brilliant philosophy, the likes of which have yet to be repeated in this cosmos. For the next two thousand years, Aristotelian thought suffused the Arab world, natural philosophy, and Christian theology. Aristotle's theory of air and spheres surrounding the earth did not topple until his vision of an egglike universe was broken and discarded by upstarts who were not afraid to challenge the established giants or to anger the authorities. One such formidable thinker was Paracelsus.

A Doctor of Hot Air

I was strange and no one could cope with me.

—PARACELSUS

A somewhat mad and lonely genius who loved to quarrel, drink, and rub the authorities the wrong way, Paracelsus is recognized today as one of the founders of modern medicine. His real name was Philipus Theophrastus Bombastus von Hohemheim. But he gave himself a nickname. *Paracelsus* meant "super Celsus," or "better than Celsus." Aulus Celsus had been a renowned encyclopedist back in the first century A.D.

Like Aristotle, Paracelsus envisioned a grand system, but his was no bedrock for the church. Paracelsus blended magic with minerals, alchemy with chemistry. An alchemist, a homeopath, a chemist, and a mystic, he accepted the four fundamental elements, but considered water the most important. He called water the mother element. In his system, all four elements were wombs and all things emerged from them. Paracelsus also adopted ideas from Chinese folklore, including the notion that soulless but mortal creatures inhabited each element. In air he saw *sylphs,* a word he coined to describe airy essences in the shapes of beautiful young girls. In the earth tunneled gnomes, another creature his inventive mind came up with. In water swam *undines,* or water nymphs. And through fire slithered salamanders. Paracelsus liked to pontificate about his system and virtually everything else, including the stupidity of conventional doctors and how much he charged the rich compared to what he charged the poor. He burned the textbooks of physicians he disagreed with, including Galen, whose grip on medicine during the early sixteenth century rivaled that of Aristotle on science. Paracelsus also moved around a lot, a testament to his conviction that "the universities do not teach all things, so a doctor must seek out old wives, gypsies, sorcerers, wandering tribes, old robbers, and such outlaws and take lessons from them. A doctor must be a traveler."

Paracelsus's travels began in 1502, when he was nine. He had been born in a rural inn and educated by his father, a roadside physician in an era when virtually anyone could hang out a shingle advertising medical services. After his mother committed suicide (she leaped off a bridge into the wild Siehl River within sight of the family inn), the young boy moved with his father to Villach, a busy city in the Tyrolean Alps. In Villach, Paracelsus helped his father with patients. Father and son hiked into the mountains and gathered herbs with the advice of peasants who had used them for centuries. Paracelsus spent time in the local mines famous for their lead, iron, and zinc. He observed metals being transmuted in vats of *aqua regia*, nitric acid and ammonia, and *aqua fortis*, nitric acid. Both aqua regia and aqua fortis dissolved minerals and separated out impurities by the poorly understood process of precipitation. At fourteen, Paracelsus went on the road, showing a proclivity for wandering he would retain his entire life. Like the students he encountered, Paracelsus drank, sold potions, and begged. He refined his rhetoric, his skill at debating. In the first decade of the sixteenth century, a mastery of rhetoric was a ticket to an intellectual future. The young man's tongue, his intelligence, and his acquaintance with the rector Joachim von Waadt, of the University of Vienna, got Paracelsus into the prestigious school in 1509. There he took his era's premed track. He enrolled in the high arts: arithmetic, geometry, music, and astrology. A thorough knowledge of Aristotle's heavenly spheres and their relationship to the earth, together with the stars' influence on the destiny of man and the planets' power over affairs inside the body, went under the astrology curriculum.

But Paracelsus soon rejected the conventional views of stars and planets shaping human health. He dropped out of the university and set off on his own path.

For the rest of his short life—Paracelsus died at forty-nine—he traveled constantly, wrote sporadically, treated patients everywhere, fought the authorities, and found the time for a seven-year odyssey that saw him engage in even more adventures. He cured the king's mother of melancholia in Denmark, operated on Teutonic knights along the coast of the Baltic, and reveled throughout the night with charlatans, mystics, and adventurers in Basil's court in Moscow. In 1524, he traveled back to his father's house in Villach. Wearing an enormous sword rumored to hold the elixir of life in the pommel, or maybe a dose of laudanum, Paracelsus attacked the medical establishment and its so-called doctors, who seldom touched patients, put an ear to their breaths, or asked what they ate. "All they can do is gaze at piss," he said, mocking a popular diagnostic procedure, urine inspection. And he was incensed by the whole idea of the four humors—phlegm, blood, green bile, and yellow bile—which conventional doctors attributed to some kind of stains on a patient's character and treated with bloodletting, sweating, and vomiting, if you could afford the treatments. Few poor people could, one more reason Paracelsus lost his temper with the medical world. He didn't think people should have to pay a lot for the wrong diagnosis and some leeches.

Paracelsus claimed that the real culprits of disease were foreign particles that entered the body through air. Once inside the flesh, they took possession of bodily organs, imposed their own rules, and threatened life. Here was a crude and colorful but correct description of the parasitic concept of disease that dominates medical thinking today. For treatment, Paracelsus devised a system based on salt, sulfur, and mercury, establishing an outline of chemistry's role in fighting sickness. He also prescribed tinctures and herbs, basic ingredients for homeopathy. Although Paracelsus disdained dissection ("You will learn nothing from the anatomy of the dead," he wrote, "it fails to show the true nature, its working, its essence, quality, being, and power") and ignored the popular "dirt pharmacy" of eggshells, toadstools, ostrich feathers, and unicorn horns—really the shavings from narwhal teeth— he did profess a certain fondness for mummy powder. Especially if attained either from a dead saint or from the body of a healthy young person recently drowned or killed in a fall. "Having died in full vigor," he declared, "their flesh, still radiating life power, might regenerate the failing spirits of the sick."

Spirits fit well in Paracelsus's airy view of the world. He believed ghosts, together with sylphs, lived in air. But the ghosts were incorporeal and soulless. He called them *evestrum*. Air also contained a mysterious and vital something that nourished life. He couldn't quite put his finger on what it might be, this *spiritus vitae,* but a lack of it could make you ill. During his lifetime, air fell out of favor as an element in the minds of many of his contemporaries. Paracelsus himself referred to air by its older name, *chaos*. He often said air acted like a vast structure, a temple, a big house. It contained the earthbound realities of life and separated them from the eternal in heaven.

One of earth's realities that Paracelsus noticed in the sick were dark clots of vile matter. They reminded him of similar-looking globs he saw clinging to the sides of empty wine vats. He called the clots *pathological tartars* and gave them a moral dimension. They symbolized "the pain of hell and its curse in the life of man." The ideal way to get rid of the clots was with a tartar solvent. Said to be the best was *liquor alcahest*. But *liquor alcahest,* like the philosopher's stone and the alchemist code for transmuting metal into gold, could not be found in any apothecary next to the dirt pharmacy. Himself a thwarted seeker of the secret of *liquor alcahest,* Paracelsus made do with wine as a spirit-enhancing substitute, although he abstained from sex and wasn't much of a fashion plate, either. He wore expensive clothes until they stank so badly that even the peasants he occasionally preached to wouldn't take them when offered.

In the 1530s, as he preached in mining villages in the mountains surrounding Villach, Paracelsus discovered that many miners suffered from serious respiratory problems. They had dry coughs, were listless, breathless. Paracelsus told the sick miners that they weren't being punished for sins they had committed against the mountain spirits, as many believed, nor were they inhaling evil vapors and spitting them out as dark chunks of phlegm. They were simply breathing dirty air. Get out of the mines and get better. That was his advice.

Paracelsus died in 1541. The impact of his career, like a meteorite splashing down unexpectedly in the collective conscious, began to ripple outward. His manuscripts were suppressed for years because he had angered so many powerful people. But they began to appear in the 1570s: *Opus Paramirum, Paragranum, Philosophia sagax, On nymphs, gnomes, etc.* Like the man himself, the books were hard to decipher and occasionally bewildering, but they helped put order in "a Tower of Babel run by a medley of maguses," as the science historian Hugh Kearney said, describing the worlds of science and medicine in the sixteenth century.

Above Europe at the century's close, the sky remained home to what Paracelsus had called *magia naturalis,* natural magic. Heaven was up there. Astronomers tracked the night skies with astrolabes and quadrants, labeling stars and trying unsuccessfully to calculate their distances from the earth. In 1616, Copernicus's *De Revolutionibus,* which explained the logic of a sun-centered solar system, was added to the index of forbidden books of the Catholic Church. Celestial ether remained the divine glue that kept the stars from falling. For the church, celestial ether satisfied both the spirit and the mind in a world where magic and nature remained unified for most people, where God was seen as both a magician who pulled sunshine out of his hat and an enigmatic scrooge counting up your sins. Religion and science still cohabited. There was friction, but hardly enough to generate heat. Except, of course, when an outraged and outrageous genius, such as Paracelsus, appeared. Or Copernicus. Or Galileo.

But a new movement was brewing. Some natural philosophers began putting their faith in reason as well as in God. They conducted rudimentary experiments and posed new theories. Air, its properties and mysteries, played a major role in this newly emerging world. Natural philosophers wanted to know exactly what air was, wanted to take it apart. Could you tease a beautiful sylph out of a balmy breeze? Could you capture toxic particles that came from distant stars before they became pathological tartars in the body? Could you identify what it was in air that made fire, rusted metals, went in and out of your nose, and was vital to your very existence?

Early Scientists of Air

For want of a name, I have called that vapour, Gas, being not far sev-ered from the Chaos of the Auntients.... Gas is a far more subtle or fine thing than a vapour, mist or distilled Oylinesses, although as yet, it is many times thicker than air.

—JEAN BAPTISTA VAN HELMONT

Robert Boyle, an upper-crust, God-fearing Englishman, was the first sci-entist to experiment extensively with air. Boyle was no mystic. He believed God had created nature like a giant clock and it was man's duty to wind and take care of the clock. As a scientific caretaker for a clock-maker God, Boyle felt that the Creator had built nature based on laws—laws that the new field of chemistry, still a sidebar to alchemy, together with an inquiry into the nature of air, might help him discover.

In pursuit of his goals, Boyle had some things going for him. The thir-teenth child of the Earl of Cork, he was rich. And he was smart enough to hire good help. At Oxford University, where he kept a lab in the 1650s, Boyle's assistant was the inimitable Robert Hooke. The two made quite a pair. In surviving engravings, Boyle looks like the long-haired lead singer of Led Zeppelin, Robert Plant. Hooke, a hunchbacked and irritable hypochondriac (he suffered from a congenitally twisted spine), resembles Toulouse-Lautrec. Boyle's essays would, a century after his death, inspire Jonathan Swift to write *Gulliver's Travels,* adventures during which the narrator spots an island floating in air, "a vast opaque body between me and the sun...about two miles high...the bottom flat, smooth, and shining very bright...inhabited by men." Hooke was bound for the history books as an astronomer, an artist, an instrument maker, a writer, and the first curator of the Royal Society of London, of which he and Boyle were founding members.

As the curator of the Royal Society, in a famous experiment, Hooke once slit the thorax of a dog, inserted a pump down its throat, poked holes in the lungs, and kept the animal alive for an attentive audience. "The Dog, as I expected, lay still, as before, his eyes being all the time very quick, and his heart beating very regularly," Hooke wrote. "But, upon ceasing this blast, and suffering the lungs to fall and lye still, the Dog would immediately fall into Dying convulsive fits; but he as soon reviv'd again by the renewing the fulness of his lungs with the constant blast of fresh Air." Hooke concluded the dog's lungs were only a pump, and "it was not the subsiding or movelessness of the Lungs, that was the immediate cause of Death, or the stopping of the Circulation of the Blood through the Lungs, but the want of a sufficient supply of fresh Air."

Hooke's "first and last Mistress—mechanics," provided Boyle with the technical means to isolate and measure "airs," as gases were then called. Hooke improved the air pump design of the showman von Guericke, who had used his pump to draw air out of a metal sphere, and the advanced pump enabled Boyle to systematically investigate airs and to discover some basics. Boyle showed that air was crucial to life and fire. He quantified the elasticity of air, its potential to be compressed and to expand, and stated the results in Boyle's law: The volume of a gas varies inversely with the pressure. In a famous experiment demonstrating "the spring of air," Boyle and Hooke used a large barometer similar to the one Torricelli had built only twenty years earlier. Placing it in a wooden box almost as tall as Hooke, they tilted the glass tube so that trapped air slipped by the mercury until the mercury balanced on air. Then they poured more mercury into the open end, reducing the trapped air to half its volume and the tube broke. They repeated the experiment with a new barometer tube of "pretty bigness," and with more precautions. "The greater the weight is that leans upon the air," wrote Boyle, "the more forcible is its endeavor of dilation and consequently its power of resistance."

Boyle wrote books with memorable titles: *Occasional Reflections upon Several Subjects; The Skeptical Chymist*, in which he criticized Aristotle's fuzzy logic and Paracelsus's mysticism; *Touching the Spring of the Air and Its Effects;* and *The Christian Virtuoso*. The text seldom delivered the promise of the titles, however. Reading Boyle, except apparently to Jonathan Swift, usually felt like a protracted wade through a gentleman scientist's painfully detailed descriptions of his experiments. In contrast, Hooke, who never seemed to have enough time to accomplish half his goals, wrote just *Micrographia*. A veritable bouquet of scientific insight and observation from a shriveled, odd little genius, it remains a genuine masterpiece.

During their lifetimes, neither Boyle nor Hooke worked with "gases," because air was not yet recognized as a blend of invisible substances that were different from one another. Air was air. Early in the seventeenth century, Jean Baptista van Helmont had coined the word *gas,* but it never gained common usage until much later. Of course, the term *scientist* was not yet popular, either. The word *scientist,* meaning "a learned man," was probably adapted from the Italian *scienziato* and only became commonplace in the 1840s. Until then, most men of science were known as natural philosophers.

John Mayow, another Englishman of wealth and privilege, was a contemporary of Boyle's and Hooke's. Mayow studied and experimented with respiration. He concluded that breathing brought fine nitrous particles from the air into your blood. Once in the blood, the particles joined sulfur particles. Their interaction caused a gentle fermentation that kept you alive. The nitrous particles also fired the heart, Mayow decided. They united with the animal spirits already circulating in the blood in a fresh and novel way, setting off controlled miniexplosions, a sort of inner fireworks of the heart that made it go. Mayow's theory had its precedents in Paracelsus's idea that something mysterious but vital, a *spiritus vitae,* existed in air, and in the more recent work of Alexander Seton and his follower Michael Sendivogius. Just as sulfur and niter (potassium nitrate) could be mixed to make gunpowder, went the reasoning of Seton and

Sendivogius, so could a spirit of sulfur and a spirit of niter combine and cause the little explosions that Mayow related to human vitality. The ideas of both Seton and Sendivogius, and of Mayow, were in turn linked with Aristotle's ancient theory of the earth's exhalations. Real sulfur and niter might be used for making gunpowder on earth, but it was the spirits of both that caused earthquakes beneath the ground and microscopic blasts in the heart fibers, making it pump.

This was a breakthrough idea. Wrong, but imaginative, a fusing of physiological, geological, and mystical thinking. And some kind of combustion was responsible for muscle contraction, although it would be more than a century before oxygen's role would be known. Mayow, who pushed the understanding of respiration long before anyone else, also figured that death did not come from an absence of the nitrous particles so important to air and life. Death came because the animal spirits in the bloodstream couldn't make their rounds once the heart stopped. When that happened, life lost its "effervescence," wrote Mayow. And it was over.

Of the scientists experimenting with air, breath, and fire in the late seventeenth century, one of the most quirky was a preacher named Stephen Hales. His main focus was vegetables, though he occasionally experimented with dogs, oyster shells, decomposed blood, and himself. Another mechanist, Hales put great faith in science as a glimpse into "a beautiful and well-regulated world" shaped by God. He wasn't afraid of getting his hands dirty and, from what others said about him, was almost insufferably serene. Horace Walpole, the English writer, described Hales as "a poor, good, primitive creature." One can imagine Hales, after he had enclosed some hapless puppy in a bell jar for an experiment that almost took its life, consoling the little critter afterward and giving it a snack.

Hales had his lab by his church in Teddington. His experiments were remarkable for their ingenuity. In a series of rebreathing experiments, he used himself as a guinea pig. He fit a homemade mask, complete with valves, tubes, and a bladder, over his nose and mouth to see if by inhaling his own breath he'd keel over like a puppy under a bell jar. Wobbling around after a minute, he had his answer. He got up off the floor. Deciding that his breathing had "vitiated" the air in the mask, robbing it of something vital, its mysterious *spiritus vitae,* Hales modified the mask by inserting several diaphragms. Each diaphragm was made of a piece of flannel stretched over wire and saturated with salt of tartar, probably a blend of grape juice from wine casks and yeast, which made a dark reddish sediment. Hales put the primitive mask over his face again. This time he stayed on his feet for almost eight and a half minutes. Hales knew his Mayow and believed that sulfurous fumes, a by-product of combustion going on inside the body, had to leave with the breath. He deduced that the salt-saturated flannel diaphragms filtered the sulfur fumes as they left his mouth. This helped the air in the bladder remain free and elastic, which was to say healthy and good, untainted by sulfurous waste. The real explanation, of course, was that Hales had overdosed twice on his own carbon dioxide. But on the second occasion, the salt of tartar, which absorbed some of the gas, kept the carbon dioxide volume low for a while. Only after seven or eight minutes did the gas accumulate enough to poison the preacher and cause alarm in the breath command center of his medulla.

An anomaly even by the standards of the early eighteenth century, Hales wrote nothing about his experiments, which ranged from the trivial to the brilliant and were often confused, until he was fifty. His book *Analysis of Air* had a major impact on chemistry. He mapped the similarities between blood circulation in people and sap flow in plants. He proved that plants breathed. From experiments with electricity, he theorized that the animal spirits in man, which powered the muscles, including the heart, carried an electrical charge.

Isaac Newton was Hales's hero, as was "the excellent Mr. Boyle." Newton had left Cambridge University in 1696, the year Hales entered. As a theology student, Hales probably watched Newton be knighted by Queen Anne in the King's College chapel of Trinity College in 1705. It was at Cambridge that Hales made a lead cast of a dog's lungs with a fellow student, a future doctor, not long before he took up his position in Teddington. There, he tended his congregation, plants, test animals, inventive apparatus, and fertile imagination for the rest of his long life. He died in 1761, at the age of eighty-four. It was Newton's laws of attraction and repulsion that Hales later applied to the electrical properties of air. An atomist, Hales believed invisible particles in air gave it the spring Boyle had identified, made air elastic. The particles' forces of attraction overcame those of repulsion to shrink air, and those of repulsion overcame those of attraction for the volume to expand.

Like Robert Hooke, Thomas Hales was also a masterful tinkerer. He designed flasks to capture and measure air. His pedestal apparatus, basically a table in a tub with a large bell jar over it to seal the interior, proved ideal for focusing sunlight through a magnifying glass and burning things, or for isolating a test animal in a controlled environment from which Hales could siphon away the air. He also invented the pneumatic trough, which would be modified and used by chemists everywhere for several centuries. The trough was a bell jar with a long tube directed down into a basin of water and was connected to an "inverted chymical receiver," also filled with water. Gases produced from burning or fermentation in the bell jar passed through the tube and through the water into the receiver. There the gases condensed as they cooled, and subsequently could be analyzed. The ingenuity Hales brought to an experiment was important. Theories were great, but he was living in the dawn of the age of proof. Once a successful experiment is over, it is too easy to think of its logic and sequence as almost preordained. But often the sequence would not have been possible without a special and new piece of apparatus. Like Hooke's improved air pump, Hales's pneumatic trough, in particular, provided his contemporaries and successors with the tools to manipulate gases in ways previously unknown. Without these tools, the making of new theories would have stalled.

Thomas Hales didn't do much analyzing of the airs he collected from saltpeter, horn scrapings, human bladder stones, and on and on. He did, however, use his new apparatus for exploring ideas he had about air's being elastic in the atmosphere, but "fixed" in dense matter, ideas that supported Georg Ernst Stahl's infamous phlogiston theory of air and fire.

Phlogiston: A Wacky Theory with a Long Shelf Life

The phlogiston theory was by no means bad science—it was simply wrong.

—LEONARD C. BRUNO, *TRADITIONS OF SCIENCE*

Boyle, Hooke, Mayow, and Hales were all British. But the foremost late-seventeenth-century scientist of air was German.

Georg Ernst Stahl was a court physician at Weimar before he joined the faculty of the University of Halle, a showcase of learning established by Frederick I of Prussia, in 1694. Intolerant, misanthropic, and smart, Stahl liked obscure words and convoluted explanations. For his theory explaining the age-old mystery of fire, he chose the word *phlogiston* from the Greek, which meant "to set on fire." For Stahl, phlogiston was something real, not just a quality, like warmth or light. He erected his theory of phlogiston from research done by Johann Becker, a contemporary who said that all substances contained three types of air: the vitrifiable, the mercurial, and the combustible.

Stahl also defined something he called the *anima,* a force he claimed got you up in the morning, gave you purpose, and steered your movements. He said your movements were a result of the mechanical functioning of particles in the body in obedience to the *anima's* commands. A formidable physiological idea, Stahl's *anima* seemed to harbor the soul. The *anima* was an immaterial driver in a material body, a ghost of sorts in a mechanical contrivance with momentum and direction, if no real map, or at least not one Stahl was able to identify. Looking back from the early twenty-first century, we can see that Stahl's *anima* has greater relevance than his more famous phlogiston theory. A formidable, if shaky, physiological umbrella, under *anima* we can line up mind and body controls, from respiration to digestion to cognitive functions and decision making. All of them are now gathered under neuroscience.

For Stahl, the *anima* was really a secondary theory, though. His big interest was in phlogiston. Today, phlogiston theory is one of those ideas of the past that makes us scratch our heads about science. Yet once scientists had decided that both respiration and fire were combustion, it made sense to look for a common substance in all combustible things, rather than for a common process, oxidation, which only became understood much later. In 1700, phlogiston theory sounded plausible. Clinging to it were some unanswered questions, but Stahl brushed them off as unimportant—the experimental method had not gotten *that* rigorous yet.

If you'd had phlogiston theory explained to you in front of a hearth at the University of Halle three hundred years ago, it might have gone something like this: The flame you see in the hearth, the dancing, flickering tongues of yellow tipped with blue—that's the essence of phlogiston. The phlogiston is in the wood. It's in all materials that burn. Phlogiston is set free through combustion. It leaves behind ash. Ash is the residue of a combusted material after its phlogiston has moved into the air. The flames you see are the visible signs of invisible phlogiston at work.

Here, grip this sword. Hold the blade over the flames. See the black powdery substance collecting on the blade? That's close to pure phlogiston. The air all around us assists the flame's motion and aids phlogiston's escape. The air holds phlogiston, as you are holding my sword. It's phlogiston's accomplice in a way because it takes up the phlogiston and makes it possible for flames to continue. Without air, phlogiston can't get free and there's no fire.

The theory did a better, if no less accurate, job of explaining the different shades of blood. Since combustion was thought to take place either in the heart or in the blood vessels because of friction (the faster your heart beat, the more heat the blood created), the observation that blood in arteries was bright and in veins darker validated phlogiston theory nicely. Arterial blood contained phlogiston, and venous blood didn't, because its phlogiston had somehow been burned inside the body

In our era of scientific peer review, a half-baked theory like Stahl's wouldn't pass muster. But in the early eighteenth century, it was a blockbuster, clever and complex and intellectually satisfying. Better than any theory before it, phlogiston explained many aspects of one of the oldest mysteries, fire. It was so persuasive that it stalled advances on other theories of combustion and respiration for most of the century. Ironically, phlogiston theory was the exact reverse of the modern theory of combustion. Oxygen is taken up by things that burn, rather than phlogiston given off. Yet Stahl was right about one crucial thing. There was a transfer going on, a concept he grasped and one soon associated with all chemical reactions. Fortunately, the theory left a lot of unanswered questions. For instance, how did phlogiston get into wood in the first place? If too much phlogiston in the air put out a fire, why didn't a huge forest fire suffocate itself? Inconsistencies and gray areas eventually added phlogiston to the junk heap of science. Gray areas stimulate a theory's deconstruction and help answer the question What's really going on here? For answers, scientists must do new experiments, devise alternative hypotheses, and postulate new theories. If the theory being challenged is well established, like phlogiston or an earth-centered universe, those who doubt it endure a lot of criticism. Challenging the status quo demands courage. It's a slippery slope up which the newcomer often makes his or her way alone. But in the evolution of thought, both the giants and the pip-squeaks alike can take solace in the personal motto of Georg Ernst Stahl: *E rebus quantumcumque dubiis quicquid maxima sententium turba defendit, error est:* Where there is doubt, whatever the greatest mass of opinion maintains ... is wrong.

*

Georg Stahl, Stephen Hales, John Mayow, Robert Hooke, Robert Boyle—all were explorers in an age of exploration. Not explorers of the seas, like Sir Francis Drake, or of the stars, like Johannes Kepler, but of the invisible. Of air, with its ghosts of dead babies, spirits, particles, phlogiston, and celestial ether. With its godly light and its evil vapors. They faced questions as difficult to answer as the parallax of stars and the longitude of ships at sea. With primitive equipment and private hunches, they experimented and philosophized in a rough and stumbling manner toward common agreement.

Continually they asked, What is this stuff I can swing my hand through with ease yet on which birds can glide and butterflies rise? What is it in the air that flows through leaves and lungs effortlessly, filling their circulation systems and sustaining life? What is given off when a person exhales, as Hales had into a homemade contraption, and almost passes out?

By the early eighteenth century, some of air's mysteries had been cracked. Boyle's law explained the pressure of gases. Air's sensitivity, its ability to refract light, its expandability—all were fairly well understood. God himself, not the "God particle" for which physicists would abandon the hunt three hundred years later, remained intricately related to air and light, to the atmosphere's vast scale and tantalizing mysteries.

In Aristotle's day, the Greeks had thought breath was the essence of life. *Pneuma,* the root of *pneumonia* and *pneumatic,* came from the Greek, meaning "spirit." By the eighteenth century, scientists were sure air was the essence of life. Air may not have gushed from the naval of the man with a thousand heads and a thousand eyes, as was once believed in India, but *where* did it come from? Where did it go? How did it work? What was it?

Three centuries later, we know the molecular design of oxygen. We know oxygen carries an electrical charge that helps it bind easily with the molecules in plants, minerals, fluids, and gases in a constant, dancing exchange of energy potentials. We know, as the physicist George Leonard wrote in *The Silent Pulse,* "At the ultimate heart of the body, at the heart of the world, there is no solidity. There is only the dance." We know that what Hales breathed back in through his mask, collected in his pneumatic trough, and called "fixed air" was carbon dioxide. We know the Big Bang created our universe, setting adrift particles and debris that eventually begat oxygen, life, and humans, about fourteen and a half billion years ago. As for what preceded the bang, or even if it was a bang—couldn't it have been a whimper, or a wail, like a newborn child exercising her lungs after that first breath, which demands an incredible jolt of energy?—we're often as deeply in the dark as the scientists of the seventeenth century.

What drove them to dismantle Aristotle's world of marvels despite the risks of being ostracized and attacked, and to replace it with one of mechanics, remains unchanged, however. It's what always moves scientists to postulate and dream, whether about sylphs brushing the backs of their necks or about rays that can see bones inside the body or about the loss of protective ozone in the stratosphere. They don't know what's happening. And they want to know.

5

Air, God, and the Guillotine

The man who first really understood air was beheaded. With the drop of the blade all his respiratory functions ceased. Or almost. It took a few milliseconds for the sensors in the brain stem of Antoine Lavoisier to register their imminent demise. Suddenly, no oxygen, the element he had named in 1779, was entering his bloodstream. A whole set of disastrous events was set off on the molecular level. Lavoisier's brain, relying on oxygen and glucose, was quickly exhausted. ATP, the energy building block found in every cell, decreased. Glutamate, a neurotransmitter, increased rapidly in his brain, becoming toxic at higher concentrations. No one knows how glutamate does that, but if God is an engineer it seems a major design flaw. At any rate, Lavoisier's blood pressure fell fast, and his blood flow slowed to a trickle. The Greeks would have said his *pneuma*, his spirit, had ceased movement. The judge who had tried Lavoisier that morning would probably have said good riddance to bad rubbish. Lavoisier's friend, the astronomer Joseph-Louis Lagrange, disagreed. "It took them only an instant to cut off that head," he said, "and a hundred years may not produce another one like it."

Marie Paulze and Antoine Lavoisier during a rare relaxed moment.
The father of modern chemistry, Antoine Lavoisier owed much to
his charming wife, whom he rescued from marriage to an old count
when she was fifteen. Marie Paulze sketched many of Lavoisier's
famous experiments and was his English translator.

Air Rivals

I have discovered an air five or six times as good as common air.

—JOSEPH PRIESTLEY

WHEN ANTOINE LAVOISIER was executed in Paris during the Reign of Terror, his rival, an English theologian, freethinker, and scientist named Joseph Priestley, was sailing to America on the *Samson.* It was not a happy voyage. Recently burned in effigy, caricatured in newspapers, kicked out of the Royal Society, and castigated from pulpits across England, Priestley was on the run. He'd been on the run, more or less, for three years, ever since his lab, house, and church, the ultraliberal New Meetinghouse in Birmingham, had been sacked and burned on July 14, 1791, two years after the fall of the Bastille. The persecution had little to do with Priestley the scientist and much to do with Priestley the philosophical theologian who attacked the Anglican Church and continued to support the French Revolution long after his fellow countrymen thought its high ideals drowned in bloodshed. Now sixty-one, with a wife and three sons, Priestley was going into exile. He was headed for rural Pennsylvania, for a region that Samuel Taylor Coleridge, then a twenty-something romantic poet with opium-stoked dreams of himself escaping to America, mistakenly called "peaceful Freedom's undivided dale."

Antoine Lavoisier and Joseph Priestley were dramatically different types of men but with a common interest in the mysteries of air. Lavoisier was the quintessential insider. An aristocrat with money, he enjoyed prestige and was the envy of his French colleagues. Priestley was the quintessential outsider. A liberal driven by passion and single-mindedness, he put his faith in God and always managed to find the courage to speak out against oppression and to champion free expression.

The two men were born a decade apart: Lavoisier in Paris in 1743, and

Priestley in Yorkshire in 1733. Lavoisier was a child of the French Enlightenment, an era enamored with philosophy, the glory of the individual,
and the power of reason. Priestley was a sickly boy raised by his spinster
aunt Sarah after age six, when his mother died. Sarah's home was a gathering place for Dissenting preachers, many of whom saw themselves as
an enlightened elite in an oppressive nation and whose orations from pulpits mesmerized audiences across England.

When Lavoisier turned ten, the indefatigable encyclopedist Denis
Diderot proclaimed that chemistry and experimental physics, two relatively new branches of science, were where the action was. Lavoisier
soon called himself a young physicist. He may have dreamed of being
another Ben Franklin, a pioneer of the new physics, or of coming up
with "a theory of the earth," an explanation of the forces that shaped the
planet. A theory of the earth was *the* hot intellectual puzzle of the era,
the equivalent to figuring out and irrefutably explaining global warming
two centuries later. Lavoisier attended Mazarin College, an institution
famous for the high caliber of its instructors, and studied the classics, literature, the sciences, mathematics, and philosophy.

Across the English Channel, Priestley studied in local parish schools,
supplementing what he learned with readings at home in natural philosophy, mathematics, and ancient languages. At nineteen, Priestley
enrolled at the Dissenting Academy at Davantry, a school for noncon-
formists who rejected the Anglican Church. There he refined his interests
in utilitarian theology and the experimental sciences. He graduated with
strong opinions against slavery and for women's rights. Priestley's first
position in Suffolk was short-lived, because of his opinions. But his second assignment to a liberal church in Cheshire proved a better fit. There,
Priestley began scientific experiments with air and electricity. In 1761, he
shifted again to the Dissenting Academy at Warrington, where he was
ordained a minister and married Mary Wilkinson, the daughter of a
Welsh ironmaster.

On trips to London in the mid-1760s, Priestley had the good fortune
to meet Benjamin Franklin and to begin a relationship that Lavoisier

might have envied. The intense young minister and the crusty diplomat, then in England to argue the interests of the Pennsylvania colony before Parliament, hit it off. Franklin loaned Priestley books. The American statesman encouraged Priestley to write *The History and Present State of Electricity*, the first book to mention Franklin's experiments with kites in thundershowers. Shortly after the book's publication, though, Priestley lost his interest in electricity and focused on the chemistry of various airs.

The switch may have had something to do with stench. The air in London, as in Paris, Rome, Madrid, and many European cities, was polluted by the rank odors of its inhabitants to a degree unfathomable to people today. In *Perfume,* Patrick Suskind's novel about a Parisian murderer in the eighteenth century who collected the scents of young women in a dissection process that would have been at home in the anatomy theater of Padua during the late Renaissance, the author presented a vivid list of offensive odors: "The streets stank of manure, the courtyards of urine, the staircases stank of moldering wood and rat droppings, the kitchens of spoiled cabbage and mutton fat; the unaired parlors stank of stale dust, the bedrooms of greasy sheets, damp featherbeds, and the pungently sweet aroma of chamber pots." Peasants lived in dark, dingy, smoky houses with fewer windows than animals. The upper classes, although they had more windows and higher ceilings, doused themselves liberally with perfume to mask the smell of unwashed armpits, greasy hair, stinky shoes, and rotten teeth. The royals were no better. Louis XIV's servants washed his clothes in spices and released doves drenched in perfumes at the king's dinner parties; they flew around the heads of his smelly guests. Streets and boulevards in European cities were widened to allow the miasma of odors and fumes to rise above the rooftops instead of clinging to the cobblestones and brick of narrow lanes and allies.

In Paris, Lavoisier grew up in a world of perfume and sweat that Priestley encountered primarily on his trips to London. Whether the odors swayed Lavoisier, who normally doused his clothes with perfume, is not known for certain. But during the 1760s, he rejected his father's entreaties to pursue a career in law and officially became a new scientist whose interests were all over the map. They included botany, chemistry, and geology. In 1767, Lavoisier accompanied the great French geologist, Jean Guettard, on a long trek through the Vosges to collect rocks and make observations for a mineralogical atlas of France. By this time, geology was telling a far different story about the origins of the planet than what Genesis told. In fact, some geologists said that if you wanted an accurate picture of the past, you should read rocks, not the Bible. They said the world had not begun on October 23, 4004 B.C., as the British Archbishop James Ussher had declared a century earlier. And these new geologists argued that the Great Flood had not been caused by a passing comet on November 18, 2349 B.C., as argued passionately by Newton's successor at Cambridge, William Whiston. Dismissing scriptural geology as charming but unscientific, geologists like Guettard claimed that the earth was a stratified library of knowledge to be dug into, analyzed, and read, just like a book.

Lavoisier, though he liked Guettard, found hiking in the mountains and gathering rocks less than satisfying to his urban tastes. Back in Paris his short-lived enthusiasm for geology was replaced by a new fixation. He got hooked on chemistry. In particular, he was mesmerized by "the matter of fire."

Like many before him, Lavoisier found himself drawn to Aristotle's four fundamental elements—fire, air, earth, and water—even if they didn't seem quite as fundamental as they had been. By the mid–eighteenth century earth had been subdivided into "earths," which included the seven sacred metals; some bastard aberration metals; non-metals such as sulfur, which arose from the underworld out of volcanoes; carbon, which appeared on charred wood; and acids and bases. Water, which would fascinate Lavoisier all his life, was still considered to be fundamental. Water might be the universal solvent, Lavoisier thought, the substance continually reshaping the planet, the true key to the theory of the earth. Air also remained fundamental, but with water vapor and gases mysteriously locked in it. Of the four classic elements, fire retained the most mystery. Phlogiston theory, still the popular explanation of fire, did not satisfy Lavoisier's growing need for scientific proof, however.

By 1772, Lavoisier's experiments with fire boosted his already considerable confidence to the point that he predicted a revolution in physics and chemistry, with himself as its leader. In England, Priestley was well established as his most serious competitor for such leadership, even though the men knew little of each other. In the narrow world of eighteenth-century science, there were good journals, along with reports of lectures, books on new topics, and encyclopedias that described pivotal experiments, but all too often they were not available for months, even years, after the fact. France and England both had academies of science, the French one funded by the state (a perfumed Louis XIV had founded it), its English counterpart paid for by its members. Still, there was no easy way to stay abreast of the latest developments outside one's circle of associates. Word about exciting work often got around first through the grapevine, by letters or visits of foreign scientists talking about what was hot back home. One hot item in 1772 was *Directions for Impregnating Water with Fixed Air, in Order to Communicate to it the Peculiar Spirit ande Virtue of Pyrmont Water,* a pamphlet Priestley published to boost his reputation. As far as advancing the knowledge of air, it wasn't much, but today, when you pop open a can of Coke, a bottle of quinine water, or any other carbonated soft drink, you can raise a toast to freethinking Joseph Priestley as you hear the fizz of bubbles escaping. The carbonated soft drink industry can be traced back to his experiments with the air wafting off a brewery next door to his vicarage in Leeds.

Basically, Priestley captured the pungent gas, called fixed air, rising off the devil's brew and combined it with normal water, producing carbonation. The same year that he published his pamphlet on fizz chemistry, he also read from a larger work in progress at the Royal Society in London. *Experiments and Observations on Different Kinds of Air* went far beyond the origins of Coke and Perrier. It detailed Priestley's years isolating and identifying new gases and predicted photosynthesis. Always intrigued by "good air," Priestley made a prophetic linkage between the breathing of animals and plants and its impact on good air, theorizing that the "injury which is continually done to the atmosphere" by animals was "repaired" by plants.

Before Priestley's experiments, only three gases—common air, fixed air or carbon dioxide, and hydrogen, which had been discovered by Henry Cavendish—were known. One of Priestley's new finds was laughing gas, or nitrous oxide. No record exists of whether his first whiffs made Priestley laugh, but I hope so. With the weight of societal disapproval accumulating on his nonconformist shoulders, he needed it.

Some idea of the alternating currents that charged his life is suggested by events in 1772. First, Priestley was asked by Captain James Cook, the great circumnavigator, to join his crew for a trip around the world as the ship's astronomer. Onboard he might have analyzed another new air, the one given off by the vitamin C–rich sauerkraut that Cook hauled along in barrels to fight scurvy. But Anglican power brokers, irked by Priestley's radical theology, bumped him from the job. Despite the snub, Priestley advised Cook to impregnate the ship's water with fixed air to make it bubbly and more palatable as a means of washing down all that sauerkraut. Then, in December, Priestley received another invitation. This one was to join the staff of the Earl of Shelburne as the in-house intellectual at Bowood, the Shelburne estate in Wiltshire, with minor duties as a tutor for the children. Priestley accepted. So instead of high adventure on the sea, he suddenly had a patron, enough leisurely time to work at great length on *Experiments and Observations on Different Kinds of Air,* which eventually numbered six volumes, and a well-equipped lab with ceramic crucibles, mortars, pneumatic troughs, and the latest burning lenses ground by William Parker and Sons of London.

The experiment that elevated Priestley into the pantheon of famous chemists occurred on August 8, 1774, during his tenure with the Earl of Shelburne. Like an inordinate number of science breakthroughs, it came about almost by accident. Priestley took a twelve-inch burning lens and focused the rays from the summer sun on red mercury oxide, which floated on top of mercury in a closed vessel that he'd flipped and placed in a pneumatic trough, a device that allowed various gases to be released from heated materials and collected separately for analysis. Priestley expected to discover another new gas, and he did. "But what surprised me more than I can yet well express," he wrote, "was that a candle burned in this air with a remarkably vigorous flame . . . and a piece of red-hot wood sparkled in it." The new gas was "five or six times as good as common air," a "good air," an air that not only made a candle blaze more brightly than normal but allowed a mouse to survive longer breathing it than breathing regular air.

A qualitative researcher more interested in observations than theories, Priestley was used to surprises. But this one stumped him. At first he thought the gas was nitrous air, which he'd already identified. Later he called the new gas *dephlogisticated air,* reasoning that this air contained no phlogiston. That was why it allowed an intense flame to continue for so long; the air had room to absorb greater volumes of the invisible essence of combustion.

Some marvelously cockeyed thinking was going on. Yet it was sensible, given the popular explanation of combustion, the phlogiston theory developed by the German doctor Georg Stahl less than a century ago. Burning released the invisible phlogiston and deposited the substance in the air. Flickering tongues of blue-tipped flame were the essence of phlogiston escaping, a sign of the invisible at work. According to Stahl's theory, air assisted phlogiston's escape, accepted it until nothing was left behind but ash. Without air there could be no fire, Stahl had said, because without air phlogiston was homeless.

Priestley still bought Stahl's half-baked theory, as did most of the scientific community, though not Lavoisier. While testing his new, possibly dephlogisticated air, Priestley did something few chemists do anymore. He sniffed his new discovery through his nose and reported, "The feeling of it to my lungs was not sensibly different from that of common air, but I fancied that my breast felt peculiarly light and easy for some time afterwards. Who can tell but what, in time, this pure air may become a fashionable article in luxury. Hitherto only two mice and myself have had the privilege of breathing it."

Worry about future addicts of pure air was not as far-fetched as it may sound. Late-eighteenth-century England had a growing drug problem, mostly from cheap opium imported from Turkey. Coleridge imbibed his share, as did his friend Thomas DeQuincy, who wrote *Confessions of an Opium Eater*. Opium tinctures—the one Coleridge and DeQuincy preferred was laudanum—could be bought at the corner apothecary; they were cheap and totally legal. Wary that his new air, which made him feel "light and easy for some time," certainly far from his normal state, might result in addictions, Priestley penned a cautionary note. "Though pure dephlogisticated air might be very useful as a medicine," he warned, "it might not be so proper for use in the usual healthy state of the body; for, as a candle burns out much faster in dephlogisticated than in common air, so we might, as may be said, live out too fast, and the animal powers be too soon exhausted in this pure kind of air. A moralist, at least, may say, that the air which nature has provided for us is as good as we deserve."

A few months after the discovery of the unusual air, which both pleased and perplexed him, Priestley went on a trip to the continent with the Earl of Shelburne. In Paris he met Antoine Lavoisier.

Air and the New Chemistry

Phlogiston is imaginary and its existence . . . a baseless supposition.

—ANTOINE LAVOISIER

For the last two years, Lavoisier had been reading everything he could about air. His wife Marie Paulze, whom Lavoisier had married when she was fifteen, rescuing the girl from the clutches of a lascivious old count, was his assistant. Fluent in English, she translated reports on the latest experiments taking place in England, where much of the action in chemistry was happening. So Lavoisier knew about Henry Cavendish and the inflammable air he'd discovered. He knew about Joseph Black, whose most notorious experiment involved fifteen hundred Glasgow churchgoers who allowed their collective breaths to be gathered high overhead by an air duct during a ten-hour devotional. The results, Black demonstrated, was lots of fixed air, or carbon dioxide, the same gas given off by the fermenting beer next to Priestley's former vicarage in Leeds. And Lavoisier knew about Priestley, the recent recipient of the prestigious Copley medal from the Royal Society, an honor bestowed in other years on such notables as Ben Franklin, Cavendish, and Captain Cook. But Lavoisier did not know about Priestley's latest find, an air five or six times as good as common air.

Lately, the matter of fire had been keeping Lavoisier especially busy. But he was always busy. Too busy, his critics said. In addition to his chemistry experiments, which Lavoisier usually conducted before breakfast or after dinner, he worked as an accountant, a social scientist, and a functionary of the General Farm, a money-collecting service that pleased the king but was "a million-headed leech" to farmers and peasants bearing the taxes. His latest reading and experiments had heightened Lavoisier's understanding of fire and increased his skepticism about phlogiston. How, for instance, could chalk be turned into lime with intense heat and the residue weigh more than the original chalk if phlogiston were given off during burning? *A body cannot become heavier unless it joins itself to some perceptible matter.* Fire did weaken the powers in most materials, Lavoisier had concluded, and the vapors given off during combustion were considerably elastic. As for the air people breathed, he had a hunch that inside the lungs it transformed into "a special kind of fluid combined with the matter of fire."

At a dinner party in Paris, Lavoisier, the intense showman with the young, fiery-eyed wife and big dreams of revolutionizing chemistry, met Priestley, the nonconformist outsider mulling over the mystery gas he'd isolated only months before. No record exists of their conversation. But as Thomas Kuhn puts it tactfully in his famous *The Structure of Scientific Revolutions,* "possibly as a result of a hint from Priestley," Lavoisier soon isolated oxygen, then launched himself on the revolutionary path he'd been contemplating for years.

Historians today still argue about who actually discovered oxygen: Priestley, Lavoisier, or Karl Wilhelm Scheele, a Swedish apothecary. Scheele, an obsessed workaholic who lived to experiment and died young (he may have poisoned himself with his habit of inhaling and tasting his own concoctions), actually isolated oxygen first, in 1771, three years before Priestley. But Scheele never wrote about his discovery to make it known in the science community. When Priestley separated oxygen from mercury oxide and isolated it in his pneumatic trough, he misunderstood and mislabeled it. After 1775, having isolated oxygen and identified it, Lavoisier laid claim to having codiscovered the gas. That was a bit of a stretch, but given other stretches he made, such as his claim that he discovered the composition of water when Henry Cavendish beat him to it, the boast fit the man. Kuhn argued that Lavoisier's appropriation of the mantle, or comantle, of the discovery of oxygen, has a basis in truth because he knew what he'd discovered, not just that he'd discovered *something.*

Whether by dint or by accident, when Priestley mentioned his latest discovery to Lavoisier at the dinner party in Paris, the Englishman gave his alert colleague the equivalent of a missing key. It was not the key to a theory of the earth, which geology seemed close to finding, but for a coherent theory of chemistry, a field still associated with alchemy and magic.

Lavoisier wasted little time in duplicating Priestley's experiment. In March 1775, he produced "the dephlogisticated air of M. Priesley [sic]." He theorized it was "the purest part of the air." Soon he realized air was a blend of gases. In 1779, Lavoisier named the purest part of the atmosphere *oxygen* on the mistaken conviction that the gas was fundamental to the creation of acids. In French, *oxygene* means "begetter of acids."

Back in England, Priestley continued to experiment with gases but receded from the forefront of chemistry. He never devised a grand scheme to explain his discoveries or to place them in a larger context. Meanwhile, Lavoisier honed his systematic thinking and assumed center stage. Or, more accurately, center arsenal.

Lavoisier's lab was located in the Paris Gunpowder Arsenal, where he and Marie lived for free because he now had yet another job; he had become a commissioner of the Gunpowder Administration. Lavoisier built a lab in the arsenal and filled it with the most up-to-date apparatus. Weekly dinner parties hosted by the intense and wiry chemist and his red-haired and vivacious partner attracted Parisians and visitors, including scientists from around Europe. After dinner Lavoisier would indulge the guests with a new kind of entertainment, the flamboyantly executed experiment.

It was the height of France's involvement in the scientific revolution, and the French clergy, nobility, and merchant classes flocked to futuristic experiments as zealously as to theater. In the right hands, science *was* theater, a little arcane but thrilling. Its props were furnaces and glassware and odd-looking instruments. Reason supplied science with the power to forecast and explain. Jean Jacques Rousseau, the writer and philosopher, was offended by most intellectuals' willingness to get down on their knees before science and its logic while discarding passion and the intuition. But Lavoisier rejected Rousseau in favor of Voltaire. Reason, the accuracy of scales, the conservation of matter—they were Lavoisier's guides. And drama was his drug. Lavoisier loved to perform as valets took notes for their masters, perfumed poodles dozed on the laps of ladies, and the sound of inlaid snuff boxes opening could be heard as gentlemen took nasal hits of stimulating particulates. In the Gunpowder Arsenal, before his select audiences, Lavoisier, a performer with shoulder-length hair and a hint of a smile at the corners of his lips, did his magic, sometimes alone, sometimes with assistants. Off to the side, taking notes and making sketches, his wife blazed like selenite with her red hair and bright blue eyes.

Lavoisier's meticulous handling of materials, his willingness to interpret what he found in fresh ways, his vision of a revolutionary new chemistry, and his flair in an era of high drama all buttressed his scientific work. But what really made it stand out was his subdued ferociousness. He was a ferret at heart, relentless, impatient, willing to go for the jugular at the opportune moment. And he was tireless. Only months before meeting Priestley, for example, he had upended alchemy's colorful lies about transmuting water into earth with fire. He'd boiled water for a hundred straight days in a pelican, a glass recipient shaped like a bird, and demonstrated that the residue inside the glass was not water transformed into earth by fire but merely the dissolved inner surface of the pelican. His initial try at isolating a gas from mercury oxide continued for twelve days and twelve nights. He put a small amount of mercury in a retort, a curved glass container whose long neck was connected to a bell jar immersed in a trough of mercury. He gently heated the mercury day after day. Gradually, it turned to ash. The level of mercury in the bell jar rose, indicating that the air in the jar was being reduced. When the mercury was completely burned, Lavoisier measured the air consumed, which was around 20 percent. Then he meticulously collected the red powder from the burned mercury, reheated it intensely in a closed system, and collected the gas it gave off. The gas was the 20 percent of the air that had been consumed in the first stage of the experiment. In other words, it was oxygen.

Lavoisier's experiments grew in magnitude as he become more adept at isolating and handling gases. In June 1782, he hauled a pneumatic chest filled with *oxygene* to the Academy of Science, aimed it into a concave lump of charcoal, ignited the flow, and melted platinum, a new metal. Seventy-six-year-old Benjamin Franklin, now in Paris negotiating treaties for the new United States and in the audience, was impressed. The Frenchman's flame-thrower produced the "strongest fire we yet know," Franklin said.

A larger, more complex apparatus with two pneumatic chests connected to a single hose was soon ready for testing. With it Lavoisier hoped to create an even hotter flame with which he could synthesize an acid from oxygen and hydrogen. He believed the synthesis was possible, because he was convinced all acids contained his recently named element. But before the inauguration of the advanced flame device, a visitor to Paris, the assistant of Henry Cavendish, told Lavoisier that Cavendish had synthesized water. He'd done it by detonating hydrogen and oxygen, scuttling yet another of Aristotle's elements by proving conclusively that water, like air, was not fundamental at all. It was a compound. Amazingly enough, the compound was two invisible, odorless gases, both of which could be found in the atmosphere.

Like Lavoisier, Cavendish's star was on the rise. Arguably the strangest air scientist of them all, Cavendish made the typical English eccentric seem normal. He had inherited a fortune at forty and lived a sequestered life, handing notes to his housekeeper, talking to hardly anyone, and pulling on an old, violet suit he neglected to have ironed when he had to deal with the public or go to lunch at the Royal Society, where he did talk with peers. Aldous Huxley borrowed Cavendish as a model for a character in *Point Counter Point,* which was published in the 1920s. Informed by his banker that he's not properly investing his money, the character snaps, "If it is any trouble to you, I will take it out of your hands!" Cavendish's discoveries included hydrogen, the nature of water, and ingenious work on carbon dioxide, which helped scientists a century later detect argon, helium, xenon, and other inert gases.

Timing in science may not be everything. But almost. Joseph Priestley had dropped the right hint to Lavoisier at the right time to help him isolate oxygen. Then Lavoisier learned about Henry Cavendish's detonations of hydrogen and oxygen to produce water just as he was building a device that would replicate the work. Freshly enlightened, the Frenchman fired up his new flame maker and painstakingly synthesized a small volume of water. All of which was very interesting but far less interesting to most Parisians than what was going on in the sky.

In June 1783, the Montgolfier brothers, paper magnates from Vidalon, had sent three unsuspecting passengers—a sheep, a rooster, and a duck—aloft in a basket attached to a balloon inflated by hot air rising from a fire fueled by straw and wool scraps. Then in Paris, the Marquis d'Arlaudes and Jean-François Pilâtre de Rozier became the first people to fly, again in a balloon designed by the Montgolfiers. A scientist watching the event might have said that the matter of fire weakened the electrical attraction between the particles of air and increased the elasticity, which swelled the balloon while making the air inside lighter—so up it went. A curious bystander, watching the brothers making a fire of straw and wool scraps in the basket, might have said these guys were nuts. But all of Paris looked up when a freelance physics professor, Jacques-Alexandre Charles, with the Robert brothers, two fellow aeronauts, took off in a balloon of rubber-coated silk from the Tuilleries on December 1, 1783, after they first filled it with inflammable air, or hydrogen, which is about one-tenth as heavy as normal air. This time both the scientist and the bystander might have commented on the ingenuity of the fuel. But both observers might also have wondered where Charles and the Robert brothers collected enough of the gas discovered by Cavendish, and isolated in small volumes by Lavoisier, so that it filled a balloon twenty-six feet in diameter and took Charles, on a later solo flight, so high that he became but a dot.

Charles's ride put the daring young physicist in the midst of the troposphere, the level of air closest to the earth and most responsible for weather and energy transfer from the sun's radiation. It also made him, given the rate of his ascent, extremely cognizant of a tropospheric phenomenon: The higher you go, the colder it gets. In December, at ten thousand feet over forty-eight degrees of latitude, it gets cold indeed. For minutes up there, peering down at Paris, Charles may have wondered why he hadn't sent the sheep up or at least had a real fire in the basket.

Lavoisier said the flights showed "one can rise up to the clouds to study the causes of meteors." Some of his contemporaries imagined such balloons exploring the mysterious atmosphere; others wanted them to spy behind enemy lines during battles. The French Academy of Sciences urged its members to invent a process to make lots of cheap, light gas so that balloons could be sent aloft in large numbers.

Lavoisier responded to the academy's call. With an assistant, he produced eighty-two pints of hydrogen by decomposing water, drop by drop. He did it by passing the water slowly through a luminously hot gun barrel. Then he had an instrument maker build him several big gas containers and organized a two-day experiment at the Gunpowder Arsenal. Members of the academy and invited guests watched as Lavoisier and the assistant, with Madame Lavoisier jotting notes, mounted a gun barrel filled with iron rings to a pneumatic trough holding bell jars. Water was percolated through the red-hot gun barrel, and the resulting gases, oxygen and hydrogen, were collected in the bell jars and then transferred to the large gas holders. The experiment fascinated the audience, pleased the military, and was acknowledged even 150 years later "to have routed all earlier chemistry work in the perfection of the equipment used, the scale upon which the work was carried out, and the importance of the conclusions to be derived from the result." No longer would aeronauts, as the first balloon jockeys were called, have to go aloft with burning straw. They could use hydrogen.

Over the next few years, before the outbreak of the French Revolution in 1789, Lavoisier and three key associates, with Madame Lavoisier, forged what can best be called "the anti-phlogistic task force." Its purposes were clear, to torpedo Stahl's phlogiston theory once and for all, and to trumpet the new chemistry. Gradually, a rough scaffolding for a table of elements was erected (eighty years later, the first complete periodic table would be designed by the Russian chemist Dmitri Ivonovich Mendeleev). Standing apart from past theories and girded with those of Lavoisier and his colleagues, the table contained fifty-five substances. Hydrogen, nitrogen, and oxygen were among the gases. The "matter of fire" was included. So was light. Despite the newness of the scheme, the matter of fire retained some phlogiston-like characteristics. Weightless and mysterious both, it might as well have been a quark or a neutrino from quantum physics, a field yet to be discovered. Lavoisier and his circle held on to the matter of fire because they could measure its temperature and figure out how much heat was produced over time. Basically, though, as it had for millennia, fire remained an intoxicant for the mind, a destroyer of matter, and a fundamental aspect of many chemical experiments, not to mention the means of cooking dinner and heating the house.

Combining what he'd learned about fire and air, Lavoisier wrote *Réflexions sur le phlogistique.* A later biographer, biased toward his subject, claimed it was "one of the most notable documents in the history of chemistry." Maybe. It did sink its target, phlogiston theory, completely, leaving it a defunct but colorful memory, and established Antoine Lavoisier, then forty-three, as king of the new chemistry.

But his reign was short.

The regime of the weak Louis XVI and Marie Antoinette, his party-loving wife, was already sweeping Lavoisier and France toward destruction. Though much better at science than politics, Lavoisier became involved in the revolution as a money man. He assumed the impossible job of balancing revolutionary finances, which proved far more complicated than the conservation of matter. The radical doctor, journalist, and sometime scientist, Jean-Paul Marat, attacked Lavoisier as an aristocrat and an opportunist whose "toadies laud him to the skies" and as the mastermind behind the wall encircling Paris. The wall, suggested by Lavoisier when he worked for the General Farm, had been erected during the 1780s as a means of taxing imports and reducing contraband entering the city. It infuriated Marat. He said the wall stopped air from circulating through the streets. It was a cruel irony. In the final assessment of the nasty relationship between the men, Marat, who wore as many hats as Lavoisier, may just have been envious of his more famous and successful contemporary. The notoriously jealous, like theorists, sometimes must win no matter what. In this case, both men lost. Marat was stabbed in the heart by Charlotte Corday, a twenty-six-year-old revolutionary who entered his bath under the pretext of betraying others during the Reign of Terror.

Lavoisier was arrested during the reign. He and Marie had already been kicked out of the Gunpowder Arsenal. The Academy of Sciences had been dissolved. Right up to his incarceration, Lavoisier had continued his experiments. One claim has him monitoring primitive breathing gear attached to an exercising assistant named Armand Seguin when the gendarmes rode up outside. Lavoisier wanted to know more about oxygen and the body, which seemed to burn the gas almost like fire. "Breathing, rusting, burning, it's all one," he said to his father-in-law, who preceded him to the guillotine. With his head clamped into the notch on the scaffolding and the rabble on hand for the bloody drama, Lavoisier listened to the blade jerk skyward and watched a bubble of air expand from the lips of another man's head in the basket directly below.

6

Mythic Gods of Air

Before there was an Antoine Lavoisier or a Joseph Priestley, before there was something called oxygen, even before there were four fundamental elements, one of which was air, there was the wondrous blue sky. Both before and after the new chemistry, most people in the world worshiped the sky. That's because they lived by myth, not by science. And in virtually all societies suffused with myth, a sky god reigned supreme. Usually a he, though on occasion, as with the Egyptians, a she, a sky god was divine, inaccessible, often violent, and eternally shaping things on land. Sussistinako of the Sia, Nzambi of the Bantu, Ruwa of the Wachaggas, Iho of the Maoris, Oke of the Iroquois, Thor of the Vikings, and on and on, these gods lived in the air, usually for eternity. They knew how to handle lightning, rain, and wind. They were mythical versions of the heavenly father who gave Christians the Lord's Prayer, the most popular prayer in the world, one that begins, "Our father who art in heaven."

Astolfo, the hero of Orlando Furioso, *a sixteenth-century tale of a daring flight to the moon written by Lodovico Ariosto, approaches "yonder moon . . . nearest planet to our early poles," a destination Astolfo believed had rivers, valleys, cities, and castles. The illustration is by Gustav Dore, for a nineteenth-century edition of the story.*

OF ALL THE myth-loving cultures, the ancient Greeks were probably as sky fixated as any. After borrowing liberally from the older cosmologies of Mesopotamia and Egypt, the Greeks couldn't seem to get enough gods into their airy pantheon. They did seem, however, a little ambivalent about the very first one.

In the Greek myth describing the origins of the world, Ouranos emerges from Gaia, the earth mother, wearing a crown of stars. Once Gaia spreads her male counterpart over her in complete union, she declares that he is "her equal in grandeur." Ouranos is trouble, though. His offspring soon include the Centimanes (monsters with hundreds of limbs) and the one-eyed Cyclops. Ouranos is castrated by his son Cronus so that he can't produce any more children. His genitals are thrown into the sea, creating a white foam out of which rises Aphrodite, the goddess of love. Then the son who did the castrating begins eating his own offspring, all except Zeus. Zeus escapes his father's appetite, usurps his father's power, and banishes Cronus to the depths, into the bowels of the earth. One consequence of this bizarre sequence of patrimony, beginning with the earth mother, is that the Greeks never created a cult honoring Ouranos, only one for Gaia.

In most of the myths we know about (many have been lost), the union of earth and sky, like the lovemaking of us mere mortals, brings wholeness to division. The image and symbolism of the earth and the sky having sex, of the planet and the atmosphere making love on a gigantic scale and with incredible consequences, appears again and again. This mythical tale is mirrored in the latest scientific thinking about the creation of the earth, with air rising out of the fertile, molten globe, to create one continuous entity. Today, scientists see air as an extension of the earth, an outer sphere of gases, liquids, and airborne solids interacting with the solids, liquids, and gases of the planet's surface. Oxygen, nitrogen, carbon dioxide, and other gases emerged from the mother just as

Ouranos came from Gaia, creating conditions favorable to life. It's a classic story: the wet, seeding sky joining with the fecund soil, male and female, sperm and egg. "I am the Earth," cries an Egyptian song. "Your wives are to you as fields," proclaims the *Koran*. The sky constantly seeds the furrowed earth, almost in ribald merriment, with lust and hunger, with the longing to be eternal.

Sky and earth unions in some myths resound with violence or with references to children, occasionally desperate children. For example, Rangi and Papa, sky and earth gods of the Maoris in New Zealand, were so tightly wrapped together as a couple that the children couldn't get out of the womb. So the children, unstable and crawling around in the dark, put their breaths together in a conspiracy "to let the heaven stand far above us, and the earth lie under our feet. Let the sky become a stranger to us, but the earth remain close to us as our nursing mother." Interestingly enough, it was the child destined to be the father of the forests who finally tore the parents apart. "His head is now firmly planted on his mother the earth, his feet he raises up and rests against his father the skies, he strains his back and limbs with mighty effort. Now are rent apart Rangi and Papa, and with cries and groans of woe they shriek aloud, 'Wherefore slay you thus your parents? Why commit you so dreadful a crime as to slay us, as to rend your parents apart?'" But the future god of forests ignores the screams and cries. "Far, far, beneath him he presses down the earth; far, far above him, he thrusts up the sky." And so, notes the historian of religion, Mircea Eliade, the children "cut the cords binding heaven to earth and pushed their father higher and higher until Rangi was thrust up into the air and light appeared in the world."

Other myths perceive the world as an emanation out of the chaos, just as the early philosophers of the Ionian school saw worlds in bubbles floating out of the *aperion*. Some myths describe a bridge linking heaven to earth. As you might expect, crossing isn't easy. Suddenly, the bridge may morph into a narrow vine with an edge as sharp as a knife. Only shamans, wise men, or holy women who know the right incantations can pass. Other seekers of the sky find themselves involved with something both ancient and very modern, a vagina monologue with the *vagina dentata,* or toothed vagina, another mythical portal to heaven. Any mortal contemplating a journey into the sacred realm of the sky, whether by bridge, ladder, or vagina, is on the verge of something "infinitely dangerous," Eliade said. Such a trip is not the equivalent of a balloon flight over Paris in the 1780s or a continental jaunt by jet two centuries later.

When the people we mistakenly call "primitive" danced and chanted, or took drugs and enacted initiation rites to cross mythical borders toward the stars, they weren't thinking about rising up through the stratosphere and the need for oxygen to stay alive. They had something both simpler and grander in mind. They sought *illo tempore,* the paradise that existed before humans screwed thing up. They were soul mates of sorts of Blaise Pascal and his concept of *deus absconditus,* and of Paracelsus and his drunken rants tearing apart the rational mind, not of Robert Boyle and his clock-maker god or of Lavoisier and his dramatic experiments proving the conservation of matter. They wanted into the mystical because it promised an end to the painful, tragic life they found on earth and a new beginning in paradise. No earthly paradise could quite compete with the stately pleasure dome of the great blue sky.

For such people, myths about the sky weren't an escape into fantasy any more than Lavoisier and Priestley's codiscovery of oxygen was salvation through science—even if Lavoisier may have felt otherwise. If air is vastly more than its components, so are humans. A human being has character and personality; air has the sky and the heavens. Myth came before science and will outlast it. "There is no final system for the interpretation of myths, and there will never be any such thing," Joseph Campbell wrote in *The Hero with a Thousand Faces.* Myths talk not about air but about the sky, not about evolution but about birth. Today, in an era dominated by science, people look up and think sky, not air. Air is too amorphous, too big, too much most of the time to really give it its due, to acknowledge that it is as important as wine and bread.

The sky, on the other hand, is a place. And has been since time immemorial, whether put up there by desperate children or wrenched from the innards of a mythical giant such as Yggdrasill, source of all things in the great Viking myths that begat Thor, Odin, and Valhalla. It seems that the keepers of myths, the shamans, warriors, and priestesses of dwindling tribes and shattered civilizations, remain today as guardians of an alternative understanding of the universe. A purely scientific take leaves too little room for the alternative mind. The relentless demands for proof and the reductionist method of getting at the truth can put any vision of an atmosphere filled with clamorous gods beyond our grasp, out there somewhere on the periphery of consciousness, and leave us simply gasping for air like Galen's gladiators or Legallois's baby bunnies. We are now best able to access the rich ancient terrain of myth through fantasy, through films, books, art, and dream. Yet the gods of myth have inhabited human consciousness for thousands of years, as vivid and real as the superstars of today until only the recent past. To reject this truth is to become a victim of science rather than a person in tune with not only the wonders of mitochondria but also the vanished intensity that our forebears felt as they stared upward in a state of wonder at the sky, a state we have largely lost.

Great artists know this. Science may have tossed myth into the dustbin of history as irrelevant, claiming that tales about sky gods are too organic, too metaphysical, too like all that prescientific revolution hocus-pocus of the alchemists and seers. Yet myths endure. They are still revered, not only in scattered societies yet to be completely overrun by science and rationalism, but in films, in books, and on the stage. Iho, Wakan, Nzambi, and other sky gods rise out of the dormant human psyche and into the modern imagination as heroic figures in entertainment spectacles such as *Star Wars* and *Lord of the Rings*. Audiences are starved for the mythic, for its reassuring whispers—*there is a heaven, there is a sacred place; it's up there, in the air, in the sweep of a humbling sky*. Watching the films, we care not a wit that God is dead, as Nietzsche once proclaimed, but embrace what the philosopher uttered another time: "Live as though the day were here." We couldn't care less whether George Lucas, Stephen Spielberg, or J. R. R. Tolkien raided primitive treasure chests for symbolic, godlike beings of air to give meaning to their dramas.

The Western Sky

Myths get thought in man unbeknownst to him.

—CLAUDE LEVI-STRAUSS, *MYTH AND MEANING*

The sky gods in Western mythology, sky gods that Americans and Europeans still relate to on occasion, are Zeus, Jupiter, Thor, and Yahweh. Like their counterparts in Asia, the Arctic, and Africa, these gods control storms and thunderbolts, are short-tempered, and like to pound other gods and giants into oblivion. Profligate, they also like to venture down to the earth itself and lust after both goddesses and mortals, leaving both with offspring and often treating them later with complete indifference. But it's air's extremes—torrential storms, raging winds, fire-spewing lightning bolts with blasts of thunder—that best define these gods at their extremes.

For a violent Titan, fighting was preferable to anything. The Western gods fought with thunderbolts, winds, and, in the case of the red-bearded Thor, a hammer gripped in a hand sheathed in an iron glove. An air god might whirl in a tornado, pelt an enemy with hailstones the size of boulders, or flood a valley so that there was no escape for terrestrial giants with vulnerable respiratory systems. Their lungs filled with water and they drowned.

Violent business is often the way of the sky and so it was with the sky gods. From Mount Olympus, where Zeus made his home, he hurled down thunderbolts. For ten years, he flung them at a group of rebelling Titans. During that time, the earth burned, forests flared, seas boiled. Finally defeated, the Titans were chained and sent deep into the earth, "buried by the will of the king of heaven," wrote Hesiod in the *Theogony,*

a poem from the eighth century B.C. commemorating the mythic struggle. In contrast to the Greeks, whose poets and artists memorialized their
heroes with lyrics, paintings, and sculptures, the pagan Vikings were too
busy swinging the long sword, whirling the firebrand, and busting the
heads of Christians to write their exploits down or paint them on walls.
Thankfully, an Icelandic historian of the thirteenth century, Snorri
Sturluson, did set down their imaginative world of myth and glory in the
Edda. Sturluson recounted the beginnings of the Viking world as a surreal birthing of the giant hermaphrodite Ymir beneath the branches of
the cosmic tree, a gigantic ash called Yggdrasill. The ash tree spans
heaven, its roots anchor deep in the earth. In Yggdrasill's crown perches
an eagle. The flapping of the eagle's wings makes the wind. Beneath
Yggdrasill's roots live gods, giants, and goddesses who gather daily at the
Well of Urd, the sacred spring that determines fate. The gods arrive on
horseback after crossing a rainbow bridge glowing with fire. All except
Thor, that is. Thor wades to the spring through its connecting rivers.

A hammer-throwing, red-bearded, gluttonous pagan with a massive
belt, Thor once ate a whole ox and eight salmon and washed them down
with mead in the halls of Thrym. Thrym was a giant who had stupidly
stolen Thor's hammer. He paid for it once the god had slaked his
appetite. Thor got around the sky aboard a goat-drawn chariot whose
passing wake left a cascade of thunder. Thor was also virile. Big surprise.
He mated with the golden-haired Sif, whose tresses suggested corn in
another symbolic melding of sky and soil. Ultimately, it was Thor's
breath that raised him into the upper pantheon of the sky gods, though.
God of the perilous journey, Thor could blow a ship across the sea or
onto the rocks; he could pummel a vessel with hail or bless it with sunshine and fair breezes. When Thor blew downward mightily into his
beard, he turned the hairs into red bristles and a gale sprang up. If
Vikings on land spotted an enemy approaching by sea, they often
beseeched the red-bearded one overhead to bristle his growth and scatter
the foe.

In the *Edda,* Thor is descended from Ymir. The giant hermaphrodite also gave birth to the first Viking man and woman. The couple emerged from Ymir's left hand, according to some interpretations, and from beneath his left armpit in others, while he slept. His sweaty feet joined in a union that produced a son. Then came a cow from whose udders flowed streams of milk that Ymir drank. Ymir is slaughtered and cut up by Odin. His flesh became the land, his sweat the sea, his hair the trees, his skull the sky, and his brains "the bitter-mooded clouds." The sky fashioned from Ymir's skull is a vast dome kept aloft by four dwarves. The *Edda* described the four dwarves holding the sky up very high, a funny image if you think about it. But in a kingdom of giants with killing and plunder on their minds, and with bloodthirsty hankerings to carve up father figures for mythical offerings of sacrifice and growth, if dwarves are able to survive and keep the sky aloft, so be it.

Valhalla, the heaven of the Vikings, lay above the dome held up by the dwarves. In this way, it was like the quintessence, as the quintessence for Christians was above and beyond the ether, a place inaccessible to anything material or corporeal, a very special place reserved for God, or gods. In Valhalla, one-eyed Odin held court and dead warriors could do what they liked best of all: drink and fight. Vikings would go to incredible lengths to reach Valhalla, validating the intoxicating power of myths in the lives of these mortal men. Ritualistic self-slaughters took place regularly at a temple in Uppsala through the eleventh century. The Vikings even hung their horses from trees surrounding the temple. All their possessions, including saddles and clothes, were burned, so that the smoke could rise to Valhalla. A Norse scholar, Hilda Davidson, related an account written by an Arab traveler, Ibn Fadlan, about a stabbing, hanging, and burning sacrifice in a Swedish settlement alongside the Volga in the tenth century. A slave girl was strangled and stabbed simultaneously by the Angel of Death, an old woman with assistants, and then burned on a pyre, a blazing ship, next to her master. As the fiery ship sucked oxygen from the surrounding air and created a powerful updraft, Fadlan heard one of the pagans say that his lord has sent the wind to carry the smoke aloft to paradise as a gesture of love for the dead master.

Like Thor and Zeus, Yahweh, the supreme sky god of the Hebrews, enjoyed throwing thunderbolts. And he was even more intent on monopolizing the sky than his Viking and Greek counterparts. Yahweh's air power displays, in which his voice is thunder and his arrows lightning, provoked revelations, said Mircea Eliade, whose portrait of this god made him even more of a single-minded power monger than the others.

Indra, the king of heaven and lord of rain for the Hindus, would ride down to earth on the back of a large white elephant named Airavata. Another Hindu god, Shiva, was a clear contrast to the Western gods. Female, with a drum of creation in one hand and the flame of destruction in the other, Shiva herself was nothingness; she was the void. The Greek and Roman myths feared the void, the vacuum. And remember, Christianity rejected it well into the seventeenth century, long after Torricelli, Pascal, and Boyle demonstrated its existence. The Western myths never gained any power in India, where the culture actually embraced emptiness instead of repudiating it. In fact, the Hindus recognized something called the *atman*. For the Hindus, the atman resided in all beings, not just humans. "Smaller than the smallest atom, greater than the vast spaces," it was spirit, immortal, an essence of the universe. At death, the atman left the body, transmigrated through the air, and then reincarnated in another person. The circular pattern went on forever. Or almost forever. Buddhists came to believe that reincarnation could be ended by attaining enlightenment. And the route to enlightenment was through the breath.

7

Is There Air in Heaven?

Crossing the skies, Stormfield faints, only to awake to "loveliest sunshine and the balmiest, fragrantest air. . . ." He soon spots millions of people in the sky of heaven. "What a roar they made, rushing through the air! The ground was as thick as ants with people, too—billions of them, I judge."

—MARK TWAIN,
EXTRACT FROM CAPTAIN STORMFIELD'S VISIT TO HEAVEN

A priest hears the beguiling call of a creature from heaven, possibly telling him of singing angels, rivers of wine, and luscious forests—all signs that air exists in the afterlife just as it does on Earth.

IS THERE AIR IN HEAVEN?

As a riddle, the question begs an answer. My answer? I sure hope so because if I get there, I want to be able to take a big sigh of relief.

As a scientific question, Is there air in heaven? is meaningless. Science dismisses heaven, just as it dismisses myth. But looked at metaphysically, the question becomes rather provocative. It rises to another dimension. For if God is inscrutable, as many saints and theologians have proclaimed, then His dwelling place must be beyond knowing as well. If you can't know God, can you know anything about where He lives? Yet beyond knowing does not mean beyond existence. Even science agrees with that. Some bits of subatomic matter, such as the graviton, are said to exist, but physicists have not detected their existence yet. Maybe heaven is like the graviton; it exists but we just haven't found it yet.

At any rate, visions of both heaven and the afterlife come from varied sources: the Bible, folklore, poetry, art, dreams. About the only thing that they all agree on is that heaven is up. And up is better.

Admittedly, heaven is a place notoriously tough to pinpoint, no matter which way you're looking. And, ironically, for artists, heaven seems much less attractive than hell. Heaven is too nice. All those halos and pearly gates. In contrast, hell is "oddly fleshly, with tortures that hurt and an atmosphere that is . . . excessively gross," wrote Alice Turner in *The History of Hell*. Sleazy and noirish, on some levels even romantic, hell has carved out a much higher audience rating than heaven has in the modern mind. Still, most people don't want to go there. In a Gallup poll conducted in the early 1990s, 60 percent of Americans said they believed in hell, compared to 52 percent in 1953, but only 4 percent said they wanted to visit.

You won't find heaven or hell on your global positioning system, although medieval mapmakers regularly identified both places. Five hundred years ago, when few people traveled farther than a day's walk from

their village or town, heaven often seemed just beyond the horizon. After
the Garden of Eden, perhaps. Or past the Elysian fields.

Some theologians in the thirteenth century insisted heaven was a real
place, with corporeality, juice, dimension, and an atmosphere, not some
immaterial *ens* up in the celestial ether. The religious philosopher Albert
the Great argued that heaven was not only real, but even had property
lines. In an odd bow to Aristotle and the four elements, Albert the Great
envisioned a heaven made of fire. Property lines of fire—now that took
some imagination.

For centuries the English went beyond the call of duty to give heaven
special attributes and appealing features. Most societies used the same
word for sky and heaven, but Old English honored the *heofonheall,* or
heavenly court, in the *uplyft,* or sky. *Heofronrice* was the kingdom up in
heaven, a tangible place for the faithful to contemplate. *Wlite* and *wuldor,*
beauty and glory, were attributes of a God who occupied a *heofonsetl,* a
throne, up in *wlitig,* another word for heaven.

Climbing up to heaven often required strength and courage, just as it
did to face the *vagina dentata* or to cross a knife-sharp vine over an abyss.
A climb, of course, needed muscles. Muscles burn oxygen, so climbing to
heaven was an oxygen-reliant activity. Jack scaling the beanstalk, the
priest king Ur clamoring up a ziggurat, Siberian shamans ascending steps
hacked into beech trees, Jews scrambling up the ladder that Jacob saw
hanging down from heaven in a dream—all needed oxygen for the ascent.
So does it make sense that oxygen would be absent once they got there?

Plato's theory of forms, which said that all reality is a suggestion, an
imperfect rendition, of some ideal form, supported heaven's having air.
The theory, which powerfully influenced the Christian belief that Jesus in
his human form was the ideal man, and the Virgin Mary the ideal
woman, also suggested that earth was an ideal of heaven. Like man and
woman, the earth was imperfect, thus understandably sinful, yet a preview of a better place where sin would not exist.

Mystics bringing visions of heaven back from trips on the wind,
prophets accompanying angels, those returning from near-death experi-

ences—all have plied the ethereal and returned with vivid stories about it. Once they returned to the earth, the travelers usually described things we associate with air: birds, trees, gardens. They recalled rivers overflowing with wine, milk, and honey. They saw golden gates and thrones. Enoch, the Old Testament maverick, on travels aloft with the angel Uriel, claimed to behold tongues of fire and a marble building, "its ceiling like the path of the stars. And I saw countless angels...encircling the house." Whether the angels were speaking Enochian, Enoch didn't say. But flying, they needed air. And so did the flames.

Despite Enoch, the Old Testament props up no single view of heaven and the New Testament doesn't straighten us out on this important matter, either. Possibly that's because the apostles believed the Second Coming was imminent. When Jesus came back, he'd have more details. Or better yet, the earth would become heaven. Or there'd be heaven on earth. It would all be a great big, grand, airy paradise without sin, fear, or cellulite.

As we're still waiting, some impatient Christians have taken the opportunity to fill in the blanks. Take the Mormons. Some years ago, I went on a tour of the Mormon Tabernacle in Salt Lake City with a group of architectural historians. We listened to the choir in the Mormon Tabernacle, we strolled through the square, we watched people being baptized in huge baptismal fonts that sat atop the backs of ceramic bulls. Then we were shown three different versions of the afterlife. To me, each one appeared to up the lushness of its predecessor, as though we were being asked to ponder whether we preferred spending eternity in a room at a Motel 6, a bigger room at a Ramada Inn, or a suite at the Ritz. The tour guide told us that when a Mormon died, the person checked into the version of the afterlife appropriate to how good he or she had been while alive. Or that's how I interpreted it. Invited to ask questions, I did. I asked if the room you were assigned had any relationship to the money you had tithed to the Mormon Church. I was told absolutely not.

The exchange, I'm afraid, revealed my ignorance of the multiple heavens available to those who believed Joseph Smith to have been a savior, a savior who happened, I feel obligated to say here, to have been born just a stone's throw from my sister's farm in Vermont. Quite a while after that exchange, I discovered that the apostle Paul had also seen three versions of heaven. That shook me up a little. Paul described them in *Visions of Paul.* Led by an angel, Paul went back and forth between heaven and hell. He wasn't holding his breath, so he must have been breathing, right? With his angelic guide, Paul is shown heaven number one, which turns out to be a kind of earthly paradise. Three thousand angels are singing, rivers of wine are flowing, there's honey, there's a forest with leaves that must be taking in carbon dioxide and giving off oxygen. Sinners are sheltered beneath the trees, waiting between death and resurrection. When Paul's angel shows him heaven number three, the apostle feasts his astonished eyes on a golden gate and pillars. "Blessed are you if you enter by these gates," the angel intones. The Biblical scholar Jeffrey Burton Russell said that Dante, writing in *The Inferno,* flipped the phrase, telling those on the verge of entering hell, "Abandon all hope, you who enter."

Saint Augustine, who lived in the fifth century, apparently believed there was air in heaven. In *The City of God,* he described a celestial Jerusalem where bodies exist in the flesh but are no longer tempted by sin, as he had been as a young man. In the afterlife, according to Saint Augustine, you can breathe easy. It doesn't matter whether you were a missionary eaten by a cannibal, a knight drawn and quartered during the Crusades, or a village churchgoer for seventy years who died peacefully in your sleep—the power of God to perform miracles insures the ascent of all the pieces of your body because He wants you to be whole. And in your prime. Saint Augustine thought you'd be around thirty and stay that age for eternity. You would even be able to transubstantiate, or travel through space, with a wish.

Closer to our times, the French poet Pierre Reverdy, a disciple of Rimbaud and a fierce fan of *A Season in Hell,* claimed you can spot someone who's been to heaven. They're crazy and wild. They're overflowing with the nectar of eternity. "Stars get all tangled up in their hair," Reverdy maintained. In "Heaven," a song written and sung by David

Byrne, he croons about heaven's being "a place where nothing, nothing ever happens." Which, as I interpret it, could mean there is no need for air. But if nothing ever happens, you might just as well be a cosmic fossil, or a meteor. Not very appealing.

Wherever heaven is, most cultures have painted it as blissful, a place worth striving for. Religion usually attempts to somehow embrace it. Fantasy often takes visions of heaven over the top, with hundreds of virgins draping themselves over heroes and Hollywood-like fights ending with wounds that evaporate. You could say, conceptually, of course, that our atmosphere is heaven. Certainly, without it, we wouldn't have the brainpower to conjure up the idea of heaven. We wouldn't be able to appreciate the sky as "a miraculous achievement" or as a structure of "magnificence," as Lewis Thomas did.

Logical, rational types tend to doubt that heaven, even one located in a structure of magnificence, is a tangible place. Of course, a state of bliss is not a place, either. It's a mood. So if heaven is bliss, which is not a place, would bliss need air to be sustained? Probably not. But in heaven, even a conceptual one, aren't there blissful saints walking about? And if a saint is part of the body of Christ, and his body stands for "in the deepest sense . . . the earthly body of Jesus, the consecrated Eucharist," as Jeffrey Russell contends, then we've come full circle. We're right back to an affirmation of air in heaven by circumstantial evidence, by Plato's theory of forms. Jesus' earthly body breathed, so therefore his heavenly one breathed because all the redeemed stay themselves, according to the Bible. This is confirmed by Saint Augustine. In such an afterlife, you couldn't shed your lungs even if you wanted to.

Of course, there's a kicker. If God can do whatever he wants, he could create everything in a vacuum—the flowing rivers, the golden thrones on white clouds, the fluttering angels—and have things function as though air was present, because he decreed it.

As for air in hell, flames are a dead giveaway. The air of damnation is well supplied with the pure air five or six times as good as common air. That explains the bright "red wheels of fire" encircling the eyes of the

white-haired boatman Charon, who ferries Dante and Virgil across the
River Styx. Shoving off from shore, Charon warns his passengers:

> *Woe to you, wicked souls! Give up the thought*
> *Of Heaven! I come to ferry you across*
> *Into eternal dark on the opposite side,*
> *Into fire and ice!*

One of the more vivid, even breathtaking, visions of hell came from
the quill of an Irish monk who wrote *The Vision of Tundal* in the twelfth
century. It was a highwater mark for lucid, factual renditions of a place
whose power reached its zenith between the fall of the Roman empire
and the onset of the Renaissance. Hell sort of topped out with Tundal's
trip there. As historian Alice Turner recalled this medieval horror story,
Tundal is a likable sort whose lapses include a little debauchery, minor
theft, and a habit of spending too much money on jesters and musicians
instead of tithing to the church. Over dinner, Tundal has a sudden stroke
during an argument. He falls into a coma and immediately feels his soul
leaving his body. Fiends claw at his flesh, screaming, "Where are the
good times now? Where are the pretty girls?" Tundal's soul is rescued by
his long-ignored guardian angel who wants it to have a good long tour of
the future awaiting a wayward, unrepentant knight.

What follows is a trip through a hell heavily dependent on air for spectacular effects. "Tundal is shown murderers sizzling over an iron grate laid
across a whole valleyful of stinking coals," Turner wrote, "then a mountain with fire on one side, ice and snow on the other, and hailstones in
between, where fiends with iron hooks and forks chivy unbelievers and
heretics from one torture to another." Tundal must cross a deep valley on
a thousand-foot plank a foot wide, off which the parsimonious and the
proud fall into oblivion. Next, Tundal enters the maw of the giant
Acheron and finds himself packed together with the greedy. "Bitten by
frenzied lions, mad dogs, and serpents, burned by fire, bitten by cold,
suffocated by stench, and clubbed by devils until the angel reappears,"

Tundal is then tended by his angel just in time to face a two-mile bridge as wide as his hand and spiked with nails. Once across, he has to lead a wild cow because "you stole a cow from one of your friends," the angel reminds him. "But I gave it back," he objects. "Only because you couldn't keep it," she says. "But because you did, you won't suffer much."

After the cow incident, his guardian angel leads Tundal to a dwelling full of gluttons and fornicators. The place is like an oven. Then there's a gigantic bird with an iron bill. The bird "eats unchaste nuns and priests and defecates them into a frozen lake where both men and women give birth to serpents."

The battered and exhausted Tundal comes now to the Valley of Fires. He's grabbed by demons and tossed into a raging furnace with twenty sinners, and all of them are smelted together. Removed from the heat by the demons, the sinners are hammered into a solid molten mass, then tossed into the air. The angel again appears and extracts Tundal from this horrible eternity.

Finally, Tundal sees Lucifer. Once "the most beautiful, the most powerful creature God made," now Lucifer is evil. He's the devil. He's manlike, with a beak, tail, and thousands of hands, "each of which had twenty fingernails longer than knights' lances, with feet and toenails much the same, and all of them squeezing unhappy souls," says the original vision. Then comes the most heinous image of respiration in all of Christendom. Lucifer "lay bound with chains on an iron gridiron above a bed of fiery coals. Around him were a great throng of demons. And whenever he exhaled he ejected the squeezed unhappy souls upward into Hell's torments. And when he inhaled, he sucked them back in to chew them up again."

Following a quick exit through purgatory, Tundal feels his soul again "clothed in his body." And, of course, is changed for eternity. "He opens his eyes, asks for Communion, gives all he has to the poor, orders the sign of the cross sewn on his clothes, and begins to preach the word of God."

Another "I've been there and returned to tell about it" tale of the twelfth century, *The Vision of Alberic* also adds to the substantial evidence proving that there is air in hell. Alberic describes burning gases, a river and lake of fire, sulfurous fumes, a forest of thorny trees. A century later an oxygen-burning oven was conjured into the imagination by *Dreams of Hell*, an allegory written by Raoul de Houdenc in which the author dines with Pontius Pilate and Beelzebub on roasted heretics in garlic sauce. In this satirical trip the pilgrim has already witnessed the river of gluttony and the mountains of despair. While munching on heretics, he reads aloud to the king of hell from a book about evil minstrels.

Questioning the existence of hell during the Middle Ages could get you in even deeper trouble than could doubting the truth of heaven. One skeptic, the ninth-century theologian Johannes Scotus Erigena, was stabbed to death by his own pen-wielding students. According to medieval legend, they were carrying out justice after Erigena had called hell into question and been labeled a heretic.

A literal hell meant air. Or, in the Middle Ages, the vaporous substance identified with the atmosphere. The word *atmosphere* originated in Greece, combining *atmos*, or vapor, and *spharia*, or sphere. And, of course, hell needed fire, fire hot enough to congeal the evil into molten balls that demons could play with forever.

The marvelous, mysterious atmosphere of the Middle Ages provided an invisible skeleton of sorts on which those in power hung symbolic tapestries, one of which was an assured afterlife that it was best not to publicly doubt. In an era when theologians spent hours, weeks, even years, pondering the question How many angels can dance on the head of a pin? it was no great leap to imagine them applying equal due diligence to the question Is there air in heaven? Is there air up in the celestial sphere, where dragons attacked angels, sweeping them unmercifully down into Satan's lake of fire?

A moldy, dank, smoky, apocalyptic hell and a bright, angelic, sunny heaven both need air as much as moist, fleshy, blood-filled human lungs need it. Neither place, given popular suspicions and beliefs, could exist in a void. Heaven and hell both need air.

Part 3

Hold Your Breath

Below them in the paling light smoldered the plains of San Augustin stretching away to the northeast, and earth floating off in a long curve silent under looms of smoke from underground coal deposits burning there a thousand years.

—CORMAC MCCARTHY, *BLOOD MERIDIAN*

8

The Little Atmospheric Disruptions of Man

Standing by a cemetery in Centralia, Pennsylvania, I watch smoke seep from the ground between gravestones. It rises into the blue sky from small tongues of flame, and I doubt the smoke will find its way to heaven. Maybe it's Lucifer's breath as he exhales unhappy souls, though I doubt it. To my right stands a sign. It warns of dangerous gases, the ground suddenly collapsing, injury, death. Beneath my feet—how far, I'm uncertain—coal burns in unmined seams abandoned and left to smolder. Photos of this thrown-away town in Appalachia taken from a satellite suggest that the coal burning underground might be jewels. Black jewels, actually, black carbonated jewels captured from the sky and now aflame, just as they were beneath the plains of San Augustin in the 1850s, beneath mountains in Australia thousands of years before that. And now are aflame in a long arc that stretches three thousand miles across China and dumps a voluminous amount of carbon back into the sky every day.

Manchester, England's textile center during the Industrial Revolution, bragged of having the worst air in the country. The air, though needled with smokestacks and thick with soot, carried the scent of money. The city made a fine vista for a Victorian picnic in a painting by William Wylde in 1851.

SMOKE IS ODD, complex, nasty, marvelous. It's older than bacteria, as modern as smog, as common as fire. Smoke shrouded the primal fire of the Big Bang. You could say that the gases and debris of the cocoon nebula, which orbited a young sun four and a half billion years ago, gave birth to earth. You could say that everything on earth is a result of coagulated smoke worked on by sunlight. Ultimately, you could say that the human race is nothing but the remnants of the debris of smoke animated by a mysterious spark of life.

Today, when we think of disruptions of the atmosphere, we seldom think of smoke in all its mind-boggling dimensions and phases. We think of smog and ozone holes and global warming—along with radioactivity, bioterrorism, and the occasional volcanic eruption. Smoke sounds almost quaint, even a little passé. A visit to a place like Centralia, a spooky town peopled only with ghosts and memories since the U.S. government bought all the houses and buildings and shifted everyone out, makes you contemplate smoke a little more intently. For there it is, coming up out of the ground. You look around, and the scene suggests hell with the lid still on. Or the morning after on a battlefield bearing the scars of a scorched-earth policy.

I bend over and press my palm against the blackened ground. I detect only a faint fever. Nothing real alarming. In the distance, the air is clearer. I hear a bird song, the notes propelled from the bird's syrinx to my eardrums, which play a tune on hammer, anvil, and stirrup. The smoke rising in front of me shimmies. It wiggles. It almost seems friendly, waving at me innocently, appealingly in a macabre yet timeless sort of way.

This is not the smoke that would accompany a bolide hit, I think. It's not the smoke of an end game, which the scientists at Princeton say we don't have to worry about for a while. This is not the smoke of incinerated forests, tsunamis clearing coastlines, seismic shocks dropping skyscrapers, ash settling.

This is not the smoke of a volcanic eruption, either. Volcanic eruptions are the earth's smoking guns. Leonardo de Vinci compared the
blood, heat, and breath of people to volcanic eruptions and the flow of
lava. The word *volcano* comes from Vulcan, the Roman god of fire who
tended a forge on Mount Olympus. In our time, daily, any number of
more than fifteen hundred active volcanoes around the world grumble
and spew; they send up grayish exhalations of carbon dioxide, sulfur
dioxide, and water vapor, along with what volcanologists call *blocks* and
bombs, molten fragments of different sizes ejected thousands of feet up
into the air. Among volcanologists there is a pseudoclub—The Bomb
Behind Club—and membership is limited to those who have had molten
bombs and rock fall behind them while watching an eruption.

This is not the smoke of a forest fire, I think. Historically, another common disruption of the atmosphere, smoke from a vast conflagration of
trees, brush, and debris, can become so thick that it deflects sunlight, fouls
respiration systems, disrupts photosynthesis, and blankets lakes, ponds,
streams, and rivers with soot. Nor is this the smoke of biomass burning,
which created a brown cloud over Asia in 2002. Two miles thick, a blend of
ash, soot, aerosols, and industrial chemicals, the cloud darkened cities and
killed an estimated half million people in India alone, then managed to
creep offshore hundreds of miles to the pristine Maldives, burning lungs
and polluting water there as indifferently as on the mainland.

This isn't that smoke. And this smoke isn't smog.

I know smog. Or knew it. Smog was my first real bad-air experience.
In the early 1970s, my girlfriend and I used to walk in winter on the
mostly empty sand of Hermosa Beach, California. Off to the north, sticking out over the coast off Santa Monica was this garish, mustard-colored
tongue. Not yellow mustard, like French's, but grayish yellow, the color
of Dijon. The tongue was smog. Sulfur lent it the yellowish hue. As
spring rolled around, the tongue seemed to gradually disappear. It hadn't
disappeared. Rising temperatures had simply swollen the tongue until it
engulfed Hermosa Beach, along with most of the Los Angeles basin, and
we were all breathing smog twenty-three thousand times a day.

Later, I lived in Quito, Ecuador, where the particulates from diesel trucks and buses flecked my tongue if I was crazy enough to open my mouth in the narrow city streets. I tasted the oily slime of Prague's air on winter mornings, pondered the perplexing concept of "permissible dose" while staring up at the impressively huge cooling towers of Temelin, which was designed by the Russians on the Chernobyl model and meant to be the world's largest nuclear reactor. I felt ozone burn while jogging on Maine beaches because air pollution had been swept hundreds of miles from its sources to the coast by trade wind patterns. I rode around Washington, D.C., on Interstate 495 during August, when air quality kept anyone with respiratory problems inside while U.S. congressmen spent their summer recess in less harmful locales.

The atmospheric disruptions I've experienced have never, as far as I know, seriously affected my health. I was always glad to get away from them. A Dijon sky, lung burn, diesel tongue, an excess dose of rads—no human voluntarily chooses to lather himself or herself with these abnormal conditions, which ultimately distill down to molecular, biological threat amid the cellular warren of the body. Even less, of course, do most of us (there are exceptions) want to hear of a bolide hit that incinerates people like crickets in a firestorm. Nor do most of us want to be very close to an erupting volcano. Or get caught in a forest fire. But bolides, volcanoes, and forest fires, as dramatic as they are, aren't the kind of atmospheric disruptions the vast majority of us have to worry about. The disruptions we have to worry about are mostly collective emissions from hundreds and thousands and millions of small sources—tailpipes, chimneys, smokestacks, factory vents, backyard incinerators, smoking dumps, airplane engines, slag heaps of radioactive waste, and others that emit nitrogen oxides, carbon dioxide, sulfur dioxide, dioxin, benzene, alpha and gamma rays, and so on, that accumulate in the air and the soil and in our bodies over time: days, weeks, years, decades. The disruptions caused by humans are nuanced, subtle, and pervasive. And they began innocently enough, with our first fires.

Smoke

The noun *smoke* comes from the Old English *smoca,* is akin to the
Old English *smocian* (variant: *smeocan*). Compare it with Middle
High German *smouch,* German *Schmauch,* and German
schmauchen, to emit a thick smoke. Also to the Medieval Dutch
smooc, Dutch (and Low German) *smook,* and also to the Greek
smukheim, to smoulder, and the Lithuanian *smaugti,* to suffocate.

—ERIC PARTRIDGE, *A SHORT ETYMOLOGICAL DICTIONARY*

We do not know the exact date—probably around 10,000 years ago—or
the exact location—possibly a cave in what is now southwestern
France—but the first atmospheric disruptions caused by man began when
he managed to light a fire. Neither apes nor chimpanzees, from which
Homo sapiens evolved via *Homo erectus,* the upright primates that stalked
around Africa on two legs more than three million years ago, could
do that.

Cave dwellers in Paleolithic times, the early Stone Age, left not only
art on their walls but soot on their ceilings. They tended badly vented
fires. They may have loved the blankets of smoke, which kept them warm
and gave their meat a special flavor, as much as twenty-first-century
Americans love their sport utility vehicles, huge, beetle-like shells of steel
holding cavernous volumes of air. Stone Age people defended their
smoky lairs, although the enclosed shelters fouled their lungs and made
some of them go blind. Fetuses brought to term in caves suffered from
the secondary effects of smoke inhaled by their mothers. And, if born in a
smoky cave, a baby inhaled soot with its first loud cries and gasps.

The first human air pollution of any real scale, however, waited for
the fabrication of tools and coins, around five thousand years ago. One of
the oldest known smelters—a Silicon Valley of the early Bronze Age, a
reporter called it—was found in the early 1990s by a team of archaeologists from the University of California. On a mesa south of the Dead Sea,
the archaeologists uncovered copper coins, axes, and other tools. Scat-

tered through seventy well-preserved rooms were crucibles, copper-working tools, ingots, and slag. Khirbat Hamra Ifdan, as the metallurgy center was called, had been destroyed by an earthquake around 2700 B.C. Similar smelting operations built elsewhere by the Greeks and the Romans also made copper coins, lead tools, and silver. Terribly inefficient (about 15 percent of the copper, for instance, swirled up the flue), but spectacularly impressive, the smelters were run by refiners who learned to cover their mouths with fabrics. "Otherwise," wrote Pliny the Elder in A.D. 77, "the noxious and deadly vapor of the lead furnace is inhaled."

Depositions hauled up from deep in Greenland's ice sheet have shown that the copper emissions of 2,500 years ago were not exceeded until the industrial revolution. Ice laid down between 500 B.C. and A.D. 300 contains lots of lead, the element of choice for Roman pipes, roofs, cisterns, wine goblets, and cooking pots. The finding helped confirm a long-standing hypothesis that the Romans poisoned themselves with lead made in smelters located far from cities and tended by disposable slaves. The Romans were apparently aware of the damages caused by lead—nervous disorders, insomnia, convulsions, and paralysis—even from the small amounts they consumed drinking out of lead goblets and from lead water pipes and eating from lead plates. The knowledge didn't persuade them to change their ways, however, ways that helped poison and collapse a civilization.

After the fall of the Roman empire, the next blip in the atmospheric record found in ice cores points to smelter activity during the eleventh and twelfth centuries. That was when the Song dynasty in China began minting coins and sending copper detritus aloft with high furnace temperatures and powerful updrafts. The trace metals rode the trade winds across oceans and continents, just as the smoke and the iridium from the bolide that exterminated the dinosaurs had ridden the winds. The copper sprinkled down through the atmosphere, settling on the snow fields of Greenland and compressing into half-inch-thick ice sheets over time from the weight of snow accumulating above. Glaciologists have found traces of Song dynasty copper above the copper and lead deposits left in ice sheets by the Romans, the Greeks, and the Bronze Age tool-and-coin craftsmen once working at Khirbat Hamra Ifdan.

It was during the eleventh and twelfth centuries that smoke from wood, dung, and coal, which was used for the intense heat required to make tiles and bricks, began leaving visible signatures on walls and roofs of cities around the Mediterranean basin and in West Africa. Maimonides, the philosophical physician and author of the infamous *Guide for the Perplexed,* an explanation of religious and metaphysical questions that had baffled the curious for centuries, traveled widely through some of the Mediterranean's smokiest dens. Maimonides stayed in Cordoba, the Muslim capital of Spain, a vibrant place of palaces, mosques, and a great library. He visited Cairo, an imperial walled city of nearly half a million and capital of the Mamluk empire, which controlled Egypt and the entire Fertile Crescent. On his travels, Maimonides said he breathed city air that was often "stagnant, turbid, thick, misty, and foggy." He concluded that such air affected the brain, causing "dullness of understanding, failure of intelligence and defect of memory."

It took almost five hundred more years for the first antismoke firebrand of note to appear. His name was John Evelyn. A contemporary of Robert Boyle and Robert Hooke (and the treasurer of the newly formed Royal Society), Evelyn lived in London in the mid-1600s, when coal was becoming a popular fuel on an island losing its forests but amply supplied with coal beds laid down during the Carboniferous period. Evelyn railed about the smoke blackening London's air. Going for a walk by Scotland Yard, he said he couldn't even see the palace of Charles II; it was too wrapped up in dark scarves of smoke twisting out of chimneys and ruining street-level visibility. Much of London was often shrouded in "a cloud of sea-coal," Evelyn lamented. "If there be a resemblance of hell upon earth, it is this volcano on a foggy day: this pestilent smoak corrodes the very yron, and spoils all the moveables, leaving soot on all things that it lights: and so totally seizing on the lungs of the inhabitants, that cough and consumption spare no man."

But, as Evelyn knew, coal had a lot going for it as a fuel. It burned hotter than wood, and longer. It was easier to transport. Found in veins that kept whole towns digging away for decades, coal was an energy sink close to the surface of the land and ready to be exploited. How much coal existed was beyond the ability of either industrialists or miners to calculate yet. They didn't know that the amount of carbon deposited in the earth's crust was proportionate to the amount of oxygen in the air. For every 100 tons of carbon dioxide that natural processes, mostly photosynthesis, had broken down over time, the result was 27 tons of carbon and 73 tons of oxygen. That's because a molecule of carbon dioxide is 12 parts carbon (roughly 27 percent) and 32 parts oxygen (roughly 73 percent). The air surrounding the earth contains approximately 1,000 trillion tons of oxygen, which means the earth's crust holds about 375 trillion tons of carbon, much of it coal. You can get some idea of how much carbon that is by considering that in an average year in the 1990s, the whole world burned around 7 billion tons of carbon in coal, oil, and natural gas reserves. That's a fractional amount, 0.0000185 percent, of the carbon in the earth, of the energy storehouse left from billions of years of carbon extraction from the sky as our atmosphere evolved. Back in the seventeenth century, no one knew any of this. People would probably have been astonished to hear that the deposits of fossil fuels in the earth were absolutely gigantic, veritable seas and rivers of carbon beneath the planet's skin.

Coal's by-products undermined some of the early enthusiasm for it, however. And especially for sea coal, which burned extra smoky because it burned wet. Sea coal not only offended John Evelyn, who wrote about it in *Fumifugium,* or *The Inconveniencie of the Aer and the Smoak of London Dissipated* (1661), but was also blamed for staining the roof and sides of Old Saint Paul's Cathedral by one of Evelyn's heroes, the Archbishop William Laud. Laud got so irritated at the brew masters in Westminster for creating the damaging smoke that he tried to fine them, intending to use the money for repairs on Saint Paul's. The brew masters told Laud to forget it, that they had a special royal exemption to produce smoke. The issue escalated and eventually reached Parliament, which sided with the brew masters. Laud, never a popular or tactful church administrator, soon found himself accused of treason and was executed. Whether the archbishop ended up on the gallows because of his audacious attack of those producing smoke that darkened the house of God is a little unclear. He went down in history as one of the first proponents of making polluters pay for damages they caused. His demise surely seems prophetic. Challenging smoke makers and making them pay would be viewed as heresy well into the twentieth century in a nation wedded to industrialism. John Evelyn avoided a similar fate by staying on the good side of the king. Evelyn relied on invective and print, rather than confrontation and fines, to make his point. As for remedies for "hellish and dismal" smoke, Evelyn didn't recommend fining brew masters. Invective was one thing; getting into it with those who placated the multitudes, as Laud had, was another altogether. Evelyn recommended reducing smoke by planting sweet-scented flowers and shrubs, by burying the dead in the country rather than in London's backyards, and by banishing especially smoky trades to other nations.

"Great Stinking Fogs," as the astronomer John Gadbury called them, scythed people down in London like the Grim Reaper during the late seventeenth and early eighteenth centuries. And in 1733, an English doctor named John Arbuthnot published *An Essay Concerning the Effects of Air on Human Bodies,* which claimed that the "unfriendliness" of city air killed many infants born and brought up in it. *A Philosophical Estimate of the Cause, Effects and Cure of Unwholesome Air in Large Cities,* published by A. Walker in 1777, refuted some of the vileness attributed to London's air and argued that lungs raised on city air could be irritated by too much exposure to fresh country air. Fifty years later, in 1845, the official keeper of Britain's death register challenged Walker's allegations. Walker had the situation backward, he declared, insisting that "smoke is irritating to the air-passages, injurious to health, and one of the causes of death, to which the inhabitants of towns are more exposed than the inhabitants of the country."

At the time, proving that smoke killed anyone, despite such assertions, was medically impossible. In London, it was easier to convince people that vitalism, a metaphysical remnant from the days of Paracelsus and alchemy, a long-cherished theory that human life possessed some ineffable spark, some godlike quality beyond the understanding of human beings or their new savior, science, was as likely to kill them as the smoky air they breathed. In *Bleak House,* which was published to great acclaim in the 1850s, Charles Dickens etched an indelible portrait of vitalism gone awry. In the novel, Mr. Krook, a slum landlord and the owner of a tenement called Tom's-all-Alone in London, sleeps in a chair in a ground-floor flat, a bottle of gin and a glass nearby, the air about him "unwholesome" and "stained." A melding of the greedy capitalists Dickens caricatured so vividly, Krook suddenly bursts into flame. He spontaneously combusts in a scene blending tragedy and comedy, a scene of vitalism gone wrong.

Upstairs, in another flat of Tom's-all-Alone, a boarder named Mr. Guppy, and his friend, Mr. Weevle, notice some awful black soot in the air. It's worse than the black grit common in the slums.

"Confound the stuff," gripes Guppy, "it won't blow off—smears, like black fat!"

Black fat, indeed.

When the twosome edge downstairs and into Krook's quarters shortly after the stroke of midnight—this is Dickens at his most moralistic and melodramatic—they can't find the sotted rat. Instead, there's "a smouldering suffocating vapour," and "a dark greasy coating on the walls."

Vitalism noir had scrubbed the evil Krook out of existence. He died, Dickens wrote, "the death of all Lord Chancellors in all courts, and of all authorities in all places under all names soever, where false pretenses are made, and where injustice is done."

Vitalism aside, in a city such as London, how much smoke was bad? And just how bad was it? These were hard questions. Yet answers were needed to put smoke, the scarf of progress, the symbol of money, in perspective. A century later, the same questions would be heard during debates over the health dangers from smog in Los Angeles, and then later over toxics such as benzene emitted from gas pumps. Even then, the establishment of cause-and-effect relationships between air pollutants and health would be an ongoing challenge and even tougher to prove. In comparison, the 1850s were a respiratory dark ages.

Not that research into respiration had stalled in Dickens's era. Since the death of Antoine Lavoisier a half century earlier (Lavoisier had been feverishly experimenting with oxygen consumption in the body at the time of his arrest, observing an assistant breathe through a primitive mask for hours while trying unsuccessfully to calculate just how much oxygen his lungs burned), respiration research had moved forward, but in fits and starts. The problem was that respiration was complicated. Gradually, scientists would understand that breathing and staying alive in a world of air involved three complex and interrelated functions—the mechanics of the breath, the circulation of blood to the cells, and the creation of energy inside the cells, or cellular metabolism. But in the early 1800s, the new field of physiology, a branch of biology that dealt with function, was just beginning to piece respiration together. And a huge hurdle was lung tissue itself. How did air pass through something that looked like liver but functioned in a dramatically different way? It was only in the late 1830s that a physics professor in Berlin, Gustav Magnus, proved that oxygen and carbon dioxide could diffuse simultaneously back and forth through the lungs. For his experiments, Magnus relied on "common people who for a modest sum permitted themselves to be bled," and a blood-gas analyzer that he devised to detect oxygen and carbonic acid, which indicated carbon dioxide in the blood. By midcentury, physiologists and chemists were closing in on the mysteries of hemoglobin, the blood pigment that carried oxygen to cells and hauled away carbon dioxide. But how cells burned oxygen, once it reached them via the lungs and the blood, and emitted carbon dioxide, remained a puzzle.

Of the promising work in respiration taking place at this time, the research by John Hutchinson, a physician at London's Hospital for Consumption at Brampton and an adviser for the Britannic Life Assurance Company, stood out because of its breakthroughs in mechanics.

Hutchinson was artistic—he painted in oils and played the violin. And he was restless. He spent countless hours walking the streets of downtown London, conducting field work during the early years of the reign of the monstrous dwarf, Queen Victoria. Hutchinson convinced more than two thousand volunteers—they included Thames Police, Royal Horse Guards, Chatham Recruits, Woolwich Marines, boxers, wrestlers, giants, dwarfs, and gentlemen—to blow into a primitive spirometer that he lugged around with him so that he could measure their vital capacity. (As mentioned earlier, vital capacity is the amount of air you can exhale after taking in a deep breath.) Administering what were basically on-the-street pulmonary function tests, Hutchinson learned that vital capacity decreased with age, weight, and history of lung disease. And that lung size was proportionate to height. He became so attuned to listening to people breathe that he often diagnosed lung ailments on the spot. Before sailing to Australia in 1852 to prospect for gold (he never returned, dying in Fiji in 1861 at the age of fifty), Hutchinson became the first physician to map lung function. He identified four distinct regions of the lung: one for tidal air, or the main breath; a second for an inhalation reserve; a third for a small exhalation reserve; and a fourth for residual air that stayed in the lungs between breaths.

An Empire of Smoke

The pattern . . . pioneered in Manchester was to be repeated, with variations, all over the world.

—MARK GIROUARD, CITIES & PEOPLE

At midcentury, when volunteers were blowing into John Hutchinson's spirometer and readers gasping over the demise of the evil Mr. Krook, London was the busiest city in the world, and England the world's first industrial juggernaut. London was not the hub of the nation's industrial might, however. That honor belonged to the Midlands. And anchoring the Midlands was Birmingham, a city of church spires and smokestacks and the former haunt of Joseph Priestley, who had preached at the New Meeting House in the 1780s and associated with James Watt, inventor of the steam engine. To the west of Birmingham rose the mountains of Wales, with their veins of coal ripe for exploitation. Off to the north loomed Manchester, or "shock city," as people called it. England's textile behemoth and a model of industrial growth, Manchester had gotten its nickname for the impression it often made on first-time visitors to this thriving, smoky, polluted version of the future.

Arrayed then in a relatively tight geographical region in the center of England, a small country with enormous military power and imperialistic ambitions, were the basic ingredients of the industrial revolution, which had already begun to disrupt the atmosphere in ways no one yet understood. There were carbon-storing mountains supplying fuel; there were carbon-burning cities providing an energetic mix of muscle and bone, organizational brain, and entrepreneurial drive; there were steam-powered railroads linking everything together. Rivers, secondary to the railroads, ferried coal-fired barges and ships, along with industrial wastes from factories to wherever they could be dumped. Manchester alone was served by three rivers and six railroads. The Liverpool and Manchester Railroad, opened since 1830, ran from shock city to the seaport. Liverpool's air was described as "a putrid miasma" above "the most unhealthy town in England."

By the mid-1850s, the air over many cities and towns across England was bad. But people liked it. It was the air of progress: dirty, gritty, oily. It carried the scent of money. In the competition for the worst air in England, London was a serious contender; however, the press gave the prize to Manchester. And it had only recently captured the sooty crown from Liverpool.

Manchester's crown, its skyline, was needled with smokestacks often puncturing a shroud of various hues. A testament to diversity, just like the number and variety of ethnic neighborhoods that drew the ambitious and the hungry off farms, away from villages, and from foreign countries, the city's smoke was an atmospheric signature. As the twentieth century would show, polluted air can be red, gray, coppery, green, and other colors, depending on emissions. As a result, cities around the world scrawl unique overhead signatures honoring their key industries. For instance, the grain mills of Minneapolis leave a different atmospheric signature than do the tailpipes of Los Angeles, and Pittsburgh's steel foundries loft different particles into the sky than do Mexico City's smoldering dumps. Signature air can have a taste: greasy from the dance of oils, grainy from the roasting of grains, sharp from the percolation of acids, acrid from traces of dyes.

Manchester amazed, intrigued, and frightened first-time visitors. Many arrived aboard elevated trains that chugged past soot-blackened tenements, eight-story mills, and vast, stained warehouses. Manchester was a "crude necessity" to some, wrote the Scottish historian Thomas Carlyle, "as fearful, as unimaginable, as the oldest Salem or prophetic city." The reactions to Manchester would be repeated by those appraising Athens, Delhi, Pittsburgh, and other cities where industry thrived and coal burned all day and all night. Visiting Manchester in 1858, Hippolyte Taine wrote in *Notes on England* about "a sky turned coppery red by the setting sun; a cloud, strangely shaped resting upon the plain and under this motionless cover a bristling of chimneys by the hundreds, as tall as obelisks." Marxist theorist Frederick Engels once described the dye-clogged Irk River as "a long string of the most disgusting, blackish-green, slime pools . . . from the depths of which bubbles of miasmic gas constantly arise and give forth a stench unendurable even on the bridge forty or fifty feet above the surface of the stream." The author of *The Conditions of the Working Class in England* managed to bend the ear of a local tycoon and ask about the grievous conditions for laborers in the mills. But the tycoon cut Engels short, declaring, "There is a lot of money being made here. Good day, sir."

Money, jobs, and promise made Manchester exciting and complex. In a mythical sense, the city inspired the gods. Not the ancient sky gods who would have choked and gagged and been hard pressed to pick out targets for their lightning bolts, or objects for their insatiable lust, but the new gods, those of industrial power and production and progress. Manchester also paid homage to underworld gods, to the earth mothers whose veins of energy fired its needs. Maybe the city's mythical qualities, its hellishness overlaying its economic heavenliness, were what combined to get the place under most people's skin. Whether a visitor shuddered and thought of hell or cried out, "This is the most beautiful city in the world," he or she seldom left town indifferent to Manchester.

Mostly, though, Manchester was admired and copied. In the growing number of industrial cities of the world, factories were praised, and smoke and other air woes ignored or denied. Jobs, wages, customers, and profits meant more to the average person than air, soil, water, and health—at least until he or she got sick. Denial that smoke, fumes, and coal dust hurt anybody in the most beautiful city in the world, and its wannabes in other nations, was relatively easy. Science had figured out air's role in fire, but remained perplexed by respiration and what damaged it. Common law, which ruled on nuisances in England, as well as the United States, didn't know what to make of smoke. Besides, in the mid–nineteenth century, the courts were wed ideologically to industry. In virtually all the industrial nations, the philosophy was the same: Strangling a person with smoke was OK; strangling the economy by forcing industrialists to deal with smoke was not.

Air Fights: First Rounds

Here were a few, perhaps a hundred gentlemen connected with the different furnaces in London, who wished to make two million of their fellow inhabitants swallow the smoke. . . .

—LORD PALMERSTON

Despite the daunting odds facing them, in England, some reformers began challenging the smoke makers over health and stench issues. They quoted Macbeth: "No pure air in the cities." They damned the "dark

Satanic mills" of William Blake. They harped on public officials and pushed for safety rules, ventilation systems, hospitals, and open spaces, especially after epidemics associated with miasma, still the popular culprit for a region's dank and smoky winds. Social critics insisted that people had the right to live in so-called new Jerusalems of light and cleanliness, rather than be forced to exist in evil Babylons of smoke and dust.

In 1842, seeing too much of Babylon in Manchester, the vicar of Rochdale founded one of the first antismoke groups. Called the Manchester Association for the Prevention of Smoke, the group's laments fell on few sympathetic ears in smokestack city. In 1853, the first antismoke ordinance appeared in London. The statesman Lord Henry John Temple Palmerston took on "these smoke-producing monopolists," as he called them, "a hundred gentlemen connected with the different furnaces in London, who wished to make two million of their fellow inhabitants swallow the smoke." Enforcement of the ordinance was left up to the metropolitan police, few of whom knew much about smoke. They came down hard on furnace-thick Southwick, but let other industrial neighborhoods smolder. Fines were numerous but small, a nuisance more than a force of change.

Ten years later, as prime minister, Lord Palmerston again took on the smoke makers, and this time, he was allied with the Earl of Derby. The two elderly statesmen were a formidable pair. By 1863, Palmerston had been in government service continually for fifty years. He'd joined the War Office during the Napoleonic era and then escaped an assassination attempt as secretary of state. In this encounter, a mad army lieutenant (the officer had cut off his own penis and then applied for a disability pension, which Palmerston's department denied) tried to shoot the secretary as he was climbing the stairs to his own office. Unfazed, Palmerston insisted he was "too busy" to have a doctor examine him. Later, he consented to an examination by his personal physician, who found burns and a bruise where the bullet had grazed him.

The Earl of Derby was famous for his efforts to abolish slavery in Britain in the 1830s, and like Palmerston, probably also drew the ire of much of the conservative gentry. Teaming up, the two aristocrats engineered passage through Parliament of the Alkali Act of 1863.

The facts behind the passage of the act, which became the first effective air quality law and was one that survived into the next century, were instructive. In the early 1860s, the Earl of Derby had gotten sick of seeing trees dying around his manor, Knowsley Hall, and was offended by the smell of rotten eggs wafting his way, along with smoke, from nearby Saint Helens, a town with a number of salt-glazing pottery works. John Doulton, owner of one of the pottery works whose stink irked the olfactories of the earl, said he thought smoke was beautiful. Factory fumes that smelled like rotten eggs were not acceptable effluvia, the earl replied. And lots of smoke only made things worse. Agreeing with the Earl of Derby, Lord Palmerston joined him to usher an omnibus antismoke bill through Parliament. The ensuing fight over smoke pitted aristocrats against industrialists, and the bill emerged a shadow of what its supporters had hoped for. It ignored the bigger issue of smoke and fumes across the landscape and focused on a single, easily identifiable pollutant, alkali. Alkali was the culprit associated with the earl's original complaint about a rotten-egg stench smothering the air around Knowsley Hall. Iron, copper, and chemical manufacturers emitting smoke escaped the admonishing hand of the law, at least for now.

The Alkali Act of 1863 brought antismoke forces a surprisingly easy, though modest, victory. One with several lessons. The first was that a skirmish over a select, known pollutant was a lot easier to fight than a war against ubiquitous smoke. And alkali was a mighty big pollutant in its own right. It had acquired a reputation decades earlier, when the largest chemical operation in the world—Charles Tennant's sprawling, hundred-acre Saint Rollox soda works of Glasgow—spewed sulfuric acid and clouds of hydrochloric gas over the Scottish countryside. Aggravating the outrage was a 455-foot smokestack that Tennant had built to send the alkali by-products higher into the sky. Taller than any cathedral in Britain, the Saint Rollox soda works became visible for fifty miles around, a sign of industrial progress few could miss or ignore.

The second lesson of the Alkali Act was a corollary of the first: Air was too big for an all-out war over pollutants. A third lesson was that the resources needed to slow industrial pollution from its headlong gallop were beyond any nation's willingness to spend, its treasury to supply, or its people to support. As bigger air battles would soon show, the air pollution in the skies over England and elsewhere was a complex, amorphous, and often invisible entity beyond the grasp of science to understand or the common law to control. In addition, the opposition to reform was powerful, united, and favored by the courts. In 1863, the narrowness of focusing just on alkali producers as the cause of air pollution may have disappointed the Earl of Derby and Lord Palmerston. But the industrialists probably did the two aristocrats a favor. The bill's opponents turned a potentially protracted and indecisive battle over smoke into a focused skirmish over alkali, an industrial product whose two by-products, sulfuric acid and hydrochloric acid, were air pollutants whose chemistry was understood. Going after the two acids promised measurable results. Going after smoke would have been futile.

Production of alkali then was a simple, two-step process. During the first step, salt was combined with sulfuric acid, which produced sodium sulfate, with hydrochloric gas as an unwanted by-product. In the second step, sodium sulfate was burned with coal and chalk, resulting in soda (sodium carbonate), which was used in soap production and glassmaking, and in another undesirable by-product, calcium sulfide. When the soda operations ran strong, the by-products mixed in the air, causing a rotten-egg smell (hydrogen sulfide) that spread far and wide if the winds were cooperating.

The new act mandated that all alkali producers condense a minimum of 95 percent of their hydrochloric acid gas. It set up a pool of inspectors with scientific training, a new precedent since the ranks of Victorian inspection departments were typically filled with illiterate, untrained laborers more comfortable downing ale in doss-houses than testifying before judges in courts. The act was effective in large part because of the tact of the inspectors; they used persuasion rather than litigation. Alkali producers grumbled, but most installed cleaning towers based on a model designed thirty years earlier by William Gossage. Gossage had cleaned up his stink works in Stoke Prior, then patented his invention—a tower packed with brush, down which water trickled, absorbing much of the hydrochloric gas emitted during the manufacture of soda. Sales of Gossage towers had been slow at first, since there was no market for hydrochloric acid. But the new law forced alkali producers to install towers, cutting their emissions but leaving them with a lot of hydrochloric acid on their hands. Many of them just dumped it into nearby ponds and streams.

The rotten-egg stench wafting over Saint Helens and other alkali-producing towns didn't die down for long. The demand for alkali boomed because of the brisk sales of soap and other cleaning products, as well as glass. The undesired by-products increased proportionately. In addition, the crude, low-tech towers were not as effective at capturing hydrochloric gas as thought, nor were the measuring tools of the Alkali Inspectorate all that accurate; they allowed more of the gas to escape the towers than was reported.

Rotten-egg smells, however, were but one part of a growing air-quality mess in the richest nation in the world. In addition to the alkali-related stench, odors also rose from pig farms and bone-boiling shops making glue, from tallow melters, from leather tanners, from slaughter-houses, and from dumps. Rank smells, smoke, and dust prowled the land, or worse, just hung in the air. On calm days the mix could lid a city with a rank and gritty atmospheric stew that people could almost eat. In 1873, a Royal Committee of Noxious Vapours was appointed to appraise the worsening situation. Committee members found themselves pretty overwhelmed and, at times, almost asphyxiated. Checking out a dunghill in Plymouth in 1874, one member reported finding "a vast heap of reeking filth which I never remember to have seen equaled for character and quality, either at home or abroad during my whole life."

In England, the Public Health Act of 1875 officially condemned a long list of bad smells and identified what made them. In 1881, the Alkali Act was broadened to cover more onerous chemicals. One chemical was sulfuric acid, an oily liquid destined to become the most popular chemical in the industrial world and a main ingredient of acid rain. Another was ammonia, a colorless and pungent gas popular for fertilizers and named after the Ammonians, a civilization that had worshiped the Egyptian deity Amun (the Ammonians used ammonium carbonate, or smelling salts, in their rituals and ceremonies). The revised act still had holes big enough to let plenty of stink through. One hole was an allowable defense called "best practicable means." Best practicable means meant what it said. If a manufacturer did the best he could and it was practical, he had a defense. Still, as a pioneering effort in regulating air pollution in laissez-faire Britain, the acts pointed fingers, trained inspectors, and tactfully wrestled the biggest problem of all, the enforcement of laws that applied subjective standards that courts seldom upheld.

In England, those who couldn't stand smoky Babylons, or bad smells, could move to less smoky Jerusalems. They could relocate to villages in Cornwall or Shropshire. Not many did. They couldn't afford it or worried about finding jobs.

But there was a new option: the suburbs. In the late nineteenth century, suburbanization began in earnest. Linked to city centers by trolley tracks, suburbs grew not only in England but in other countries followings its lead: the United States, Germany, Russia. A growth pattern emerged based on a precedent first established and expanded in Manchester as the city had boomed. Following the pattern, a city center, where industry usually had begun alongside a river, became the service hub for all the growth around it. Land prices in the center rose, and congestion increased as everyone came there to shop, bank, and do business. Streets were widened, roads and rail links built to outlying towns, and neighborhoods seeded where the city abutted the retreating countryside. The scale of the city center kept increasing: bigger banks, train depots, hotels, markets, and department stores, along with theaters, opera houses, and parks. If the scale became big enough, soon a metropolitan area surrounded the original core of the city. The pattern was analogous to the growth of communities of bacteria that had transformed the Archaean world billions of years earlier, but faster.

The growth pattern transformed Manchester, and it became the template for dozens of other cities. There was little containment, no membrane separating city from countryside. The city swallowed the country relentlessly, inexorably, with little time for second thoughts about lost villages and farmland. By 1900, Manchester was one big metropolitan sprawl. In 1910, its population topped two million. Air problems grew proportionately. Once pristine suburbs offered less relief from smoky air. So people moved further out into the country, if they could afford it. It was way out in the country, on the estate of Clifford Chatterly, that two of D. H. Lawrence's characters in *Lady Chatterly's Lover* discussed the state of the atmosphere in early twentieth-century England:

> "It's amazing," said Connie, Lord Chatterly's wife, "how different one feels when there's a really fresh fine day. Usually one feels the very air is half dead. People are killing the very air."
>
> "Do you think people are doing it?" he asked.
>
> "I do. The steam of so much boredom, and discontent and anger out of all the people, just kills the vitality in the air. I'm sure of it."
>
> "Perhaps some condition of the atmosphere lowers the vitality of the people," he said.
>
> "No, it's man that poisons the universe," she asserted.
>
> "Fouls his own nest," remarked Clifford.

The fourth child of an illiterate coal miner, Lawrence knew what he was writing about. He'd been born and raised in Nottinghamshire in the East Midlands, and pulmonary disease had invaded his lungs at an early age. His initial attack of pneumonia, in his late teens, cost Lawrence his first job. He soon became a writer. He spent many days of his life in some of the world's more pristine places, wheezing and coughing, and died at forty-five from tuberculosis.

Lawrence shared none of the ambiguity toward industry that many reformers felt. He reviled and steadfastly denounced it, along with the conditions that industrialism bred. "Ours is essentially a tragic age," he wrote in the opening lines of *Lady Chatterly's Lover*, "so we refuse to take it tragically. The cataclysm has happened, we are among the ruins, we start to build up new little habitats, to have new little hopes. It is rather hard work: there is now no smooth road into the future: but we go round, or scramble over the obstacles. We've got to live, no matter how many skies have fallen."

The morality of the dispersion of air pollutants became a conspicuous feature of industrialism during the twentieth century. The affluent usually could get out of harm's way. The better neighborhoods were situated upwind of factories in most industrial cities. For the poor, hard work and more income held out the hope of moving away from smoke and squalor, common urban partners. Some workers pulled it off. Most stayed put. That the poor often lived in grime, dust, and soot close to their jobs merely proved their inferiority to many of the well-off, who had the capital to do as they chose.

"Hell is a city much like London," Percy Shelley had written at the outset of the nineteenth century. Robert Southey, his friend, added that London was "a compound of fen fog, chimney smoke, smuts, and pulverized horse dung." By 1863, when Henry Coxwell and James Glaisher went high enough over Birmingham to have a grand, albeit short-lived, view of the new industrial paradigm below, fen fog and smoke had increased immeasurably. The noxious haze blanketed not only London but also many other English cities and some of the green countryside as well. By the century's end, more than 100,000 steam engines burned coal and most Englishmen thought it was a God-given right to be able to sit in front of an open hearth, smoking a pipe, sipping tea, and watching smoke rise off the glowing coals. Around the world, a world following England's lead, coal consumption had increased a hundredfold since 1800.

Babylons ruled. Life went on. Impacts of smoke, fumes, and dust on lungs and plants and surfaces of all sorts were dramatically worse than they had been, but remained impossible to quantify. No instruments could yet easily identify, never mind quantify, levels of arsenic, cadmium, nitrous oxides, carbon monoxide, or sulfur dioxide, all of which came from burning coal. Smoky industrial regions made goods in Belgium and Germany, the Ukraine, the Osaka region in Japan, the American Midwest, and sections of India and Australia. Like the Midlands, the regions were crisscrossed with railroad tracks and close to natural resources, especially coal. And no matter what the regions' smokestacks pumped out, went the popular defense, it could not harm the air. The allegation that smoke damaged the air was preposterous, the industrialists claimed. The sky was just too big; wind blew the smoke away. It was certainly more profitable and less bothersome to put a positive spin on smoke.

Across the Atlantic, in Chicago, an American businessman named W. P. Rend did just that. In a famous declaration made in 1892 to justify his position, Rend said, "Smoke is the incense burning on the altar of industry; it is beautiful to me."

Perils of Smoke

My strength is dried up like a potsherd; and my tongue cleaveth to my jaws; and thou has brought me into the dust of death.

—PSALM 22

Smoke was probably beautiful to W. P. Rend because he didn't have to breathe it all the time and he knew very little about respiration. In a speech to civic leaders in Chicago, the coal merchant said carbon in the lungs purified the air before it reached the blood, sort of like a charcoal filter. On another occasion, Rend went to Pittsburgh to defend smoke before a group of emancipated women.

In Pittsburgh, Chicago, Cleveland, St. Louis, and other industrial U.S.

cities, there was plenty of smoke to defend because during the last twenty years of the twentieth century, the United States competed head-to-head with England to be the world's industrial Goliath. The Midwest was America's Midlands, but spread over a much larger region. From Buffalo to Chicago, from Cleveland to Baltimore, iron works, chemical factories, steel mills, and other industries appeared and flourished. Bigger coke ovens were built to convert coal into the denser, hotter-burning carbon needed in blast furnaces. The coke made it possible to extract iron, which went into the manufacture of high-quality tools and machines, from ores. Hotter furnaces sent a wealth of by-product swirling aloft with smoke and soot. In the Midwest and the East, coal-fired locomotives hauled heavier loads on more railroad tracks. In Chicago and New York, coal-fired boilers heated the first generation of skyscrapers, ten- to fifteen-story buildings whose decorative cherubs and floral patterns didn't look so attractive layered in black.

Despite the sooty assaults on American cities, England suffered even worse air problems. British fog had achieved almost mythical status as a blurry monster with powers that could turn day into night, make people stumble into rivers, force moths to darken their camouflage to survive, and invade Parliament despite special barriers installed to keep fog out. Compared to England's dirty fog, America's dense gray days seemed almost benign.

Though not that benign. "Hell with the lid taken off" was how one visitor described Pittsburgh during this time.

America's smoke was bad. Following British precedent, some cities passed smoke ordinances. Cincinnati acted first, in 1871. Chicago followed suit in 1881. Then more than two dozen cities passed similar ordinances, most of which were paper tigers. All snarl, no bite. Merely defining smoke proved to be an exercise in futility and juvenile science, usually buttressed with the same descriptive adjectives, "dark" and "gloomy," which appeared in one ordinance after another. To initiate a complaint, you had to make sure the smoke was "dark." How dark? As dark as the greasy coating on the walls of the devitalized Mr. Krook in Tom-all-

Alone's? As dark as the soot said to have driven a newlywed in Pittsburgh to commit suicide, leaving behind a note blaming the smoke? Like white in the Arctic or green in Ireland, *dark* had more shades than ordinance writers ever imagined.

Courts weren't much help. The classic defense of smoke, alluded to for decades by defendants in cases that even made it to court, was written by the Pennsylvania Supreme Court in 1871. The opinion justified a brick maker's right to damage his neighbor's orchards with his atmospheric waste. From a mythical standpoint, by the 1870s, in the industrialized world epitomized by the United States and England, the Garden of Eden *was* secondary to a foundry. Explaining the rational for its decision, the court said that a brick factory had precedence over an orchard near a big city because of its greater benefits to more people. "The people who live in such a city or within its sphere of influence do so of choice," said one judge, writing for the majority, "and they voluntarily subject themselves to its peculiarities and its discomforts, for the greater benefit they think they derive from their residence or their business there." Echoing the reasoning in Pennsylvania, a judge in New York ruled on a similar case: "If one lives in the city he must expect to suffer the dirt, smoke, noisome odors, noise and confusion incident to city life."

Emancipated American women didn't buy it. They began to insist that smoke was synonymous with a lapse in values, a blindness to aesthetics, a disregard for the health of children. Smoke, they said, was immoral. Riding the high horse of Victorian morality and energized by a growing awareness of their own power, women organized in groups to oppose smoke and to fight for suffrage. They were called the Wednesday Club in St. Louis, the Woman's Club in Cincinnati, and the Women's Health Protective Association in Pittsburgh. The women were not naive. Nor were they opposed to progress as it was epitomized by science, technology, and medicine, three of the hot and promising currents charging the Age of Empire, the peaceful and optimistic era between the 1870s and the start of World War I in 1914. Because of their efforts to become voters in the United States, the women knew how to float petitions, conduct rallies, and seed political initiatives. They were also accustomed to defeat (suffrage was not approved until 1920, with passage of the Nineteenth Amendment). And they recognized the fecklessness of many male politicians, some of whom were more eager to admire the women's audacity than to react to their seriousness. With the smoke issue, the women lacked the grim drama of London's killer fogs to point out, so they emphasized smoke's ravages of beauty and joy, of goodness and cleanliness. Occasionally, the women linked forces with more zealous activists who threw everything but the kitchen sink at America's smoke, claiming it caused tuberculosis, bronchitis, pneumonia, and whooping cough, as well as low energy, depression, criminal behavior, and suicide. Mrs. John B. Sherwood, president of the Chicago Woman's Club, said the city's smoke was "responsible for most of the low, sordid murders and other crimes with its limits." When the newlywed in Pittsburgh, probably the smokiest city in America, committed suicide and left a note blaming smoke, some women in the antismoke crusade must have felt despair but also a little revitalized. The poor woman had just been pushed over the edge by all that soot.

For support, women often approached doctors sympathetic with their cause. They invited physicians to attend meetings and to voice support for cleaner air. At a meeting of the Engineers' Society of Western Pennsylvania in 1892, the meeting in Pittsburgh attended by W. P. Rend, the Women's Health Protection Association of Allegheny County brought along Dr. Sutton. After an engineer, a member of the engineers society, stood and declared, "there is nothing particularly unhealthy about smoke; on the contrary, it may mitigate other and worse evils," Sutton refuted that contention. He mentioned a recent study in the respected British Medical Journal linking smoke and disease.

That was Rend's cue. He stood. "Now, I am not a doctor," he said, "but if I was, I probably would differ with that gentleman."

Some of the liberated women in their high-collared gowns and elaborate hats, a few with fans by their chins, stared at Rend. His reputation had preceded him. The women knew of his speech given to the Union

League Club of Chicago the year before, when he glorified smoke as the incense of industry. Some had heard his medieval-like pronouncement that carbon in the lungs actually helped purify the air a person breathed. For his part, Rend probably wished he was facing a group of bustle-wearing beauties of a decade earlier, rather than these attentive females in their hourglass-shaped gowns with something to say and the courage to say it—even if they had brought along a doctor to give them an edge with the engineers.

Despite Rend's appearance, the engineers society decided to support the women, who soon sat down with the Pittsburgh city fathers, urging them to pass a smoke ordinance. The city officers did approve an exceptionally weak one, gerrymandered so that most industrial zones in the city avoided any controls. Steel and iron works remained virtually untouched. As a result, Rend got his way after all.

Not that a better ordinance would have made much difference in Pittsburgh, or anywhere else in America, for very long. Judicially, the country was not ready for curbs on industry. Especially controls championed by women whose lifestyles were often funded by husbands employed by the smoky enterprises their wives abhorred.

Research in respiration was some help, but not enough, despite articles in foreign journals, such as the one that Sutton had cited. Since John Hutchinson's pioneering work on respiration mechanics around 1850, advances in an understanding of the diffusion of gases in and out of the lungs and blood and of hemoglobin's role in carrying gases to and from cells had advanced considerably. In the early 1860s, the brilliant German physiologist and chemist, Felix Hoppe Seyler, explained how the blood pigment hemoglobin carries oxygen to cells and hauls carbon dioxide away, just as a barge hauls raw materials to a mill and waste away from it. Meanwhile, two professors in Bavaria, Carl von Voit and Max von Pettenkofer, teamed up. With the help of generous funding from King Maximilian II, they built a big, complicated apparatus—its parts included a steam engine, gasometers, absorbent compounds, and a sealed chamber. Inside the chamber the professors watched "the secrets of metabolism unroll before their eyes." The secrets were sweat, carbon dioxide exhalations, and other metabolic evidence collected from a human guinea pig who biked, ate, and slept inside the sealed environment for days. Von Voit and von Pettenkofer used the evidence to calculate how much fat, protein, and carbohydrates the man burned, or metabolized, in his cells. In the 1870s, Eduard F. W. Plfuger, another German physiologist, theorized that it was the difference in partial pressure between blood in the capillaries and in the tissues that diffused gases back and forth between them. He also postulated that the tissues controlled the rate of exchange and that blood was only a delivery system. The role of carbon dioxide, rather than the more plentiful oxygen, as the control gas of respiration was confirmed by F. Miescher-Rusch in the 1880s. He showed that carbon dioxide traveling in the blood as carbonic acid was the key to breath control, although how the control worked remained a mystery.

All this information was well and good. Physiologists at the end of the nineteenth century had solved most of the mysteries of respiration. In the first decade of the twentieth century, experiments by the Danish team of August Krogh, and his wife, Marie, proved that partial pressure and diffusion through tissues totally explained gas exchange. Combining lab work, mathematics, and theory, the Kroghs showed that no matter how hard the body worked, or how varied its metabolic needs, diffusion could handle the transport of oxygen through the lungs, aboard hemoglobin, and through cell membranes. The researchers further showed that carbon dioxide returned along the same route to be exhaled. The biochemistry of the primal switch, the speck of flesh in the medulla where the chemical decisions triggering respiration are made, would have to wait for more sophisticated instruments and greater understanding of neurotransmitters before giving up the mystery of its complex and fundamental role in the breath. This would not happen until well into the twentieth century, after physiologists had figured out how metabolism works inside every cell of the body.

Today, we know that the action inside each cell occurs in sequences along specific paths. A single cell in your body can have thousands of different pathways. A separate reaction can take place on each pathway simultaneously, and each reaction can repeat itself without upsetting the reactions on the adjacent pathways, everything regulated by genetic and chemical controls that are functioning inside the cell as well. It's no wonder that our cells, with so much going on inside them, have no room to store extra oxygen, extra glucose, or much more of anything else. Or that they must be fueled steadily by those masterful waiters, the hemoglobin pigments circulating continuously in a closed loop that passes through our lungs for gas exchange, our hearts for momentum, and our cells for the heavy lifting that gives us energy. If we are forced to go without food, our blood can burrow into the backup fat carried in love handles, or in little pockets of fatty tissue in our thighs, shoulders, and jowls. But without oxygen, we've got only about four minutes. There is nowhere for oxygen-hungry blood to go inside the body but to the lungs. Our cells demand oxygen right now for their repairs, their heat, their growth. We've evolved as a species that presumes a bountiful oxygen supply will be available to our lungs at all times. All we have to do is breathe.

In the 1890s, physicians knew that too much smoke could really mess up a person's respiration. But they didn't know how or why exactly. So few of them were ready to condemn industrialists or their smoke as a public health threat. There simply was no cause-and-effect relationship between smoke and health; even a doctor working full time in smoky city slums could not identify such a relationship with any reliability or proof.

When thinking about physicians and their apparently slow response to the dangers of smoke, remember that in the 1880s many doctors were just getting used to washing their hands. Germ theory, developed by the French biologist Louis Pasteur in the 1870s, had identified microorganisms called bacteria as the cause of disease. Pasteur proved that the practically invisible germs were transferred by hands, soil, water, and air. He'd made his revolutionary discovery assisted by air. Another master of the simple yet profound experiment, Pasteur simply heated meat broth in a tube, then kept the broth cut off from air. The broth did not decay. Once Pasteur exposed the broth to air, it quickly was infested with bacteria, which demonstrated germs' invisible presence virtually everywhere.

Physicians initially mocked germ theory, just as they had mocked Ignaz Semmelweis's mid-1800s recommendations that doctors should start washing their hands before examining patients. A Hungarian teaching medicine in Vienna, Semmelweis had instituted unusual policies after noticing that students in his hospital went from handling cadavers in the dissecting room to handling mothers in the delivery room without washing their hands. Motivated by the high death rate of new mothers from puerile fever and by his suspicions that unwashed hands contributed to the tragically high figure, Semmelweis began having students clean their hands thoroughly after dissecting cadavers and before entering the delivery room. Mortality from puerile fever dropped dramatically, but Semmelweis's superior at the hospital grew angry at the young doctor's methods. Soon other Viennese doctors also criticized Semmelweis and he was denied a promotion. Frustrated and angry, he returned to Budapest and instituted a regimen of cleanliness at another hospital, with positive results. Semmelweis wrote a book about his methods only to have it virtually ignored by the medical community. He suffered a nervous breakdown in 1865, managed somehow to cut a finger and get it infected, and soon died from the same infection he had fought relentlessly to eliminate from his hospital's wards.

Despite Semmelweis's tragic experience, in the mid-1860s some surgeons began to use a carbolic acid solution spray to kill germs in the operating room. And once Pasteur's germ theory emerged in the next decade, it ushered in a new era of concern for cleanliness. Germ theory also led a German doctor, Robert Koch, to discover that a rod-shaped bacterium was the cause of tuberculosis. In 1882, after eight years of work, Koch published a medical first, an etiology of a bacterium, a description of the origins of tuberculosis, one of the nineteenth century's more relentless killers. The disease was associated closely with the smoky cities that industrialists and politicians defended and that courts, both in England and America, upheld as reasonable places for people to live and work.

Not that coal smoke wasn't reeking havoc in the lungs of hundreds of thousands forced to breathe it almost constantly. And that coal dust wasn't worse.

Neither W. P. Rend nor other Midwestern smoke barons were making trips to the mining towns in West Virginia and Pennsylvania, where
coal miners could be found sitting around with club fingers and coughing
a lot. Nor did the industrialists visit the blackened backwaters and praise
the incense of industry or claim that the carbon in miners' lungs made
them healthier. Rend and his peers might have been shot. In Appalachia,
in the Ruhr, in Wales, in the coal regions of Ukraine and China, miners
didn't know much about respiration. But many knew they were dying
from the air they breathed, even as coal company doctors and officials
usually insisted it wasn't so.

A poison is a poison because of the dose. Excess carbon in the lungs
was, and is, a methodical pathogen. Coal smoke is different, less damaging in the short run than coal dust, but ruinous in the long. Breathing
coal dust initially causes bronchitis, an inflammation of the throat and
bronchial tubes. Prolonged breathing of the dust, which is mostly carbon, overwhelms the airways and damages the tissue itself, resulting in
fibrosis, or black lung. Anatomists of the nineteenth century who cut
open miners often found lungs rigid with carbon. Miners afflicted with
black lung often sought the nineteenth-century version of Paracelsus's
liquor alcahest. That is, a carbon solvent for their ailment. But neither
charlatans nor pharmacists had yet concocted a carbon solvent that could
scour the alveoli clean of the micron-small particle deposits that robbed
miners of adequate oxygen. More honest doctors than those employed
by coal companies weren't much help. Typically, they gave the same
advice Paracelsus had given miners in the Villach four centuries earlier:
Get out of here.

To get a clear picture of the magnitude of the damage going on in the
lungs of some miners, again imagine the tennis-court-sized surface that
represents the area of your lungs if they were laid out flat. Envision the
thin coat of blood spread beneath the tissue, one molecule thick and
flecked with hemoglobin. This blood respiratory pigment is eager for
oxygen molecules to bond to heme, the pigment's iron components,
which secure oxygen for the long slosh to all parts of the body. Now, with

every breath you take, think of the sheet of blood being swept away, like pages snapped out of sight in a copying machine. A tennis-court-sized copying machine. This is the scale and rate of the blood and gas exchange chemistry going on in the most inner recesses of the lungs at every breath. The blood, roughly a pint in an adult, supplies about 300 million alveoli. If 30 or 40 million of them are fouled and malfunctioning because of carbon, there's enough left to keep the oxygen and carbon dioxide exchange going. But as the carbon spreads and more alveoli black out, complications multiply because the lungs can't supply the body with enough oxygen or remove carbon dioxide efficiently.

In the mining towns of West Virginia, Wales, and elsewhere, a coal miner thought he was sick when he developed a dry cough that wouldn't go away. Or his fingers swelled up in a troubling physical condition called club fingers. How fast the disease progressed varied. One miner might take sick and die in a year. Another might survive for thirty. But the average time between the onset of the dry cough and the procession through a cemetery gate was five years. Even if a miner did escape the landscape of coal shafts, slag heaps, and the company town, the carbon load in his lungs could be lethal. If the respiratory defenses had been breached and fibrosis set in, the prospects of growing old were slim.

Nothing that had happened during the last ten thousand years upset the atmospheric balance, a balance not even vaguely understood in the 1890s, more than the increase in carbon emissions of the industrial revolution. And no group paid a heavier price for the change than did coal miners. Anyone going into the mines knew the tunnels laden with dust were bad for the lungs, but until there was proof of damages from the dust and courts were no longer ideologically allied with industry, there was little justice.

So, although the accumulating evidence that infectious germs caused tuberculosis, pneumonia, the flu, and other diseases by taking up lodging in bronchial tubes and tissues seemed to make assertions like those of Rend somewhat ridiculous, the contentious statements were welcomed by industry and reported in newspapers. Industry—coal-fired industry—

occupied the catbird seat during the Age of Empire. And so, despite the ever-increasing, mostly empirical but sometimes scientific and medical evidence of smoke's perils, coal and its smoke remained sacrosanct.

Nor were by-products seen as profit-making to a significant degree. In fact, by-products were a new concept, despite being as old as humankind. Before the industrial age, people used mostly everything. Waste was minimal. But with the high temperatures of carbon-fired enterprises, and with large-scale industrial production, wastes became voluminous. They had to be profitable to be dealt with aggressively. The hydrochloric acid collected from the Gossage towers and dumped because industry had yet to find a commercial use for the acid was a good example of a toxic by-product recaptured only to be thrown away. Gases and smoke were the main by-products of the burning of coal and coke, and manufacturers usually wanted to get rid of them; they were a nuisance.

In the United States and England, common law dealt with nuisances, both private ones and public ones. But to win a case, you had to establish a cause-and-effect relationship. That remained virtually impossible, especially as the sources of smoke and pollutants increased and made one big atmospheric blend overhead. The idea that the public had a right to clean air, to air devoid of nuisance, just as a person had a right to the pursuit of happiness and to make a dollar, was entertained in the United States during the 1890s, but never gained broad acceptance. This tenet would become popular almost a century later, in the 1960s. The challenge of dealing with smoke in the United States and around the world would have morphed into a need to face atmospheric problems so pervasive that the civilized world often did its best to deny their importance and to ignore their developmental histories.

A paradox facing clean air crusaders in the United States in the 1890s was that smoke caused no single big problem that terrorized people. For instance, it didn't stoke fears of an epidemic like a ship of rats hauled to midtown and left at a dock would have. Politicians weren't taking to the streets and pointing angrily at smoke and fumes as killers, or ushering asthmatic kids into city parks, the so-called lungs of a city, for a choreographed cough-a-thon attended by newspaper reporters. To be more than an eyesore, more than an aesthetic headache, air needed a calamity. A Pompeii. A Black Death. Then reformers, whether emancipated or just enraged, wouldn't have to convince anyone that air pollution was evil.

But as air battles would show repeatedly throughout the next century, there was no Pompeii, no Black Death—at least not yet—of the atmosphere. There would be radioactive fallout from atomic bomb testing. There would be more London killer fogs that rivaled the respiratory hell of the winter of 1879–1880. There would be Chernobyl, the Soviet nuclear reactor meltdown in 1986 that contaminated millions of acres of land and water and spread genetic damage throughout the cells of tens of thousands. There would be an ozone hole over the South Pole. There would be the distant rumbling of global warming off in the faintly discernible future. But there would be no single gigantic dramatic slaughter of people by dirty air, although the brown cloud over Indonesia, which killed thousands of people in 2002, came close to being an atmospheric sledgehammer. Instead, air's damages tended to affect people, one by one. A coker got black lung, a child developed asthma, a senior died during a stinking fog. Airborne pathogens were governed by a complexity that did not lend itself to melodrama. They rarely made big waves. Instead, they dropped people, one by one.

America's First Air Fiasco

You can't stop it!

—W. P. REND, ON SMOKE

In 1892, ten years after the city had passed America's second antismoke ordinance, Chicago became the site of the nation's first environmental fiasco over air pollution. An elite group of businessmen led by a real estate developer founded the Society for the Prevention of Smoke and got the city council's tacit endorsement to enforce the moribund antismoke ordinance. The society wanted to resurrect the ordinance because the Columbian Exposition, a celebration of the 400th anniversary of Columbus's landing in America, was coming to town. The society's goal was to protect the fair's centerpiece, an architectural cluster of huge buildings called the White City. But the White City promised to be a smoke magnet of the first order for Chicago's factories, meatpacking houses, apartment buildings, public institutions, and waterfront tugboats, all of which pumped out the incense of progress in volumes that seemed at times to give the city's air a personality of its own. "The sheer dirtiness and heavy blackness of the smoke that polluted the city was appalling," wrote business historian Christine Meisner Rosen about Chicago before the world's fair. "The most badly smoking buildings disgorged from their smokestacks columns of black smoke and greasy soot that reminded spectators of erupting volcanoes. Some black smoke was so heavy that it could barely float in the air. It often fell to the ground, creating almost solid banks of soot and steam and ash on city streets. An observer looking back on the smoke generated by the Chicago Edison plant on Adams Street noted that the smoke was 'so dense that one could almost lie on it.'"

The Society for the Prevention of Smoke had problems with many of the smoke makers from the outset. Some Chicago businessmen, including the city's biggest power broker, Marshall Field, were prosmoke and didn't want to see the ordinance resurrected. The city's daily newspapers mostly opposed the society, too. But more importantly, as things turned out, the society made the mistake of getting into a losing battle with the little guys, the tugboat captains on the waterfront.

The tugboats were basically steam engines on hulls. They burned tons of soft coal loaded with sulfur and other impurities. The boilers in tugs operated under the ideal conditions for producing tons of smoke. More coal was needed to move a barge or ship, meaning a fire suddenly smothered with new fuel. But then when a tug idled, it burned coal slowly, which also produced lots of smoke. The Society for the Prevention of Smoke went out of its way to educate the tug stokers in the nuances of smoke control. But persuading a tug captain to feed his fuel box conscientiously, like an efficient stoker in an apartment building or a factory on land, proved impossible. As for controls on their boilers to make them burn cleaner, the tug captains told engineers hired by the society to forget it. Controls didn't work, they said; the gadgets needed constant tinkering. Another recommendation made by the engineers—switching to hard coal that burned cleaner because of its high carbon and low sulfur content—failed to attract converts because hard coal was more expensive than soft.

The society's efforts on land—an educational program about smoke and ways to reduce it, various appeals to the citizenry to be socially responsible, threats to take smoke makers to court if they were intransigent—were fairly successful. Their ongoing battle with the tugboat captains, however, turned into a fiasco and public relations disaster.

Pugnacious and defiant, the tug captains lived in smoke. They joked about it. They mocked Chicago's lax ordinance, enforcement of which had been virtually ignored by overworked health department inspectors for the last ten years—until the Society for the Prevention of Smoke showed up. Tug captains occasionally dusted folks with smoke just for fun. An editorial in the *Chicago Tribune* said, "It is pretty generally agreed that the tugboat operators love to get under a bridge when it is crowded with vehicles and pedestrians and vomit forth from their stacks vast volumes of blinding smoke and chuckle with delight when they see people above them choking and gasping." These guys were not going to change fuels or add expensive controls that didn't work so that business barons could point to the White City going up in Jackson Park and say how it symbolized how wonderful it was to be middle class and white in a progressive city. What did all that have to do with Columbus and the discovery of America, anyway?

Adjacent to the waterfront in Jackson Park, the White City threatened by all the smoke numbered more than two hundred buildings. One hall was so big it could have held Winchester Cathedral, Madison Square Garden, the Great Pyramid, and the U.S. Capitol beneath its roof. Once opened, the fair drew as many as 700,000 people a day. They munched on a new treat, Cracker Jacks. They looked at villages brought by ship and rail from Egypt and Dahomey. They applauded Buffalo Bill's Wild West Show, which included hundreds of Pawnee, Sioux, and Cheyenne, along with uniformed members of their former enemy, the U.S. Cavalry. They stared in amazement at Krupp's Baby, a cannon cast by the famous munitions manufacturer in Germany and capable of firing a 2,000-pound shell through three feet of iron.

To protect the White City, the society hired a gung-ho prosecutor named Rudolph Matz to shove the smoke ordinance down a few throats. Matz took several tug owners to court. But they remained united, vocal, and angrily opposed to an ordinance they said was unfair and being enforced by an elite group of private citizens who had taken government power into their own hands—their own greedy hands since all but one of the directors of the Society for the Prevention of Smoke were also directors of the Columbian Exposition. They stood to make bundles from its success. Going on the offensive, the tugboat captains sued the society. They started a prosmoke group with a populist name, the Citizens' Protective Association.

By mid-1893, when jurors acquitted two tug owners whom Matz had brought to court for flaunting the ordinance, public opinion had turned against the antismoke society. The society was also short of cash. Meanwhile, the White City was open, luring tens of thousands of visitors and getting rave reviews, despite its patina of smoke. The exhibition halls, entertainment palaces, and midway still glowed compared with the rest of grimy Chicago. "If the voice of the people is the voice of God," wrote the editors of the *Chicago Herald*, "it is evident that the Almighty is exercising a special providence over smoke makers."

Fair goers caught a glimpse of the victorious tugboat captains when they rode the sensational new Ferris wheel, which took them three hundred feet above the White City and Jackson Park. The wheel had been designed by a young engineer from Pittsburgh, George Washington Dale Ferris. A mercurial figure, Ferris provided the Columbian Exposition with "something novel, original, daring and unique," in the words of the fair's chief architect, Daniel H. Burnham, something that could hold its own in the public imagination with the recently erected Eiffel Tower in Paris. Big like everything else in the American fair, the Ferris wheel had arrived in Chicago on 150 railroad cars, all 100,000 pieces of it. The main axle and its carrying hubs weighed seventy tons. Once erected, the wheel carried 2,160 passengers in thirty-six cars, each the size of a Pullman railroad car, each with its own snack bar. Twin, 1,000-horsepower steam engines hidden off the midway provided the rotational power, driving gigantic sprockets with a 20,000-pound chain.

In the competition to see who would provide such a ride for the fair, Ferris had beaten out some formidable, if impractical, options. One, apparently a nod to America's more humble beginnings, was to have been a log cabin perched on top of a 1,500-foot tower of logs. Another proposal had urged the fair's main committee to approve the construction of a gigantic telescope-like ride, inside of which passengers would rise toward the heavens as the telescope barrel extended, then return to safety when it retracted. A third idea came from the Chicago-Tower Spiral-Spring Ascension and Toboggan Transportation Company. It proposed a tower just under two miles high. Thrill seekers would ascend in elevators to the top of the tower, then load into toboggans, if they dared, and slide all the way to Boston, New York, Philadelphia, and elsewhere, traveling down long steel rails anchored hundreds of miles away in the ground. Arguably the wildest proposal of all, however, was a ride far ahead of its time. It was a bungee-cord-like contraption designed to get people to a 4,000-foot tower. Two hundred riders at a time would board a cabin, and then, after it was shoved into space, they would experience a thrill few would ever forget. First, they would plummet straight down. After 3,000 or so feet of free fall, an elastic band fashioned from "best rubber" and attached securely to the cabin roof would slow the descent. An emergency feature, eight feet of feather bedding, covered the ground. If the ride worked as planned, you would stop above the feathers and then shoot back up a couple thousand feet.

A ticket on the more practical Ferris wheel gave you two complete rotations to a much less exciting altitude of only 300 feet. At the apogee, you could look out over the White City and the vastness of Lake Michigan dotted with ships. You would see America's first generation of steel and concrete skyscrapers, then called "tall buildings" and epitomized by the Monadnock Building, the huge Chicago Auditorium, which covered an entire city block, and the in-process Reliance Building, a steel-framed and glass-skinned structure of sixteen stories. Some of Chicago's tugboat captains must have ridden the Ferris wheel. When it stopped for each cabin below to be loaded, and rocked slightly back and forth, it may have reminded a tug captain of being on the lake in rough water. He may have looked out toward the lake and seen gasps of smoke jetting out of the stacks of his fellow tug captains' vessels and smiled.

Members of the Society for the Prevention of Smoke weren't smiling. Their egos were bruised (though their bank accounts were not, given the success of the fair), their efforts thwarted. The momentum they had picked up during the first months of the campaign, momentum that appeared to be pulling smoke out of the air with moral suasion, minor intimidation, and little need for the courts, had crashed against the hulls of the tugboats and sunk.

In England, an earl or the prime minister might be able to propose legislation to improve air for "the lower classes," and have something come of it. Not in America. In Britain's former colony, the working class bristled at the idea of the business elite shaping its destiny. Ironically, in this fiasco, the real losers were the public, the city of Chicago, and the *idea* that clean air was attainable through mutual cooperation.

The brief tugboat war was a classic confrontation of groups with different values over something completely new: air quality. Somewhat comic, a sidebar to a successful extravaganza, it pitted self-driven capitalists gowned with the self-righteousness of the civic-minded against pugnacious little guys with the toughness of crabs. What might have been a template for dealing with smoke became another much ballyhooed tale of the toppling of the powerful, much to the delight of the crowd and the long-term detriment of practically everyone, including W. P. Rend. In his boisterous naïveté, Rend admitted that he too sometimes choked on smoke, but it didn't harm him or anyone else. In fact, he said, it might—and here, Rend echoed the tugboat captains—be good for you. Smoke toughened up your lungs and body so they could defend you against more formidable foes, such as bacteria and other germs.

What had really failed in Chicago, of course, was the smoke ordinance. Sloppily written, subjective, and unenforceable, the ordinance, when challenged, had no support in the judicial system. It proved a better weapon for the tug captains than for its enforcers. In a drama of well-intentioned people trying to make Chicago better while earning all the money they could, the outcome helped set precedents for how America would deal with air problems for the next sixty years, until Los Angeles got hit by smog after World War II. That is, the nation wouldn't deal with air problems. After a brief resurrection of antismoke momentum in the early nineteenth century, air problems would be pigeonholed as engineering challenges.

After the Columbian Exposition closed and the White City was dismantled and Jackson Park returned to the squirrels, Chicago kept on making plenty of smoke. And it cost the city. In 1911, the city's smoke inspector estimated that the damages from smoke that year were around $17 million, or maybe three times that amount—his data wasn't all that scientific, he admitted. Cleveland's Chamber of Commerce estimated its city's losses at $6 million per year around that time. Cincinnati's smoke inspector put his city's annual costs at $8 million. "In the end, however, no one could propose an economic figure for the true cost of smoke in American cities," said David Stradling, in *Smokestacks and Progressives: Environmentalists, Engineers, and Air Quality in America, 1881–1951*. "No one offered figures for the cost of shortened lives, chronic unhealthfulness, the drudgery of constant cleaning, or the dullness of dreary days. Nor could municipal leaders offer an estimate for the value of their cities' reputations."

Artists Take a Whack at Smoke

Can we imagine a Jack the Ripper story without its fog?

—PETER BRIMBLECOMBE, *THE BIG SMOKE*

When concerned businessmen and progressive aristocrats in England, as well as the new women's clubs in America, all failed to eliminate much of the vile smoke and fumes of the second half of the nineteenth century, an opportunity opened up for more creative types to have a go at it. Artists and writers gave the public examples of air terror, of air fear, of air mystery. They portrayed smoke as ugly yet mesmerizing, as characteristic of the times as money. They sometimes succeeded in giving smoke a personality of its own, even a little soul. Or as soulful as something that regularly killed animals, trees, and people could get.

The working-class Parisian painter Maximilien Luce visited Charleroi in Belgium and made distressingly beautiful canvases of smokestacks and mills. In *The Steelworks of Charleroi,* men shovel coal in the foreground. Hellish smoke tinted pink and blue lifts above gray buildings. Above the roofs loom four huge chimneys, out of which my imagination couldn't help seeing billowing smoke rising up the gallery walls when I first saw the painting in an art museum. The men shoveling are dwarfed by the factory they serve, the overall impression garish and surreal. Luce's stippled brushwork was like that of the American expatriate painter James McNeill Whistler, who lived in London about the same time that Luce painted Charleroi. Whistler's aesthetic credo, *l'art pour l'art,* influenced generations of artists. His own canvases, many of them of his adopted city, contain little fear about a bleak future contaminated by anything more threatening than moody colors.

Another American expatriate, Henry James, boasted that he loved London's smoke. "Its thick dim distances," he wrote, "are the most romantic town vistas in the world." Fog "flatters and superfuses," he claimed, "makes everything brown, rich, dim, vague."

And, at times, deadly.

In probably the bleakest tale from the great age of smoke, a Canadian-born writer, Robert Barr, slaughtered thousands. A friend of Arthur Conan Doyle (whom Barr parodied in "Detective Stories Gone Wrong: The Adventures of Sherlaw Kombs"), Barr had moved to London to edit a foreign edition of an American newspaper, but soon cofounded *The Idler*, a magazine aimed at young, hip, irreverent Victorians. In 1862, an issue printed Barr's apocalyptic vision of London's being practically annihilated by smoke. "The Doom of London" was light on characters but heavy on prophetic darkness. It was an early sample of a new genre called science fiction. Reading it in *The Idler* on the way home in a tram must have given even the young and foolish pause or caused them to glance nervously out the window at the soot and smoke.

Barr placed his story in the distant future, the mid–twentieth century. The tale begins ordinarily enough—a fog creeps into London on Friday. Nothing unusual there. But then the air gets dense and really gritty. People start to die. The narrator, who happens to have a new piece of technology, a respirator, flees through streets littered with corpses. Dead horses lie sprawled in their traces. Dead passengers dangle from the windows of trams and carriages. An eerie glow from London's new electric lights suffuses the smoke. The story attributes many of the special effects on an absence of oxygen, claiming that the lack snuffs out gas and candle lights, along with the lives of animals and people. The narrator finds a train still running, boards, and is soon careening into the countryside. Wildly out of control, the train passes through tunnels still blessed with bubbles of oxygen. The residual oxygen helps revive some passengers, but they can't survive. The narrator, saved by his respirator, does. And London slowly recovers as well, once winds blow the killer fog away. "The story, despite both literary and scientific flaws," wrote Peter Brimblecombe in *The Big Smoke*, "is fascinating because it characterizes some of the feelings of London's impending doom that appear in the nineteenth century."

Barr's friend Doyle later wrote a short story called "The Poison Belt," in which a wave of gas anesthetized most of humanity. In other stories, Doyle often relied on fog to set a mood, just like a painter. Stuck inside for days on one occasion, Watson, the loyal sidekick of Doyle's alter ego, Sherlock Holmes, sees fog lurking like a criminal just outside. Says Watson, "We saw the greasy heavy brown swirl drifting past us and condensing into oily drops upon the windowpanes."

After the turn of the century, some writers began taking their readers into space, escaping smoky cities, probing the layered atmosphere that had just been discovered, and traveling on into the void. The relentlessly inventive H. G. Wells invented a gravity-shielding substance called "cavorite," named after its inventor, Mr. Cavor, in *First Man to the Moon*. Cavorite blocked "all forms of radiant energy . . . anything like light or heat or those Roentgen rays . . . or gravitation." The French author Arnould Galopin had his Mars-bound explorers rely on "repulsite" for the same effect. Repulsite freed a bullet-shaped rocket from the earth's gravitational field in *Fantastic Adventures of Three Frenchmen on the Planet Mars*, adventures that pit the heroes against dozens of antlike creatures on the red planet. Making Mars a lot friendlier, *On a Lark to the Planets*, by Frances Trego Montgomery, claimed canals were in fact a cool place for young people jaunting around the solar system to touch down and relax. Cartoonists working in another new kind of mass publication, the comic book, sketched adventurers in space ensconced in buslike rockets with fish-eye headlights, their own heads in glass bulbs filled with oxygen.

Whether prophecies of death by fog or dreams of escaping a smoky earth into space had any effect on a second wave of clean air activism that swept through both England and the United States after 1900 is hard to say. But with the new century came revitalized spirits and fresh visions of cleaner air over cities in the two smokiest nations on the planet.

In England, a powerful voice for change was that of Lord Reginald Brabazon, the twelfth Earl of Meath. Lord Brabazon's distaste for smoke had a distinct military flavor. Imperialistic expansion dragged Britain into several wars at the turn of the century, including the Boer War of 1899 to 1902, and the pool of available soldiers was found to be dangerously shallow, in part because of smoke. "The constant inspiration of impure air," said Sheffield's smoke inspector, thinned the ranks of healthy recruits. If Pax Britannica was to endure, those sustaining it had to be able to breathe without coughing and to run with their rifles and packs without getting exhausted. Dead forests in Nottingham, ruined crops near Swansea, and pitted ironwork in the Midlands were indeed worthy of concern, but military preparedness was of real importance.

Lord Brabazon led a powerful group of royals calling itself the Coal Smoke Abatement Society. They soon learned that the Boers were easier to fight, even with coughing recruits, than smoke swirling around the homeland.

The society did take a slightly new angle on an old problem. It claimed that last century's hopes that the enlightened self-interest, education, and tactful enforcement of ordinances now on the books would clear England's skies were a Victorian pipe dream. What was needed instead was a regional approach. Brabazon's group wanted officials to get tough about smoke beyond town borders and to strictly enforce existing laws but on a larger scale because the spread of smoke insisted on it. But when some society members sat down with the London County Council, explaining their reasoning and urging the council to follow these recommendations, the council begged off. There were so many more factories, ovens, chemical plants, and so on, than there had been even a decade ago, said council members, that the problem was far beyond their power to control. So Lord Brabazon and the Coal Smoke Abatement Society tried Parliament next, making a case for smoke as the nation's problem. But Parliament also declined to act decisively. Instead, the legislators opted for a tactic that would prove hugely popular throughout the twentieth century, a tactic city officials and regional authorities and national leaders would choose again and again when faced with too many bad-air days. Parliament ordered a study.

Finished in 1904, the study was circulated by the British Foreign Service. Much ink was spilled describing what other industrialized powers were doing, and not doing, about smoke. Mostly, they were doing little, but intended to do a lot. Having initiated a study and learned that England was no better or worse than other countries—a little better, actually, since the country had broadened the Alkali Act of 1863 to cover sulfuric acid, ammonia, and other hazardous fumes and had added the Public Health Act of 1875—Parliament seemed pleased with itself. Whether dirty air was fouling the lungs of young recruits or not, no new laws or new muscle for old laws was forthcoming, much to the chagrin of Lord Brabazon and the Coal Smoke Abatement Society.

In America, the antismoke revival had a stronger impact than in England, at least until World War I. The movement got a lift from the wave of progressive thinking sweeping the nation and softening the opposition of politicians and judges to the notion that health was a public issue. Thirty years earlier, the argument that people, rich or poor, urbanites or farmers, were entitled to clean air would have been ignored as the rant of radicals and socialists. But now the populists said that children shouldn't be exploited and that neither monopolists nor trust-forming cartels should rule finance and industry. As part of the new wave, progressive city councils dusted off old smoke ordinances, added new provisions, which were hardly more scientific than those they displaced, and gave enforcers new authority to levy fines and to take violators to court. In court, judges listened more closely to the testimony of doctors and inspectors; members of the bench were less deferential to the lawyers for industrialists and other smoke producers than they used to be.

In 1906, reformers going after smoke makers received a big boost from the American Medical Association (AMA), which took a public stand against smoke. Defending its position, the AMA referred to the pioneering epidemiological studies of a German physician named Louis Ascher. Using statistics and microscopic examinations of lung tissues, Ascher had linked pulmonary disease with smoke, both in the coal-mining regions and in the industrial cities of Germany. In 1907, two American doctors, Theodore Schaefer in Kansas City and Abraham Jacobi in New York City, published articles backing up Ascher's work and confirming the harmful effects of sulfur dioxide and soot. The AMA seemed poised to connect the dots between smoke and respiratory damages. But it didn't. Most doctors were conservative. They were reluctant to take a hard stand on what was still shaky evidence, and justifiably so. Smoke was not always the monster it seemed. For instance, when some doctors learned that sunlight killed germs, they theorized that smoke, by obscuring sunlight, allowed germs to ride dust and other particles into homes and noses, causing tuberculosis. The hypothesis, though intuitive and appealing, was wrong. Smoke aggravates tuberculosis; it does not cause it.

During the progressive era, empowered by new medical evidence about the evils of smoke, enforcers in New York City even arrested some violators. Other cities followed suit to show they meant business. Inspection departments in many cities received better funding for their work.

Judges found themselves in new terrain. The ordinances they had to interpret were not much stronger than those thrown out decades earlier. But there was a new concern with "the ethics of air," with the damages air caused to health, even if the damages could not be explicitly proven. Basically, judges still had to rule based on weakly worded, subjectively enforced documents that defendants attacked as unjust. Well-meaning industrialists who wanted to clean up air pollution pointed to a dilemma; there were few if any affordable technologies to reduce smoke. Even the most liberal, pro-clean-air judge couldn't totally ignore the fact that smoke also remained the banner of economic progress.

One result of these often conflicting, unresolvable issues was a change in the law. Nuisance law in America, as well as in England, remained the judicial yardstick for litigation about smoke. But for a nuisance to be prosecuted, first it had to exist; a privy had to smell, a dump reek, or a factory emit the old standby "dark" smoke before a complaint could be made. A plaintiff then had to link the complaint to a specific source and prove it caused harm. To preempt damages before someone could complain, judges reshaped traditional nuisance law into rudimentary zoning, which began to give shape to cities and growth. Zoning was a way to separate smoky factories and industrial neighborhoods from residential ones. As a legal tool, it concentrated smoke but didn't do much to reduce it.

Industry and the railroads fought the sea change in the law, but the law had turned, becoming sympathetic to plaintiffs. Engineers and doc-

tors—not offended ladies with fans and parasols—now brought the violators to court. Some powerful oratory on air was heard. Of the memorable declarations, one by Charles Reed, a surgeon in Cincinnati, stands
out. "To breathe pure air must be reckoned among man's inalienable
rights," said Reed. "No man has any more right to contaminate the air
we breathe than he has to defile the water we drink. No man has any
more moral right to throw soot into our parlors than he has to dump
ashes into our bedrooms."

Finally, smoke-control technologies—automatic stokers, better ther-
mostats, improved grates—appeared as judicial intolerance of ordinance
violators became self-evident. Inspectors, often members of America's
burgeoning population of young and ambitious engineers, gained
respect and authority. "As cities turned toward a technological solution of
the air pollution problem," wrote David Stradling and Peter Thorsheim
in their history of air pollution controls in America and Britain during
this period, "engineers eclipsed women as the voices of reform." Labeled
"combustion experts" by *Power* magazine, the engineers, were "compe-
tent to weigh the enormity of a violation against the offender's ability to
comply." No longer a "badge of great prosperity," smoke had become
"an unnecessary nuisance," reported *Harper's Weekly*.

So, with reformers successful in cities across the country, was U.S. air
getting any cleaner in the first decade of the new century? Not much.
Engineers, liberal courts, and tougher laws helped the reform, but the
economy was still number one. To grow, it needed power. Coal—not oil, a
new fossil fuel then refined mostly for kerosene used for lighting—produced most of the power that made the nation grow. City, county, and
state governments, despite earnest efforts to back reform, still allied themselves with industry when jobs were at stake. Politicians didn't want more
smoke, but they didn't want fewer jobs, either. Fear of labor unrest caused
by job losses and threats by factory owners to pull up stakes and move elsewhere if regulations squeezed them too hard thwarted dramatic change.

Of course, no one in early twentieth-century Chicago, St. Louis, New
York, or elsewhere wanted to breathe bad air. But neither the judges nor
the politicians proved capable of making a truth—that the industrial

growth that had justified and driven the increase in smoke had worn out its self-righteousness—a fundamental plinth for concerted change. Zoning was a limited tactic. In one way it was more detrimental than helpful to air. By removing people from the smoke of enterprise, zoning shifted the issue from a local problem to an out-of-sight one, especially with the help of taller smokestacks that sent the emissions high into the winds. Out of sight, out of mind—that was the philosophy. It was ironic that the industrialists who had choked urban neighborhoods without cease were given license now to foul the regional atmosphere rather than being forced to clean up their wastes.

A bigger hurdle to clean air around 1910, though, was the numbers. Sources of smoke and fumes across America continued to soar. The country was on an economic roll. The smoke revival movement ran into a conundrum that would plague many later efforts at cleaning the air: limited regulatory success but increasing air pollution because of more and more sources. At least the second wave of reformers could celebrate some victories even as air quality continued to deteriorate. They also could point with enthusiasm to the Mellon Institute Smoke Investigations. Funded by a scion of a Pittsburgh family whose name was synonymous with smoke, the investigations looked not only at Pittsburgh's hellish air but at the conditions in foreign cities. Health data linked excess carbon intake with lung damage, tuberculosis, and pneumonia, validating the conclusions of Louis Ascher in Germany a decade earlier. The Mellon Institute studies provided a basis of comparison between industrial cities, a comparison that had been lacking before, and established a new level of sophistication in air research.

But then World War I broke out; it really torpedoed clean air reform.

The "outbreak of war has caused the committee to suspend its sittings," read an announcement from Britain's Smoke Abatement League in December 1914. In London, practically overnight, smoke the enemy became smoke the friend because it obscured city targets from German bombs. When America joined the Allies in 1917, enforcement of smoke ordinances in cities smacked of a lack of patriotism. Smoke intensified. As Stradling and Thorsheim put it, "Whether one believed that the war was a life-and-death struggle to defend democracy or a senseless act of collective self-destruction, the horrific reality of warfare robbed smoke of the symbolic weight it had earlier possessed."

★

World War I slammed the curtain down on a lot of things: the prolonged era of relative peace and stability known as the Age of Empire, a sense of optimism about man's basic goodness, the brief progressive movement in America to clean smoke from the air. In America, the clean-air movement went out the window, so to speak. It would be fifty years before serious efforts were made to do much about that. Only by then air pollution, which had been essentially an issue of coal smoke and some industrial by-products during the nineteenth century, would have risen to another level.

The war also marked the end of what I have called the little atmospheric disruptions of man. Afterward, the disruptions became larger, chemically more complex, and mythically richer in character because they wreaked havoc on people with the ferocity and indifference of the ancient sky gods. Air pollution, devious entity that it was, also became harder to see. Initially, it was just harder to see in America, where cleaner-burning oil and natural gas became popular first. But as the century progressed, the cleaner fuels became popular around the globe.

If we think of air pollution as a kind of dark figure that first stalked the industrial era of the eighteenth and nineteenth centuries, then plagued the technological era of the late twentieth century, and finally afflicted the information age that bridges the twentieth and twenty-first centuries, the hundred years ending with World War I were its adolescence. That is, a protracted era of growth, but not one of real character development. Real character development of the pollution mix now infiltrating the earth's air began with World War I. With poison gas. It was followed in quick succession by a series of heretofore unimaginable airborne terrors: radioactive fallout, biological drift, and killer smogs. Waiting in the atmospheric wings were nuclear plant meltdowns, a vast ozone hole in the stratosphere, and other anomalies such as the Bhopal gas disaster in India in 1984, the blooming spread of dust and toxics radiating from the collapse of the Twin Towers of the World Trade Center in September 2001, and the SARS (Severe Acute Respiratory Syndrome) epidemic of 2003. All were new, fast, unanticipated additions of chemicals, rays, germs, and particles to local, regional, and global air sheds, the physical theaters of the atmosphere shaped by natural forces such as topography, weather, and the seasons. The newcomers increased the scale and magnitude of atmospheric change, which was still headlined by the pervasive by-products of fossil-fuel burning from factories, power plants, vehicles, and homes. Poison gas, radiation, and new industrial toxics of the twentieth century resulted in a double-decker new reality: bad-air episodes that were swift, terrible, and either local or regional, *and* long-term ones that were gradual and global.

As a force of nature stirred by people and their inventions and emissions, air pollution in the twentieth century bifurcated. And bifurcated again. It evolved just as the lungs of mammals had evolved tens of million years ago to supply growing bodies and bigger brains with more oxygen. As this happened, for roughly the first fifty years, between 1915 and 1965, safety standards for air remained virtually unknown. Doses of poison gases that soldiers inhaled, rads that penetrated the flesh of those exposed to atomic-bomb test fallout, or concentrations of manufactured chemicals such as DDT that were sprayed on crops and worked their way up the food chain were neither measured nor measurable. The advances in technologies that built sophisticated weapons and consumer products only slowly found application in the making of tools for identifying and quantifying pollutants. Moreover, instruments capable of analyzing damages done by the pollutants to the human body, to plants, and to other materials only slowly appeared. One result of the slow transfer of knowledge was that up until the flurry of U.S. air pollution legislation during the 1960s, legislation that actually began with California's attempt in the late 1940s and throughout the 1950s to do something about smog, was people's lack of technical knowledge. Even the most concerned politician, judge, physician, or smoke inspector had limited ways of grasping a problem, and even fewer available responses, because without knowledge a response was hard to formulate and justify.

The twentieth century developed into a special chapter in the earth's long atmospheric history, a chapter we're still figuring out, and living with. It was a unique chapter, fresh and different in its own troubling but exciting ways. Maybe it signaled a pending extinction event. Maybe not. Maybe all the computer models describing the immensity of human impacts on air and climate and all the scientific hand wringing accompanying them, are misleading, though I doubt it. Of course, the twentieth century also taught us how the atmosphere goes about renewing and repairing itself, just like the living cell to which it is often compared. And how the atmosphere cleans itself, dragging acids, sea salts, particles, and anything else that water can get a hold of and precipitate out of the sky.

But I'm getting too far ahead in the story. For now, let's just say that as World War I began, smoke received a temporary reprieve as a problem, as *the* atmospheric nuisance of the last two hundred years. And, to be fair to the dark figure that had been around for a while, smoke hadn't lived up to its potential to cause catastrophic respiratory harm. Sure, London had terrible fogs. And mining towns were respiratory outposts from hell. But *Homo sapiens* had weathered the great age of smoke pretty well. Tens of thousands died from smoke-related disease but still the population had increased manyfold. Smoke was basically an urban malaise, a city thing. It left most of the countryside alone. Looked at historically, the era of great smoke couldn't hold a candle to a series of volcanic eruptions, never mind to a bolide splashdown. Compared to what awaited people throughout the twentieth century, a very special case of worsening air pollution assisted ably by prolonged denial, coal smoke had been civilization's youthful friend.

But those days were about over. Newcomers were incubating. Some would scare the bejesus out of people. Others would accumulate slowly and miles high in the sky. Collectively, the new pollutants were something that organic life had never faced before.

9

A Few Twentieth-Century Air Nightmares

On April 22, 1915, against the advice of his High Command, Kaiser Wilhelm ordered thousands of chlorine gas canisters opened along the four-mile front at Ypres. A cooperative light breeze carried the roiling, yellowish green cloud across no-man's-land. Neither barbed wire nor sandbags slowed it down, and the chlorine snaked into the Allied trenches like a sudden fog. It routed thousands of French, Algerian, and Canadian troops, their eyes, throats, and chests on fire, vomit spewing from lips, throats gagged by coughing, and blood spurting from noses. Soldiers unable to flee pushed their faces in the dirt; they shoved mud into their nostrils and throats. But to little avail. Within minutes, five thousand of them were dead or dying. Ten minutes more and the roiling cloud of poison that smelled a little like bleach—like the odor of a chlorinated swimming pool on a hot summer day—had dissipated.

One afternoon in 1954, members of the Highlands Park Optimist Club, a group of boosters with a macabre sense of humor, ate lunch in smoggy Los Angeles beneath a banner proclaiming that if the air didn't get cleaner soon, it was curtains for all of them.

THE APPEARANCE of poison gas at Ypres marked the beginning of a new era in air pollution. Not because poison gas would spread everywhere and kill men, women, and children indiscriminately, or because it was ever that great a weapon (fickle winds and gas masks saw to that), but because it made air, the membrane of life, an acceptable channel of death on a large scale. Theoretically, it was a channel limited to the battlefield. But like many changes wrought by war, the poison gas developed for combat would morph into a number of peaceful uses once the fighting was over, uses with safeguards but with inherent dangers as well.

The bottom line at Ypres was that air had been invaded by man-made chemicals that intended to kill people on a large scale. That was a first.

Of course, war intends to kill. But seldom had so many died with such little effort, and with no loss of life or even injury on the attacker's side, as they did during Germany's first surprise use of poison gas. Precedents existed of soldiers killing each other with airborne poisons, but not by the thousands. The first toxic weapons were probably refined by the ancient Greeks (the word *toxic* comes from the Greek *toxon*, which means "arrow"), whose mythical superhero Hercules slew the multi-headed hydra by using arrows dipped in pitch and lit on fire, which drew the monster out of its den. After killing it and slicing open the body, Hercules plunged the tips of his remaining arrows in the hydra's venom so he'd never be without poison missiles again. Scythian archers coated their arrowheads with manure or swabbed them in the maggot-clotted cavities of corpses, hoping the dirty weapons would spread disease behind their enemy's walls. During the French and Indian War, British soldiers eliminated entire villages of Indians friendly to the French by bearing gifts, blankets wrapped around smallpox bacillus, to unsuspecting families. But the yellow-green gas launched on orders from the kaiser at Ypres during the cruelest month of 1915 took the cake as an airborne poison with superior killing strength.

At a hospital in Boulogne, not far from the battle trenches, the physiologist Joseph Barcroft examined hundreds of survivors of the attack. Barcroft, forty-three, a Quaker with pacifist convictions, had volunteered for military duty the year before, not long after Britain had declared war on Germany. A professor at Cambridge University, he had just published *The Respiratory Function of the Blood,* a book focused on the chemistry of hemoglobin, the diffusion of oxygen in and out of the body, and the dissociation curve, the relationship between the amount of oxygen binding to hemoglobin and its concentration in the air. Barcroft's knowledge about respiration had landed him across the English Channel immediately after the attack in the hope that he could treat soldiers and help contain the panic spreading through the ranks in the wake of a warfare first.

Most of the troops, thankfully, seemed to be recovering. They could breathe OK and had little residual lung damage. Very interesting to Barcroft was the remarkable similarity of their postexposure symptoms to those experienced by mountaineers or balloonists at high altitudes. Barcroft was very familiar with this area of study. In 1910, he had received an invitation from Nathan Zuntz, a German physiologist now classified officially as an enemy, to conduct high-altitude research. Barcroft had traveled by ship to Tenerife, where he shared close quarters with his colleague, first at the 7,000-foot work station above Santa Cruz and again at the 11,000-foot Alta Vista station, where a couple of French astronomers were camped out, waiting to look at Halley's comet with a telescope they'd lugged up from sea level.

Zuntz had been intrigued by the telescope, as he was about most new technology. He was already famous in the small and competitive world of research physiologists for having built the first portable metabolism machine and for having defined *dead space.* The metabolism apparatus consisted of a backpack with tubes, a skullcap with gages, and a mouthpiece that looked like a flute. Wearing the thing, Zuntz looked more like a one-man band than a world-renowned scientist plotting the variables of human metabolism while on the move. Dead space, a discovery he'd made back in 1882, quantified a much-pondered variable of the human breath. That is, when you exhale, "used" air remains in your lungs and trachea, and when you exhale, this air is sucked back in before the fresh, oxygen-rich air the lungs really want. This "in-the-way air" filled what Zuntz labeled the dead space. He calculated that a person's dead space at rest, just sitting around and breathing normally, was about one-third of each breath. Under exertion, however, when bigger breaths were demanded, the space shrank considerably, to a fraction of each breath. What was really fascinating about dead space was the fact it never totally disappeared. So a person's lungs were never completely empty of air from their very first gasp, and the lungs never pulled in a completely fresh breath after the first one. Even then, there was a tiny dead space in a fetus, although it was filled with fluid that got secreted as the newborn squeezed through the birth canal into the world of air. After birth, throughout life, try as a person may, he or she can never completely get all of the dead air out of the lungs.

At any rate, Nathan Zuntz had become a friend of Joseph Barcroft's during the research trip to high altitude in the Canary Islands five years before the poison gas attack at Ypres. But now he was classified as Barcroft's enemy. To contact him, Barcroft occasionally relied on an intermediary, the Dane August Krogh, a friend of both men and destined to win a Nobel prize in chemistry in 1920 for his work on gas exchange in the capillaries, work he did in collaboration with his wife, Marie.

In Boulogne, working mostly alone, Barcroft may have longed for some input from his colleagues, either friend or foe. Although most of the troops were getting better, a small number had been badly damaged by the poison gas; their faces turned blue from the least physical exertion. Tests showed that the gas was no longer in their blood, but their hemoglobin was not functioning correctly. Instead of carrying oxygen to cells, the blood pigment was overloaded with excess carbon dioxide, which was not diffusing out of the lungs as it would normally and instead was interfering with oxygen supply.

After long days in the wards and various attempts at therapy, including the use of pure oxygen, which proved beneficial for some patients, Barcroft made the kind of simple but profound observation he was known for: "It is surprising how little is known about the circulation through the lungs in various chest disorders."

This lack of knowledge not only hampered the treatment of the soldiers. It had also plagued inhabitants of heavy smoke regions and coal miners for more than a century. The pathology of respiratory diseases remained largely a mystery. Chlorine gas, it would be learned later, inflamed capillaries in the airways, swelled membranes, and increased mucus, resulting in a reduced oxygen intake, or hypoxemia, and, in more serious cases, an increase in the partial pressure of carbon dioxide, or hypercapnia. In 1915, chlorine gas was a substance few chemists knew much about, except maybe the Germans responsible for it. An attack with the gas was simply an exceedingly vivid, scary, and much publicized example of an event whose symptoms startled people during wartime but were nothing new. The symptoms had been recognized not only in mountain climbers but in people living in smoky cities and dusty coal-mining regions since the start of the industrial revolution and even before. As far back as the days Paracelsus wandered through the Tyrol disguised as a preacher to avoid the authorities, symptoms of hypoxemia had been noted. The symptoms were one of the trade-offs of progress, just as poison gas attacks became one of the trade-offs of war.

Returning to Cambridge University from Boulogne, Barcroft joined a group of chemists and physiologists assembled to create Britain's response to Germany's use of chlorine gas. Working for the Trench Warfare Department, they soon settled on an insidious alternative called phosgene. Phosgene smelled faintly like new-mown hay and didn't burn the respiratory tract. Instead, the gas wended its way deep into the lungs where, reaching a lethal concentration, it hydrolyzed into hydrochloric acid. Ten times as deadly as chlorine gas, even modest amounts of phosgene would leave a soldier dead within hours, his respiratory system in shock, his lungs filled with fluids.

Poison gas was a nasty death trip. Of that Barcroft was well aware.

But one of the casualties of World War I was scientific detachment. Since the 1880s, many European scientists had joined universities and industry in the service of economic progress. They abandoned the older tradition of the independent thinker working alone and found themselves working with the latest equipment in the best-equipped labs and with other scientists in their fields. These scientists were treated like gods. Now, though, war had drawn the new gods down from their lofty perches and into the maelstrom. Scientists such as Barcroft found themselves supporting war and weapons for a variety of reasons: patriotism, excitement, fear, a need to be on the cutting edge. Divorced from myth and disdainful of heaven, the industrial world in service to war had few qualms about turning the realm of the former sky gods into an airy sewer of death. In the new world that began in 1914, a young soldier breathing in air through his nose and daydreaming about spring haying back on the farm could find himself in agony. Better reach for the gas mask and make damn sure it was strapped on tight.

After Ypres, few of the poisonous gases intently developed and tested on dogs and prisoners affected the outcome of World War I, however. The military potential of lethal molecules launched on the wind was muted by gas masks, which became standard issue, and by uniforms redesigned to limit skin exposure. Not that both sides didn't keep trying to kill each other with relentlessly imaginative chemicals they could fire in shells or, the wind willing, simply release from canisters. Joining phosgene and chlorine were arsenic powders, tear gases, and various combinations of all these toxins. Klop, a mix of bleaching powder and picric acid, a bitter, toxic, yellow crystalline acid, became popular because it could foil the standard gas mask, making the user vomit and pull the mask free. Without a mask, the victim was often exposed to phosgene, which was packed in artillery shells together with Klop.

Mustard gas, probably the best known of World War I's chemical weapons, wasn't used until late in the fighting. The Germans, who usually beat the Allies to the new weapons because of the superiority of their chemical industry, introduced mustard gas one warm summer night in July 1917. Again, the attack occurred at Ypres, where armies remained dug in two years after the first chlorine gas attack. Unlike its predecessors, mustard gas hardly smelled, and it worked slowly. Many soldiers, convinced after a while that the gas attack was over, removed their masks and tucked them away. Then their eyes began to swell shut, exposed skin started to blister, and the fabric of their uniforms began to disintegrate because the gas ate through fabric, as well as rubber hoses and leather. Rescue personnel found hundreds of troops wandering blindly out of the trenches, their eyes, noses, and throats burning, their chests tight, as their upper airways shut off their breath. Thousands died.

During the final months of the war, the Germans disguised their mustard gas, which did have a faintly detectable odor similar to horseradish, with a sweetener, a tear gas whose active ingredient, xylyl bromide, smelled like lilacs. "So it came to pass in the wartime spring," wrote historian Richard Rhodes, "that men ran in terror from a breeze scented with blossoming lilac shrub."

Before the war, there had been episodes of air terror. But none of them rivaled the power of poison gas to spread fear and death so quickly. The new era brought to an end the little atmospheric disruptions of man, and set the stage for disruptions both large and small. Some struck like poison gas and were gone in minutes; others aggregated slowly, like the CFCs (chlorofluorocarbons) that would gradually eat away part of the stratospheric ozone layer. Smoke, however, was on its way out as a palpable evil, at least in the more advanced industrialized countries. Throughout the war, industrial cities in Europe and the United States were wreathed in smoke. But afterward, smoke thinned, as oil, natural gas, and electricity (sometimes called "white coal" because it was produced at power stations alongside rapids or with dams that restrained water) increased in popularity and became synonymous with modern times. Even industries notorious for dark smoke, such as smelters and city power plants, cleaned up. Smoke meant they were old-fashioned, or worse, inefficient. Smoke meant not only incomplete combustion but squandered profits. More than a century earlier, Count Rumford had said as much. Promoting his so-called smokeless stoves in London (designs so efficient that "coal steept in Cats-piss makes not the least ill scent," a user claimed), Rumford told prospective customers that hundreds of tons of unburned coal hung over their heads, the result of incomplete combustion his stove would rectify. Rumford seemed born to hyperbole; in fact, the unburned carbon above London on any given day then was probably around one-hundredth of his unchallenged claim. But his boastful manner apparently caught the eye of Marie Paulze, Lavoisier's widow, when Rumford traveled to Paris to oversee the casting of cannon barrels for the Prussian army. The two fell in love and got married. Regrettably, as it turned out. Soon less enamored by Marie Paulze's charms than his predecessor, Count Rumford said that Antoine Lavoisier had been lucky to be guillotined. Be that as it may, the industrialist's message that in smoke there was profit survived. It became an axiom of industry after the World War I and probably did more for clean air than all the laws and antismoke campaigns of the previous decades combined. Fuel economy made sense to industrialists, who hated to be told what to do by government, but who liked to save money.

Not that air pollution went away because the world got smarter about coal. Not by a long shot. But it helped. And air pollution began to evolve.

In the United States, the period between the 1920s and the 1950s marked the decline of smoke and the birth of smog. Air quality was in the hands of engineers and technicians; they were the go-to experts for smoke control and reduction. Of course, it was a myth that they were cleaning up the air. A complete fabrication. The atmospheric poisons and particles were evolving as oil and natural gas displaced coal as the fuel of choice; they were less visible but there nevertheless.

Air still lacked a big fear, a calamity to remember, except in the history books about London. Poison gas receded as such a potential evil, at least for a while, after 140 nations signed the Geneva Protocol in 1925, prohibiting mustard gas, chlorine, phosgene, and others in the military pipeline. The Great Depression further reduced the use of coal around the world and slowed the growth in popularity of oil and natural gas. Fog as a fear stoked by pea-soup dawns seemed to enter remission. Between World Wars I and II, no memorable killer fogs, as prophesied by Robert Barr for London, struck any city. But, beginning in 1943, Los Angeles was increasingly blanketed in hazy, yellow smog, a new and nasty atmospheric phenomenon.

Smog Man

A melting pot in which the civilization of the future may be seen bubbling darkly up in a foreshadowing brew.

—BRUCE BLIVEN, ON LOS ANGELES

In 1946, America was in a postwar euphoria. Industrial juggernaut without parallel, great smoke maker and oil burner and chemical producer, nation with the world's most cars and most extensive public transit system, America could do little wrong. It had emerged from World War II physically unscathed, while Germany, Japan, France, Russia, and other former powerhouses were in ruin or partial ruin. America seemed to exist beneath a golden halo. When the halo turned to smog over Los Angeles, the nation's sunniest Shangri-La situated alongside the Pacific Ocean in a beautiful natural basin surrounded by mountains, it just didn't seem right. But there it was: dense, yellow, and irritating. Smog. This atmospheric condition that had first appeared regularly during the war was now forcing folks to pull their cars over, wipe their eyes, and then stare down boulevards lost in an opaque haze. To combat the smog, the *Los Angeles Times* invited a Midwesterner with a track record for fighting and defeating smoke out to California to study the situation and make some recommendations to clear the skies. Almost immediately, journalists labeled the stranger Smog Man. And the campaign to eliminate smog was framed as a fight.

Smog Man was Professor Raymond Tucker of Washington University in Saint Louis. And the first three days Tucker spent in Los Angeles in early December 1946 were disappointing. The weather was totally uncooperative. Blue skies. Balmy breezes. Convertibles with their tops down and snow-capped peaks to the east. Then on the fourth day, an inversion moved in; Los Angeles became capped with smog. To get a good look Tucker scaled the steps of City Hall Tower for some elevated sleuthing.

He couldn't see much. Both the plumes he'd noticed rising from factory smokestacks and the dank handkerchief-like clots of haze out near the foothills had been absorbed by the smog. Whatever the makeup of the stuff—an L.A. smog official said even a chemical engineer couldn't figure it out—the air certainly had changed since Tucker's last visit, back in 1932, when he had attended the Olympics. Despite the lack of visibility, Tucker knew some of the causes. In 1946 Los Angeles contained 4 million people, 1.5 million cars, 300,000 backyard incinerators and hundreds of factories. There were public dumps all over the place. Adding to these atmospheric insults in the winter months were a few million so-called smudge pots, rough stoves for burning old motor oil and old tires. Farmers used these smudge pots to blanket their orchards and crops with thick smoke on chilly nights. All in all, Los Angeles was a sprawl of development and industry and farmland held together with roads and trolley tracks. It was Tucker's job to identify the smog-producing ingredients of all this activity and to come up with some recommendations for controlling them.

For two weeks, Tucker visited the city's air and smog experts; he observed tests being run on human guinea pigs to check their "eye-smarting tolerances." He talked with prospective members of a Smog Committee being set up by the *Los Angeles Times* and its publisher, Norman Chandler. Tucker went on radio, gave journalists interviews, declared that smog could be licked with an "unceasing fight and an aroused citizenry." And, of course, he heard all about inversions and the history of smoke and smog in the Southland, as locals called it.

The history went all the way back to October 1542, when Juan

Rodriguez Cabrillo, a Portuguese explorer sailing for Spain, had anchored offshore and gazed at the impressive horseshoe of mountains enclosing the basin. Cabrillo wondered why smoke from fires tended by Native Americans rose in long columns and seemed to hit something, a kind of an invisible ceiling, and sprawl, and he dubbed the place La Bahia de Los Fumos, or "the bay of smokes." What Cabrillo had witnessed was L.A.'s famous inversion layer. That is, a layer of warm air, usually up a couple thousand feet, that trapped smoke and gases rising from below almost like the lid of a box. More recent anecdotes that Tucker heard about L.A.'s air quality described times when smoke had blocked the sun at midday, and the first record of smog. Tucker was certain smog had made an appearance here many times before, but it had only been officially recognized on July 26, 1943. That day, nobody—trolley operators, policemen, the mayor, or movie stars—had been able to see farther than three blocks. Workers coughed, residents complained, visitors stayed away. This was certainly not the Los Angeles loved for clear skies and fine weather. A culprit was needed, and after the desired one—Japanese submarines launching a sneak gas attack from offshore—was discounted, a more likely prospect was identified—Southern California Gas Company's Aliso Street plant. The plant made butadiene for synthetic rubber. It was quickly closed. But the smog had kept coming back.

Which was why Professor Tucker had been summoned to conduct two long weeks of research from dawn to dusk.

Finished with his fact-finding, Tucker prepared to board the train at Union Station to return to the Midwest. He was tired, a little disappointed. He had not experienced even one really spectacularly bad smog day. Nevertheless, before climbing onto the *Super Chief,* he rose to the occasion and told the gathered reporters, "I can tell you one thing. My smog report for *The Times* containing suggestions will be an exact description of what I think should be done here." He urged patience and resolve. During the fight against black smoke in St. Louis before the war, victory had needed both virtues. Along with some political muscle. In those days, he had not been Smog Man, as many of these reporters liked to call him, nor Smog Fighter, another of their favorites, but simply the assistant to the mayor of St. Louis. The mayor had thrown his considerable political clout behind an ordinance devised by Tucker that required all businesses, factories, and homeowners in the city to switch from dirty soft coal to cleaner, but more costly, hard. Or else to install expensive smoke controls. The strategy had worked. Infamous for its streetlights blazing away even in midday, so that people could see where they were going through the smoke and not drive into the Mississippi River, St. Louis then enjoyed much cleaner air, though it was hardly smoke free. No fuel, especially coal, whatever the grade, was smoke free. Some smoke was just less visible.

A month later, in late January 1947, Tucker sent his report to Norman Chandler, who ran it in the *Times* Sunday edition. Tucker made twenty-three recommendations for reducing smog. Get rid of backyard incinerators, he said. Give smoking trucks citations. He pointed an accusatory finger at factories, power plants, railroads, cars, and buses for their emissions. Ammonia, formaldehyde, acrolein, acetic acid, sulfuric acid, sulfur dioxide, hydrogen sulfite, mercaptans, hydrochloric acid, hydrofluoric acid, chlorine, nitric acid, phosgene, and organic dusts had been found in the air, he said. Various places emitted these known irritants through stacks, vents, and natural ventilation. The offenders he cited included foundries, oil refineries, chemical manufacturing plants, fish canneries, incinerators, hydroplating plants, fertilizer plants, packing plants, soap factories, and waste disposal plants. Most importantly, Tucker urged Los Angeles County to establish an air pollution board with clout. Have it replace the forty-five disjointed groups then responsible for smog abatement, he advised. Tucker also said that the city ought to take a cautious approach about passing anti-smog ordinances and rules. No single industry, plant, or group of people caused smog, he said. "Each contributes its share."

All in all, Smog Man gave his employer, the *Los Angeles Times*, and the Southland good advice. He was circumspect. He was honest. As an editorial in the *Times* said, "Prof. Tucker waved no wands. He rode in no blimps. He visited no movie studios. He kissed no babies."

He also said, "Don't expect miracles."

But this was Southern California, circa World War II, land of movie stars and Hollywood and grandiose dreams. Of swift, sweeping, positive change, often linked to technological innovation and imaginative thinking. Here, miracles *were* expected. People got paid lots of money for them.

First, though, before any miracles, the Los Angeles County Air Pollution Control District was formed. A smog czar, formerly chief of the coal division of the U.S. Bureau of Mines, was hired. His name was Louis McCabe. Once in Los Angeles, McCabe muttered the right homilies: He felt confident, progress was being made, a good law was in place, and so on. New rules went quickly into effect. Manufacturers found they had to get air pollution permits; they lost their old right to "necessary" discharges of toxic fumes at their own discretion; more inspectors with college degrees started nosing around.

Then a series of solutions to smog, which was appearing more often, came across McCabe's desk. Many read like pages torn from film treatments for futuristic thrillers. They seemed written by aspiring sci-fi screen writers rather than by scientists and engineers. One proposal urged that tunnels be drilled through the San Gabriel Mountains, and huge fans installed in them to blow the smog out over the Mojave Desert. Another wanted big holes blasted through inversions by cannons placed on Mount Wilson so that smog could rise up into the sky. A third suggested focusing sunlight on inversions with giant mirrors, a mid-twentieth-century version of the burning mirrors used by Lavoisier and Priestley centuries ago, the intent being to sear holes in the warm air blanket, through which smog could escape. One doomsayer suggested simply moving everyone out of Los Angeles because the basin wasn't suitable for the energy-hungry lifestyles of *Homo sapiens*. Plans for seeding clouds from planes to wash the night air, and for spray towers to treat the air with car-wash-like hoses and leave it sparkling each dawn, were also considered by the new air pollution control district. All impractical, though imaginative, the proposals highlighted the unreality surrounding the first campaign to eliminate smog and the inclination to reach for some single, big, impressive treatment of an amorphous, vast, and relentless phenomenon then believed to exist only in Los Angeles. Expensive, technologically unachievable, occasionally right up there with the toboggan ride from two miles high that would have whisked riders to their homes from Chicago's Columbia Exposition in 1893, the schemes were all shelved.

In real life, the control district had to make do with more mundane approaches. In addition to writing new rules for air pollution permits and rescinding the right of necessary discharge long enjoyed by factory owners, the district mandated bag houses to capture dust and fumes from certain factories and smelters. To learn more about smog, its health impacts, and what might control it, chemical engineers, meteorologists, physicians, and field inspectors were added to the district's payroll.

But smog kept increasing.

Immediately after Smog Man's visit in 1946, the L.A. assistant director of air pollution, Isador A. Deutch, had warned that smog was complex and would defy easy analysis. Deutch prophesied that the city and county faced "an immensely complex scientific task of identifying and controlling the infinitesimally small quantities of irritating substances which cause so much trouble." By the early 1950s, Deutch's assessment had been borne out. Smog proved too big and elusive for even a unified control district with a good budget to rein in. Los Angeles jabbed at smog here and there to keep it at bay. But pedestrians crossed wide, car-busy boulevards on more and more smoggy days, handkerchiefs pressed to eyes to shield them from the hazy phantom everywhere about. Drivers continued to pull over alongside palm-lined avenues because they could hardly see in front of their chrome bumpers. One afternoon, members of the Highlands Park Optimist Club, a group of boosters with a macabre sense of humor, held a much publicized lunch during which many of them wore long-snouted gas masks as they dined beneath a banner proclaiming that if the air didn't get cleaner soon, it was curtains for the bunch of them.

Ironically, cars, their exhausts the real culprit behind smog, avoided regulation. Car sales boomed. Freeways were built, trolley tracks torn up. Little was done to promote, support, or save public transit. It was the dawn of the Cold War; public transit smacked of Communism, of taking care of the proletariat. A city commission appointed by the governor in 1953, following another fog episode in London that killed four thousand people on a three-day weekend, urged the county to build a public transit system. But it didn't. "Wide snaking rivers of concrete, teeming machines, and legless air-conditioned inhabitants swarming beneath a nimbus of dirty yellow haze," were more the norm in California during the 1950s and 1960s, wrote the self-proclaimed car biologist K. T. Burger, who grew up there. "I was naturally trained as a car biologist," Burger explained. "Just as many wildlife biologists grew up in the outbacks of Idaho and Montana, I grew up in the urban wilds and freeway wilderness of L.A." As a boy, he might as well have had wheels for legs. "Walk to school, take a bus to a concert?" Burger wrote sarcastically. "You've got to be kidding."

California suburbs grew like algae crawling out of the sea and given new life by wheels to spread and multiply. The idealized Californian family lived in one suburb, worked in another, shopped in a third. A normal family owned two cars. Tying this modern family's random and excess movement together with mass transit would have been expensive and complicated. Funding would have taken money away from roads. A car-crazed California culture shaped its environment for cars only. California's great equalizer—a car, or two, for everyone, at an affordable price—had not yet turned against it. And in those days, Detroit supplied them, with fins, whitewalls, and lots of horses under the hood. Rebel without a cause James Dean died in a Porsche head-on while driving to a race, but Hollywood romanticized his death. Popular music idealized cars. The faster, the snazzier, the better. The Beach Boys were soon singing about GTOs and Little Deuce Coupes and about having "fun fun fun 'til her daddy takes the T-Bird away." Suggestions that California might curb its car and road appetites were brushed aside as naive, backward, or just plain stupid. A car critic might as well have been a seer in ancient Rome advising the brain-addled who were drinking wine from lead goblets to switch to ceramics. Being anticar in California was heretical, could get you condemned to a fate even worse than death: riding the bus, if you could find one.

Public transit continued to disappear in Los Angeles and across California throughout the 1950s and 1960s. Likewise, it was stripped from cities in the Midwest and the East as the car culture epitomized by California, and the suburban lifestyle romanticized there, spread eastward in the wake of asphalt, gas stations, and the U.S. interstate system. Initiated during the Eisenhower administration, the huge road-building project was destined to link the country together with 42,500 miles of multilane blacktop. In 1900, the United States had the best trolley car systems in the world. Europe's longest, located in Berlin, would have ranked twenty-second in America in size. The U.S. trolley network reached its zenith in the mid-1920s, then declined slowly, not just because of the popularity of the car but because of bad management and greed on the part of trolley line owners. When General Motors bought the trolley systems across the Southland in the 1950s and tore up the tracks, it was an act of gross self-interest but only one more of an ongoing continuum paying homage to the car. Los Angeles merely became the first city exposed to the atmospheric consequence, smog redolent of nitrogen oxides, hydrocarbons, and carbon monoxide pouring out of tailpipes, along with toxic vapors rising from various points along the distribution channel, from the refineries making gasoline to the trucks delivering it to the gas stations selling it. In a preview of the future of cities on every continent, the smog routinely robbed the L.A. basin of sunlight and linked it to the atmospheres of mass extinction that had covered the entire planet eons earlier, when mammals were ratlike and furtive rather than two-legged and cocky, ensconced in stylized steel shells out of which came carbon-rich compounds until recently sequestered in the planet's crust.

During the embryonic stage of smog's growth from a one-city phenomenon to a worldwide dilemma, it is not as though well-intended and deeply committed people such as Tucker, Chandler, McCabe, and others weren't doing everything they could to reduce the spread of the stuff. But they faced forces they weren't even aware of, opponents who far outnumbered their supporters, and a myth of intoxicating power: the freedom of private mobility, the freedom of the car.

The first guy to call the freedom into question, and he did it only indirectly by analyzing smog and linking it to auto tailpipes, was Arie Haagen-Smit. A dapper, European-born chemist, Haagen-Smit had a penchant for cigars and an endearing optimism that would be pummeled but never defeated as he became America's first champion of clean air. In the late 1940s, Haagen-Smit, a professor at the California Institute of Technology (Caltech), was hired by the Los Angeles Chamber of Commerce to figure out why farm crops near refineries had discolored leaves that looked bleached and ghostly. Soon, Haagen-Smit's research put him onto a chemical analysis of smog.

He knew the local version of smog was different from the toxic mix that had killed twenty people, and hospitalized six thousand, in Donora, Pennsylvania, in October 1948. The Donora smog was a factory blend: chemicals, heavy metals, and soot, but not much auto exhaust. The bad air had clobbered the mill town south of Pittsburgh on Halloween weekend. Ambulances got lost in the haze. A football game was played by phantoms in the gloom, fans in the stands covering their mouths. Donora's huge zinc smelter, a four-mile-long plant that employed six thousand, had kept pouring pollutants into the air above the river town, despite rescue personnel's lugging tanks of oxygen from house to house in the streets. Conditions turned lethal the second day of the episode. But the smelter kept running, and people started dying. Finally, it rained. The rain washed the deadly brew from the sky. By then, the twenty victims, all fifty-two and older, most with respiratory or heart problems, were dead and thousands were in local hospitals. The U.S. Public Health Service, which subsequently investigated the disaster, blamed it on an inversion that capped the town on the fatal weekend. That the zinc works, which did finally shut down after the rain, may have been discharging deadly fluoride gas on the Friday that kicked off the weekend, was suspected but never conclusively proven. U.S. Steel, which owned the smelter, sealed its records. In comparison to that airy soup, though, the smog in Los Angeles was unique, Haagen-Smit realized. L.A. smog contained some sulfur dioxide, heavy metals, and even minute traces of hydrofluoric acid, a poisonous halogen, but these chemicals weren't fundamental. What was fundamental, Haagen-Smit soon announced, was ozone, the ground-level version of the stratospheric gas that protected organic life from DNA-damaging ultraviolet light.

Once Haagen-Smit analyzed smog and linked it to the tailpipes of cars and the city's fabled sunshine, he was vilified. His chemistry was attacked, his motives questioned—not only by the automakers and their allies, but by car lovers across the Southland. How could tailpipes and sunshine cause most of the smog? they demanded. What about the chemical plants, the refineries, the backyard incinerators, the trucks, all those millions of smudge pots? Surely, they were more at fault than the finned beauty beneath the carport, or the big-tired monster under the sooty tarp protecting it from smog.

But the auto and oil guys, the motor heads, the highway lobby, and the chemists who said Haagen-Smit was wrong had picked the wrong man to attack. The Caltech professor did not back down. He might have looked like an effete intellectual, with his glasses and European suits. But the chemist had guts and principles. Chemistry was chemistry, he said, and he had the chemistry right. Ozone was produced by hydrocarbons and nitrogen oxides hit by sunlight, and ozone, as his experiments confirmed, damaged the leaves of plants, cracked rubber, and irritated people's eyes—just like smog did because ozone was smog.

Over the next several years, the charges and countercharges about the true chemistry of ozone flew through the yellow haze as it intensified. One miserable July day in 1955, Los Angeles had the highest ozone level ever recorded: 680 parts per billion of air volume (today the federal ground-level ozone standard is 120 parts per billion). City health officials identified a "smog complex." Symptoms included lung burn, watery eyes, and slow-to-improve bronchial problems. Autopsies of people who had breathed L.A.'s air for decades revealed riblike bands of soot wrapping once pink lungs, with black splotches of carbon here and there. A lifetime breathing smog, officials said, seemed more than just a nuisance, as the forces fighting Haagen-Smit's assertions claimed.

Despite its reputation in L.A., smog had its fans, just as smoke had fans in England during the nineteenth century. Smog possessed a certain appeal. It was here today, gone tomorrow. It created awesome sunsets of unnatural colors. Like wine, it had notes, bouquet, toxicity, hue, odor. But its basic chemistry remained constant: hydrocarbons and oxides acted upon by bright sunlight.

In those days, identifying ozone and measuring its concentrations was anything but easy for pollution control technicians. In the field a technician relied on a vacuum pump that drew in a measured volume of air across a rubber band. The technician, armed with a stopwatch, watched the band until it cracked. Refrigerator-sized monitors were developed and stationed around the city. The refrigerators were filled with chemicals that turned different colors according to the ozone level, but the equipment was hard to maintain. Chemicals leaked, and walls of some monitoring stations dissolved. But the big monitors had it all over a technician with a stopwatch keeping an eye on a rubber band. They provided the technological backbone for an ozone-monitoring network, which helped prove Haagen-Smit's chemistry and confirmed ozone as the core component of smog.

Official proof only came in 1957. An independent commission established especially to settle the matter sided with Haagen-Smit. Begrudgingly, even the Detroit automakers acknowledged that their products emitted hydrocarbons and nitrogen oxides, and that sunlight knocked electrons off the oxides and cooked them up with the hydrocarbons, creating ozone. But not to worry, they said reassuringly. Nothing like this can happen elsewhere. Smog is unique to Los Angeles because of its geography and sunny weather.

Basically, the car guys had adopted the Count Rumford stratagem.

That is, exaggerate, lean on hyperbole to protect your interests. The science to disprove you is still weak. And everybody loves your product.

Meanwhile, smog appeared in Los Angeles not as some isolated entity but as the result of a complex system of variables: cars and trucks, driving habits and freeways, brilliant sunlight and inversions, a box made of mountains on three sides and of a curtain of cool air rising off the ocean on the fourth. The system's variables increased and diminished, they intensified and dissipated, they looped together and transformed. Understanding such a complex system, even for an instant, was difficult. Its very essence was change. The volumes of exhaust; the intensity of the sunlight; the density of the inversions; the movement of winds; and the concentrations of gases and particles emitted from smokestacks, vents, and tailpipes were not then, and are not now, easy to predict or calculate. Each component was dependent on numerous factors and often affected by other components and conditions. Decisions made by individuals about what size vehicle to drive, where, how fast, and how often played into the system. In the evolution of any such complex system, wrote the Nobel laureate physicist Ilya Prigogine, the decisions made by each person are important, especially when millions of people are making choices. Collective action accumulating from individual choices shapes the future. "Since even small fluctuations may grow and change the overall structure," Prigogine said, "as a result, individual activity is not doomed to insignificance." Accordingly, beginning in the late 1950s, once the causes of smog were beyond doubt, Southlanders breathed more and more of it because individually they choose to drive everywhere all the time.

By that time smog was en route to hazing the skies of other cities across the country—places where cars were replacing trains and trolleys, where sidewalks were seldom built, and where walking was seen as something grandpa used to do to get to school. Smog had not yet appeared in European cities, nor in Japan, which was still rising out of its abominable ruins. But smog would come to cities on the Pacific Rim, and across Europe, soon enough. It was destined to be the new miasma,

generic bad air plaguing cities and their surroundings, as well as regions
where trade wind transport brought smog down like a curse.

What happened in Southern California, site of the first limited
engagement to clear the air of smog, proved to be a skirmish. It was a
prelude to a big and worldwide war. The war spread to other cities
around the world after the 1960s. By the year 2000, megalopolises from
Houston to New Delhi, from Mexico City to São Paulo, had joined in.
But no city on the planet came close to defeating the enemy. By most
yardsticks, smog emerged victorious in city after city. It was stalled at
times on some urban battlefields, most notably in Los Angeles. Concerted campaigns to fight smog by reducing ozone did achieve some victories. But smog reappeared in cities where defenses were not well
coordinated, where the numbers of automobiles grew into snarling clots
of traffic during rush hours, and where the will and money were lacking
to fight such a diabolical foe. Overall, the smog war of the second half of
the twentieth century wasn't really much of a contest. Smog Man
notwithstanding, there was no miracle.

That a miracle of some kind was needed, however, had been apparent from the early days of L.A.'s unsuccessful battle plan. Making it obvious that smog wasn't a problem just facing L.A., a so-called killer fog had
hit London with a vengeance in December 1952. This fog was not a traditional smoke-caused pea-souper, but a new mix. Every Englishman's
right to burn coal in his hearth certainly contributed to the incredibly bad
air, but so did factories, cars, and buses in the sprawling megalopolis.
Assisted by an absolutely dead calm, the fog cut the visibility in places to
less than three feet. The gritty, toxic haze was everywhere: up the
Thames, down the Thames, on the marshes, and on the heights. It crept
into cabooses, it hovered in riggings. It settled "in the eyes and throats of
ancient Greenwich pensioners . . . in the stem and bowl of the afternoon
pipe," just as it had in *Bleak House*. Robert Barr's sci-fi tale of doom, written in the 1890s but placed sometime in the mid–twentieth century,
seemed uncannily prophetic when the fog came in on a Friday and just
kept getting worse, and by Sunday the city's hospitals overflowed and the

dead started piling up. When the fog lifted on Tuesday, the smoke filters used by London inspectors were so clogged that their "smoke shade" ratings were a tragic joke. The city edged back toward something resembling normal, or as normal as a city could get after four thousand sudden deaths.

One consequence of the ghastly fog was England's Clean Air Act, which was passed by Parliament in 1956. But much sooner, in Los Angeles, residents were shaken up.

Could that happen here? they asked.

Damn right it could, was the answer, given the foot dragging then going on to do more about dirty air, to get rid of the smudge pots and backyard incinerators, and to get auto and oil companies to even admit they were part of the smog problem. Governor Goodwin J. Knight appointed a committee to look at additional clean air reforms in Los Angeles. It soon made some recommendations: tailpipe standards for cars, propane in trucks and buses instead of diesel fuel, a date certain for the end of backyard burning, and a rapid transit system.

It would take thirteen years for the California legislature to set the state's, and the nation's, first auto emission standards. Backyard incinerators in Los Angeles disappeared a little earlier; they were outlawed in 1958. Diesel remained the fuel of choice for trucks and buses. But rapid transit? Forget about it.

A couple of painful ironies of the early years of smog are now self-evident. First, suburban Southern California provided other places with a badly flawed template for how to fight smog. As a cultural trendsetter, California was schizophrenic. It demanded cleaner air while doing much to insure that it stayed dirty. Southern California made cars and movement synonymous with a cool lifestyle. California sprawled. It championed cheap gas and oil for the unabashed pursuit of hedonism, happiness, and endless cruising. A better model of how to pump carbon into the atmosphere is now hard to imagine.

Second, the war on smog was, in the early years, reactive rather than proactive. Los Angeles responded to smog because otherwise the air might have killed thousands, as it did in London, and have driven away those smart enough or wealthy enough—just as smoke had driven people out of Manchester, Birmingham, Pittsburgh, and other coal-burning cities in the late nineteenth and early twentieth centuries. Despite evidence that cars caused smog as early as 1950, when the embattled Haagen-Smit associated the two, Los Angeles just kept building freeways. The car-dependent lifestyle contained within itself the seeds of certain defeat in the fight against smog. The first battle with smog cried out for peace, not war—for strategies to keep smog at bay, for curbing its rise as an air hazard to begin with. But Los Angeles was determined to kill smog with no disruptions to the status quo, with no compromises, no inconveniences, no hard changes. In other words, the city underestimated its foe. And to this day, Los Angeles is being hammered senseless by it. Ultimately, the heart of the problem, the gasoline-powered car, stayed the centerpiece of life, and all the pollution control experts did their dance around it instead of shoving the thing into the Pacific and finding a cleaner way for locomotion.

At least California reacted. I do not want to totally belittle the state's pioneering efforts on behalf of clean air. They were laudable, but doomed. The California lifestyle insured an adverse outcome in the future. What's more, the state was initially alone in fighting a worldwide monster in waiting, albeit one it fought with one hand and fed with the other. California set the rules for grappling with a worldwide atmospheric adversary, but one it had also played a major role in creating. In the smog war, and in the larger clean air war of cities and regions, California played dual roles as both hero and villain. The result, unfortunately, was that the state led the world into an awful mess. There's nothing like a know-it-all to really screw things up. And in the tale of dirty air, once California had smog and began trying to eliminate it back in the late 1940s and early 1950s, the state became air's know-it-all, big time.

Since Smog Man's visit to the Southland, California has led the charge for clean air. But always while preaching conflicting messages: cars, cars, cars; suburbs, suburbs, suburbs; growth, growth, growth; and, at the same time, *clean, clean, clean.* Lifestyle changes, questions about how Californians lived and their carbon-squandering ways, their auto addiction—these issues were studiously avoided while the California way to be was promoted and promulgated across the United States and around the world as an ideal worthy of emulating. Not that the auto industry couldn't comply with most of California's basically low-tech, though capital-intensive changes to the ways of making internal-combustion engine vehicles cleaner. But this was a war. Not a collaboration. In war, you fight. That's the mentality. The auto industry mentality—*we'll strangle to death with smog before we let government tell us what to do*—was older, more deeply rooted in laissez-faire capitalism. Whatever California came up with, *it's never going to happen,* the auto guys shouted back. *Never going to happen!*

Then it did happen, and often. And the battle shifted to another technology, another health issue, and another way to handle emissions.

Never was the air as a membrane, as a living entity, put first. Made *the* priority.

It was only in 1963 that the federal government, following California's lead, finally waded into the air quality arena. By then, smog wasn't the only atmospheric miscreant on the nation's plate. There were many others. Radioactive fallout in particular, a new reality of the Cold War and of the emerging nuclear power industry selling itself as an alternative to fossil fuels, made smog seem minor league, a nuisance people contributed to but could put up with if they had to. Ironically, of course, just the opposite was true. Radioactivity was less worth worrying about than smog. But it's a peculiarity of human nature that what fascinates us most is what we can't do much about, what we can't take for granted. Air we take for granted. A fine breeze, a deep breath. Our alveoli exchanging all those gases without a moment's hesitation. Those don't fascinate us. But let a hurricane tear your roof off, a piece of chicken clog your throat, or too much tar from all those Camels give you emphysema, and then you pay attention. You stop taking the wind, a single breath, your lungs, for granted. Radiation, in contrast, is a freaky throwback to distant time.

But in the world we've created, your chances of dying from too many rads compared to too much smog are very small. That doesn't mean that radioactivity isn't dangerous and couldn't be a lot worse than smog if there were a nuclear war or a global nuclear-power industry with the kind of sloppy safeguards perfected by the Russians during the 1970s and 1980s. For the short term, however, radioactivity is a latent air nightmare, something invisible and melodramatic that sizzles the imagination. Radioactivity is also a scary reminder of how close the whole world came to disaster during the Cold War, when radioactive substances were the preferred power toys of the generals and strategic planners on both sides of the divide.

Rads and Bugs

It would simply lift a chunk of atmosphere—ten miles in diameter, something of that kind—lift it into space.

—EDWARD TELLER, ON THE POWER OF A 100-MEGATON HYDROGEN BOMB

Radioactivity has always been here. It comes through the atmosphere from the sun and from distant supernovae. The cosmic rays explored so feverishly and with such innocent glee by Auguste Piccard in the early 1930s plunge through the earth's atmosphere. Analogous to neutrons fired in a cyclotron, cosmic rays simply originate from supernovae in deep space. Those that hit carbon atoms in the atmosphere produce carbon-14, a radioactive isotope. The far more common isotope of carbon, carbon-12, is not radioactive. Both forms of carbon occur in all living things, but carbon-14 decays into carbon-12 over time. Thus, analysis of the carbon-14 content of fossils and other carbon-containing rocks can give scientists an approximate age of these materials.

Inside the earth, there is radioactive decay from unstable elements that gravitated there as the planet cooled and a crust formed. Most of the decay is harmless. It helps keep the molten interior of the planet soft and flowing. Some decay closer to the crust is bothersome, however. Radium can produce a gas, radon, which people don't like to find in their cellars. The harm from radon comes from its energetic alpha rays. Though large, as subatomic particles go, and slow, alpha rays can be inhaled, setting off tissue damage from ionization, the knack that the particles have of knocking electrons off their orbits around more stable atoms.

The telltales of radiation, electrons and nuclei shooting off from unstable atoms, were not detected by scientists until the 1890s. William Roentgen, working with cathode rays in his lab, discovered that the rays penetrated some materials, such as skin, but were blocked by others, such as bone. The strange, unknown power Roentgen observed passing through his own hands led him to the name *X-rays*.

X-rays gave scientists at the turn of the twentieth century new powers to explore the small, the quirky, and the invisible. This special form of radiation helped Henry Becquerel figure out the nature of radioactivity, and J. J. Thompson to elucidate electron theory. X-rays allowed doctors to diagnose tuberculosis earlier and to examine fractures without cutting a patient open. Toxic, even at low levels of exposure, X-rays and other forms of radiation also innocently killed patients, doctors, and scientists alike. During the first decades of people's fascination with radiation and radioactive substances, there was an ironic and tragic disconnect between research and health. More than a hundred scientists working with the new materials died; the energetic particles welcomed so warmly into their labs penetrated their brains, bodies, and bones. Marie Curie, who, with her husband, Pierre, isolated the first milligrams of radioactive material by refining tons of pitchblende hauled from mines in the Ore Mountains of Saxony, was a victim. Thomas Edison's research assistant Clarence Dally was another. Working on his boss's X-ray lamp, Dally lost his hair and felt ulcerated sores on his scalp; tumors began working their way up his arms from his hands. He soon died.

What was happening in this new world of the very small and very fast was a throwback to medieval visions. Bizarre particles zooming to earth from distant stars and rising out of the planet itself had merely been discovered, concentrated from their scattered mineral sources, and used for experiments. Physicists were especially drawn to the exploration of radioactivity: its sources, properties, and lethality. The dimensions they probed were infinitesimally small, the particles of radioactive matter not solid, as had been believed about atoms since antiquity. Like atoms of more stable elements, those of radioactive materials such as radium were found to be mostly emptiness in which extremely small particles, electrons, swirled around a nucleus and were bound to it by electrical forces. Radioactive elements possessed order and simplicity, just like regular elements. What distinguished the radioactive ones was that they were always coming apart.

Stable elements usually have about the same number of neutrons and protons in their nuclei, but unstable elements have too many protons. Protons have a positive charge. Since particles with the same charge repulse each other, protons are always trying to get apart. They don't separate in a nucleus of a stable atom because of a unique quality of neutrons. Neutrons possess an atomic power that is comparable to glue. The glue overcomes the forces of repulsion between the protons and binds them together. Up to a certain size nucleus, that is, which happens to contain eighty-four protons. When a nucleus contains more than eighty-four protons, the atomic glue doesn't hold. So elements whose nuclei contain more than eighty-four protons are always coming apart. They're decaying. They are radioactive. As they decay, they release all the energy that went into forcing them together. The energy is what we call radioactivity.

In the early 1930s, Ernest Lawrence and M. Stanley Livingston, a student of his, built the first particle accelerator, or cyclotron, at the University of California in Berkeley. The cyclotron could smash the nuclei of atoms together at high speeds, producing atoms with unnaturally large numbers of protons. For instance, when you smashed deuterium, an isotope of hydrogen with one neutron added to its nucleus, into uranium oxide, which was radioactive but relatively stable, you created plutonium, a highly unstable element with ninety-four protons. As the scientists who first isolated plutonium in 1940 discovered, the instability made the element an ideal powder keg.

In a chapter titled "Elements from Hell" in his book on molecules, science writer John Emsley recalled how Glen Seaborg, Arthur Wahl, and Joseph Kennedy captured the first atoms of radioactive plutonium, which would provide the core for the first atomic bomb, and "quickly realized they had stumbled upon a remarkable metal." Working in Berkeley, the scientists learned that plutonium atoms bombarded by neutrons in a cyclotron split and emitted other neutrons, along with energy, and that successive splitting of neutrons sparked a chain reaction. Seaborg, Wahl, and Kennedy took almost a year to make just three-millionths of a gram of plutonium, an amount just large enough to be weighed. Then they calculated that if they had nine pounds of it, enough to make an apple-sized ball, they'd have a critical mass, enough plutonium to sustain a chain reaction.

By 1945, after America had mobilized the top-secret Manhattan Project, enough plutonium existed to make two apples. Most of it had been made at a vast plutonium-manufacturing facility built in Oak Ridge, Tennessee. One apple provided the core for a test bomb that exploded in Alamogordo, New Mexico, in July 1945. Watching the blast from a shelter (by then radioactive materials had caused enough damage to justify protection, at least for senior officers) was Brigadier General Thomas Farrell. "Unprecedented, magnificent, beautiful, stupendous and terrifying," Farrell said of the explosion. "No man-made phenomenon of such tremendous power had ever occurred before. The lighting effects beggared description . . . [lighting the countryside] with a clarity and beauty that cannot be described but must be seen to be imagined. It was that beauty the great poets dreamed about but described most poorly and inadequately." A month later, the second plutonium-cored bomb vaporized much of Nagasaki and killed seventy thousand people. The bomb that had obliterated Hiroshima three days earlier was cored with uranium.

Before the blasts, there had been some serious disagreements over whether an atomic bomb might ignite the atmosphere. Robert Oppenheimer, director of the Manhattan Project, didn't think it could happen. Edward Teller, who subsequently directed the making of the hydrogen bomb, disagreed. At one meeting, Teller "proposed to the assembled luminaries the possibility that their bombs might ignite the earth's oceans or its atmosphere and burn up the world, the very result Hitler occasionally joked about with Albert Speer."

Whatever the temperatures or the atmospheric damages, Winston Churchill simply didn't want the bomb to fall into the hands of the Russians. Churchill's fear of that imminent catastrophe brought him to the campus of Westminster College in Fulton, Missouri, on March 5, 1946. There, he delivered his infamous Sinews of Peace speech to forty thousand people, with live radio broadcasts carrying his mesmerizing basso across America and around the world. Convinced that the Communists' philosophy of world domination was as threatening to the future as Hitler's Nazism, an evil just defeated, Churchill informed the world that "an iron curtain has descended." On one side he put "the Soviet sphere." On the other, the democracies, led by America and Britain. He wanted those two countries to wake up to the threat of the Russian Communists and to the need to lock up the secrets of the atomic bomb rather than sharing them with the world. The long-term impact of Churchill's speech on the atmosphere, and on the future of the world, was monumental. Its short-term influence, which shaped the long term, had more to do with the pending decision of the Truman administration about what to do with its atomic bomb know-how.

In early 1946, the Acheson-Lilienthal report proposed the sharing of that nuclear know-how. The report, a product of advisory committees chaired by Dean Acheson, soon to be secretary of state under President Harry Truman, and David Lilienthal, head of the Tennessee Valley Authority, wanted the danger of atomic weapons defused by a kind of global sharing of their control and responsibility. The report, which was radical and visionary, proposed not only broad international cooperation but also global sanctions. Churchill, however, did not see the world in the same way as the liberal-minded Americans who briefly had Truman's ear. The British leader wanted America, Britain, and their allies to monopolize atomic energy and keep its secrets, and those of the bomb, to themselves. Churchill's way of thinking soon prevailed.

A fly in the ointment, however, was the Soviet Union. With the assistance of spies, it threaded the security network set up to guard secrets and used what it stole to build an atomic bomb of its own. In August 1949, Joe I, named after Joseph Stalin, exploded in Kazakhstan.

The ensuing Cold War bomb race, during which the United States conducted 216 above-ground nuclear tests and the Soviet Union, France, and Britain carried out approximately 150 more, was bad news for the atmosphere. And in the United States, it was especially bad news for a group of people who lived in rural southwestern Utah and found themselves rained on steadily by radioactive fallout from tests in nearby Nevada. Energetic protons and electrons stacked up in flesh and bones and made a lot of people and animals deathly ill. Thus, Churchill's view was incorrect. The Soviets got the atomic bomb anyway in a context that promoted further development of nuclear weapons. The Acheson-Lilienthal proposal might have spared the atmosphere and humanity from the growth of radioactive pollution, but the cooperative approach didn't prevail.

Under the Vault of the Desert Sky

I remember the ground so hot that I couldn't stand on it, and I was just burning alive. [Until then] I was happy, full of life. Then I understood evil and was never the same.

—ROBERT CARTER, SEVENTEEN-YEAR-OLD RECRUIT
AFTER HIS FIRST ATOMIC BOMB BLAST

It may seem hard to understand now, but in the 1950s the Mormons of southwestern Utah were tagged "a low-use segment of the population" by U.S. Cold War managers and strategists. Beneath the humbling sky once sacred to the Shoshone, like a mythical tale of betrayal visited on the excessively righteous, the true believers, the Mormons got betrayed by their own government. They got betrayed by their own submissiveness to authority and to its well-spoken liars who reassured them time and again that everything was OK. Meanwhile, their children, animals, and plants withered beneath what often was some of the most beautiful blue sky on the planet. That air so pristine became the medium through which radioactivity from nearby nuclear bomb tests pierced the living with subatomic-sized spears simply elevated the wrongdoing to the status of an American fairy tale. Not the syrupy gloss manufactured by Disney Studios later in Hollywood, but a real-life fairy tale, one that would have excited the Grimm Brothers, because in this perversion of justice in the name of national security, evil was rewarded and never fully acknowledged or made to answer for its acts.

In the 1980s, a New York photojournalist, Carole Gallagher, documented the fallout years on the rural Mormon communities in Utah: Saint George, Parowan, Cedar City, and others. By then survivors were willing to talk and be photographed. The result, *American Ground Zero: The Secret Nuclear War*, told a chilling tale of God-fearing, American-loving, authority-believing people being terrorized, deceived, and dismissed as expendable. They were dealt, as one victim put it, "a very bad hand."

Early in her book, Gallagher described a scene from a propaganda film being shot by the army. In it, an actor playing a preacher reassures two nervous soldiers that there's no need to worry, they're safe, the army has taken all necessary precautions to insure they don't get hurt during the blast they're about to experience up close. When an atomic bomb detonates, the preacher tells them, "One sees a very, very bright light followed by a shock wave, and then you hear the sound of the blast. Then you look up and you see the fireball as it ascends up into heaven. It contains all of the rich colors of the rainbow, and then as it rises up into the atmosphere it assembles into the mushroom . . . a wonderful sight to behold." In real time, Gallagher said, soldiers often hustled back to their barracks after a test blast, bleeding from their eyes, ears, noses, and mouths. And radioactive fallout spread not only over nearby Utah, but across the Great Plains and the Midwest, and caught trade winds all the way to the Adirondacks in New York, where it could jiggle a Geiger counter.

Utah was "a living, breathing hyperbole of Americana," Gallagher wrote. Mormons didn't ask questions of the army, the state government, or the Atomic Energy Commission, which was in charge of testing nuclear weapons and informing people about them. For the Mormons, authority had a cast of the divine. Making a fuss about what the higher-ups were doing to the air and the soil just wasn't done. It could get a person ostracized. One woman who did eventually challenge the authorities told Gallagher that she'd been painting in Bryce Canyon when an orange-pink veil lit up the sky. It was like the sun had come up backward. The shadows of the trees were cockeyed. "I should have known then that the world was upside down, that it was wrong, but I didn't."

Some women's hair fell out, others got leukemia, a number had miscarriages. Kids developed thyroid problems. U.S. Public Health Service personnel told women who complained that they were neurotic, had "housewife syndrome." Farm agents told farmers whose lambs were born with hearts beating outside their bodies that the ewes had been zapped by too many rads and just couldn't handle them. Locals granted front-row seats on the mushroom-cloud show just to the west knew the tests were important to national security, but they didn't know they were human guinea pigs in the "most prodigiously reckless program of scientific experimentation in United States history." They didn't know that President Eisenhower had said, "We can afford to sacrifice a few thousand people out there in defense of national security." Or that Ike was also behind a top-secret congressional hideaway beneath the posh Greenbrier Resort in West Virginia. Code named Project Greek Island, it was a nuclear bunker for protecting the lives of 1,100 high-priority people "for the sake of the union." The Mormons didn't know their animals, kids, homes, crops, streets, tabernacles, and everything else were being dusted regularly by the most toxic element known, plutonium, one pound of which, if ground up and evenly dispersed and inhaled by everyone on earth—granted, this isn't very likely, as pro–plutonium users declare in support of its ongoing production—would exterminate human life everywhere on the planet.

The United States had not declared a secret war on a religious minority known for polygamy. The Soviet Union posed a real threat and was building atomic bombs as fast as it could. But, as a fallout map, a stark black-and-white affair printed in *Under the Cloud,* by Richard Miller, makes clear, if there were a single location to choose for the tests, one that guaranteed fallout from coast to coast, southeastern Nevada was it. The map looks as though a huge inkwell blew up in Nevada. Ink is smeared over Utah. You can't even see the state lines. Ink splotches, splashes, jumps to, and speckles every state to the east and flicks a few daubs to the west as well. No studies have ever been done to ascertain how much radiation remains in the atmosphere from the atomic bomb testing era. The amount is not insignificant. "Enough plutonium was scattered to the winds to ensure that we each now have a few thousand atoms in our body," wrote Emsley about the atomic bomb testing era. He added that plutonium tends to cluster in bones, has the lowest permissible dose rating of radioactive elements, and, because of its long half-life (24,100 years), is destined to be around for a long time. Adding to the atmospheric load overhead today has been a slow leakage of radioactive particles from the decommissioning of nuclear weapons and from the nuclear power industry, its waste repositories, and its occasional accidents, both big and small.

The so-called downwinders in Utah suffered from what was later labeled *delayed mutation effect.* Radiation rode out of each mushroom cloud during the aboveground blasts because even the most clever bomb design couldn't keep more than about a quarter of the radioactive atoms around long enough so that the chain reaction gobbled up the energy like some berserk and alien sky god with a tongue of multihued fire. People got sick. Couples brought into the world offspring they were shocked to see. "His face was a massive hole and they had to put all these pieces of his face back together," said Ken Pratt, a stuntman for movies being shot in southern Utah during the 1950s, recalling the birth of his son. "I could see down his throat, everything was just turned inside out, his face was curled out and it was horrible. I wanted to die. I wanted him to die." Pratt recalled "going outside the hospital, laying on the grass and just crying and sobbing." He became suicidal, was still suicidal thirty years later, couldn't hold a job, always thought of ramming buildings in his car. He told Gallagher, "We don't know our enemy, really."

Taking a comic-tragic view of his participation in two successive blasts, the 44-kiloton Smoky on August 31, 1957, at Frenchman Flat, and the 9-kiloton Galileo on September 1, 1957, at Jackass Flats, a former member of the 82nd Airborne Division named Russell Jack Dann described Smoky like a stand-up comedian:

> We wore no protective clothing or anything, no gloves, no gas masks. We were in completely open space, right on the top of the hill, like a bunch of dummies out there. At 3.8 miles the heat and the light is instant. First of all, when the bomb went off the light was like a thousand suns and the sound was like a million cannons. Then we saw this tidal wave of dirt and dust and sagebrush and rattlesnakes and wires coming after us, it could have been any damn thing out there, but it was coming and the sergeant hollered, 'Hit it!' We went down like damn bowling pins. First of all, you could see right through your arm as if it was an X-ray, the sound was just earth shattering and deafening and a tremendous roar. The wind was blowing at 150 miles an hour, peppered the hell out of us and everything went flying, everything you could hold on to. There was nothing to hold on to.

A "dirty" bomb, Smoky had been packed with coal. It was a fossil-fuel cocooned weapon meant to cause maximum damage to life, air, and water. In a few seconds it unleashed air pollution of awesome intensity.

The fireball rose straight overhead for miles. A black-looking stovepipe appeared, glowing deep purple in its center, a mushroom cap forming on top. Russell Jack Dann marched with his squad right to where the tower holding Smoky had been. But there was no tower. Only melted tanks, crystalized sand, and lots of dust kicked up by the soldiers. Geiger counters were going bonkers, clicking like cicadas. The troops hastily boarded trucks. "They simply used whisk brooms to get the dust off us, alpha particles, and that was the end of the decontamination," recalled Dann.

Yet compared to the nuclear bomb tests near Bikini in the South Pacific, Nevada's blasts were nothing, claimed Colonel Langford Harrison, a pilot in the 4926th Test Squadron, which flew missions into mushroom clouds above both the Nevada and the Bikini test sites. And it wasn't just one sweep through, then back to the base, said Harrison. The planes stitched the clouds again and again. "The scientists wanted us to get our tanks seventy-five percent full [of radioactive materials from the blast] before we left the cloud." The Bikini blasts made the job easier because of their scale. "You haven't seen an atomic bomb until you've seen one of those down in the Pacific," Harrison attested. "You'd wipe out the entire state of New York with one fell swoop. It stretches out 125 miles across, a realization of man's insanity. In Nevada the clouds got only a couple of miles across, little firecracker ones by comparison."

Aboveground testing ended in 1963, when the Nuclear Test Ban Treaty was signed. A partial ban, it left underground testing as an option.

However radiation gets into the atmosphere, once there, especially at close range, it cares as much about air as a rocket cares about a cloud, a bullet cares about the fine hairs over your heart. Alpha rays, positively charged particles ejected at high speeds from radioactive materials, cannot penetrate skin, but once inhaled into the body, they remain very energetic and can damage DNA. Beta rays, negatively charged particles more powerful than alpha waves, penetrate better because of their very high speeds, which can approach the speed of light. But it's gamma rays, a form of electromagnetic radiation that has short wavelengths, similar to those of X-rays but even more penetrating, that are particularly nasty.

They go through the skin much easier than either alpha or beta rays. Inside the body, gamma rays can "nick" DNA strands, which results in genetic recombinations. Tissue exposed to gamma rays suffers cell damage, but the cells need to divide if the damage is to spread. Yet, radioactivity suppresses cell division. This results in a delay of symptoms, a time during which a victim might feel as though he or she is not going to be sick from a high, or prolonged, exposure. Once cell division in the damaged tissue gets going, however, an individual's body informs him or her otherwise. Things begin to go wrong. White blood cells needed to fight infections aren't there, because blood-forming functions have been compromised. Add to that a radiation-caused reduction in the blood's ability to clot, and as Richard Rhodes informs us, "the outcome of these assaults [is] massive tissue death, massive hemorrhage and massive infection."

*

In a Churchillian world of peace by paranoia, Seventh-Day Adventists were also categorized as a "low-use segment" of the American population on whom tests could be run. Like the Mormons, the Adventists didn't complain much. Unlike the Mormons, however, they were conscientious objectors. Whether drafted, or enlisted, once in the military, they lived by the Old Testament injunction Thou shall not kill. They made ideal human guinea pigs.

So it came to pass that one day in early 1955, several dozen Adventists encircled a four-story-tall metallic sphere within shouting distance of an eight-story anthrax factory in rural Maryland. Inside the sphere a batch of Q-fever microbes was exploded. From one to five microns wide, the microbes traveled in mists down octopuslike tentacles attached to the sphere and through masks and into the trachea and bronchi and alveoli of the religious hosts. Monitored by army germ warfare personnel, who told them to just breath normally, don't worry, the effects are only temporary, we've got an antidote in case you get real sick, the Adventists did their duty for God and their country.

Camp Detrick, home of the Q-fever-exploding sphere called the Eight

Ball and of the eight-story anthrax incubation matrix, anchored the U.S. germ warfare program. The camp had been germ central since World War II, when a drug company president named George Merck had been put in charge of this old army base far enough away from Washington so that if the germs got out of hand, the top brass at the Pentagon and in Congress could be warned. Sleepy Fort Detrick was encircled in wire. Guard towers went up. Two hundred and fifty buildings were built. Inside them, batches of anthrax, botulism, and other biological germs were hatched, none of which the United States put to use before the end of World War II. Postwar funding for germs was cut, then restored in the early 1950s after high-altitude U-2 spy planes took photos of suspect Russian germ factories and espionage agents confirmed the sightings.

At Camp Detrick, scientists began feverish incubation of bacteria, viruses, and rickettsias. Bacterialike and halfway in size between most other bacteria and a virus, a rickettsia like Q fever, or *Coxiella burnetii*, was not really alive until it located a living host. A rickettsia needs to weave itself into something biologically more advanced in order to thrive. As a weapon for germ warfare, a rickettsia was ideal, however. Small enough to penetrate pulmonary defenses, invade cells, and then become one with its host, the microorganism synchs into the more sophisticated biochemistry and, like an uninvited visitor that refuses to go, makes life pretty miserable. The host begins to shake, feel hot, slur his or her speech, and want to lie down, which helps explain why attentive germ personnel at Fort Detrick eased the Seventh-Day Adventists down all around the Eight Ball after the volunteers had done their duty. Of course, some of the volunteers continued hallucinating and feeling hot and hurting like hell in their muscles and joints and wondering about racing hearts. But they all gradually recovered. The slow recovery of the victims of germ warfare, and their need for constant care while recuperating, was one of the goals of the designers fashioning the bugs. Not killing the enemy, only incapacitating him, making him a burden, was felt to be both smart warfare and very humane.

In July 1955, after the successful Q-fever test in Maryland, thirty-seven

Seventh-Day Adventists were flown to Utah, out to "the devil's own laboratory," as Mike Davis called Dugway Proving Ground in his jeremiad, *Dead Cities*. South of the Wendover Range in Tooele County, a place subsequently friendly to chemical-weapon storage and toxic landfills for low-radiation waste, Dugway had risen out of the sands during World War II. A hazardous weapons outpost not prophesied in the Book of Moroni but welcomed with wide arms by the descendants of the prophet Joseph Smith, Dugway was vast, mostly empty, and well guarded. A faux German village built there during World War II had been used to test incendiaries, which were firebombs for carbonizing people and ashing cities. A decade later, once they arrived at Dugway, the volunteer Adventists were led to a giant grid beneath a dazzling night sky and told to just breathe normally. A half mile away, five aerosols, each holding five ounces of Q-fever microbes, sprayed the test germs into a breeze blowing their way. Maybe the Adventists heard angels out there beneath the gilded vault stretching over them, like a reassuring testament of the atmosphere at its most glorious. Maybe compassionate military personnel whispered reassuringly into their ears that the microbes they were about to breathe weren't going to kill them, were just going to make them drop to their knees in the praying position but with heads splitting and minds awhirl and with sweat bursting from their brows unless they felt like they were freezing. If the shits got too bad or the vomiting unbearable, there was always the antidote.

Huddled together like sheep beneath the stars, did any of the volunteers hear the Q-fever microbes riding the wind, feel them entering puckered nostrils, sense the micron-small and half-alive things penetrating cells and starting to set up shop? Of course not. One of the scariest things about biological weapons is their absolute silence. Alive, half alive, barely alive—they're all eerily quiet. Unlike bullets and bombs, germs aren't showy. They don't make a lot of noise. Showtime for them is yet to come.

The aim that night in Utah was excellent. Right on target. Millions of rickettsias went straight to the center of the circular grid, at the eye of which clustered the willing victims, breathing as normally as you might expect, given the circumstances.

But how could Q fever take out an enemy division? Or incapacitate a small city? Following the successful tests at Camp Detrick and Dugway Proving Ground, these were the questions military strategists wanted answered. One answer seemed to be wider dispersion by air. To put it to the test, several F-100 Super Sabres, America's newest jet fighters, were converted into low-altitude aerosols. Nozzles were riveted to aluminum fuselages where conventional bombs normally rode. Large concentrations of Q-fever microbes were lifted off the runway to see if they could be transported at the speed of sound, then sprayed for maximum dispersion and effectiveness according to mathematical models.

Dispersion of germs was always a headache, a potentially lethal screwup. British troops might have wrapped smallpox bacillus in blankets and handed them over to an unsuspecting enemy with nary a fatality. But that was two hundred years ago. Spraying rickettsia germs from the thorax of an aluminum bug moving horizontally at a hundred miles per hour upped the complexities considerably—as did the chance that the wind might change. Nobody wanted Q-fever microbes blowing the wrong way. Then there was the issue of turbulence. Just enough turbulence beneath the belly of a Super Sabre was needed to churn the air like the blades of a mighty Mixmaster, dicing microbe clusters into fighting size. Otherwise, they would tumble down to the ground, wasted by gravity.

During the Super Sabre Q-fever tests at Dugway, animals of various size and maw, furred and feathered, clawed and hoofed, bore the blunt of the deliveries. One pilot took sick, as did three guards posted fifty miles away, keeping traffic at bay during test sprays over the desert. Both incidents of friendly contamination were given positive spins by the military because they validated the infiltration and dispersion power of Q-fever germs. The waywardness was deemed within the acceptable margin of error.

The development of Q fever and other biological agents during the 1950s can be viewed as an escalating invasion of air and as a crazed competition to sustain a precarious peace. Each sphere in the developmental program of Q fever as a weapon was bigger than the previous sphere. Each subsumed its predecessor, and in turn was subsumed by its successor. First there had been the Eight Ball, literally a sphere four stories high with hoses jutting out of it, a kind of lethal octopus. Next came a circular grid laid out on the Utah desert with a dome above it, a vault, peppered with stars. Once F-100s began spreading Q fever as moving objects describing much larger spheres, at least fifty miles in diameter, the scale of atmospheric tainting spread from the local to the regional, from germ-suffused air threatening a house, then a city, then an entire county. Each sphere demanded more of the incredibly expensive eau de *Coxiella burnetii,* a war perfume perfected and refined and put in spray applicators in homage to a dark art, an insane alchemy, allied with technology and stirred by the fear of men. With the F-100 dropping germs carried tight to its belly like a gleaming fast insect with stubby wings, the spheres of airborne terror seemed almost complete. But like the radiation delivered by Little Boy to Hiroshima and by Joe I to Kazakhstan, the spheres weren't big enough. During the Cold War, at times only an arsenal of rads and bugs capable of poisoning the entire atmosphere many times over seemed satisfactory.

Toward that end, the United States expanded its germ warfare program in 1957 after U-2 shots taken from the stratosphere revealed Russian labs and experimental buildings on an island in the Aral Sea, then earth's fourth largest inland body of water. On the island, Q fever, anthrax, glanders, plague, and other germs were being tested in the loose and dangerous fashion that the Russians had perfected and that was as environmentally ruinous as their sloppy handling of radioactive waste from nuclear power stations would be in the 1970s and 1980s. Biological drift swept germs from the remote island into and across the Aral Sea, contaminating the water, along with sections of Uzbekistan and Kazakhstan. In 1957, with American and Russian bombers constantly aloft and carrying atomic weapons, and with the hard-nosed Nikita Khrushchev poised to become premier of the Soviet Union, once American intelligence obtained photographs that even suggested Russia's germ warriors in the middle of the Aral Sea were outflanking those at Fort Detrick, a major response was warranted.

A super germ factory rose out of the Arkansas woods. Inside its well-guarded buildings, scientists incubated all the standby germs and some new ones, including VEE, Venezuelan equine encephalitis. VEE "just makes you think you want to die," said the factory's designer. "Your eyes want to pop out of your head."

The morale caveat continued to restrain America's germ program, however. Cripple, sicken, incapacitate the enemy, but don't kill him—that was the germ warrior's mantra. It irked some military leaders. It left a big window open for retaliation. What happened once a foe shook off the Q fever, was no longer hobbling to the toilet every two minutes, or vomiting—would he want to be humane? Would he be satisfied to just make American soldiers and civilians puke and shit, sweat and shake, hide from light that made the eyes burn as though they were on fire? Or would he simply want his counterpart dead? The conundrum left some strategists skeptical about so-called humane biological weapons and their ultimate effectiveness.

Still, germs were cheaper than bombs. To be safe, America made plenty of both. Large, reputable corporations such as General Electric, Monsanto, and Goodyear, along with the country's biggest cereal maker, General Mills, joined the hush-hush germ warfare program. In 1959, in light of Fidel Castro's takeover of Cuba and the nearness of Communism, a very special timed-release germ weapon was brewed in the Arkansas factory for possible dispersion over America's wayward neighbor in the Caribbean Sea. The concoction was a blend of Q fever, VEE, and staphylococcal enterotoxin B (SEB). The sophisticated biological cocktail was cleverly diabolical. Delivered by air in bombshells, then inhaled as the contents spread through the streets, it attacked the body in an orderly fashion. Within hours, SEB induced fever, headache, and muscle pain, along with coughing fits. VEE, an infection traditionally associated with the brains of horses, kicked in after incubating for a day or two, causing nausea, diarrhea, more fever, and the popped-out-eyes syndrome that made you want to die. Finally, the Q-fever microbes became one with their hosts, bringing on more body pain, headaches, and hallucinations. Anyone breathing the Cuban cocktail would be a wreck for a month, completely spent. Meanwhile, American paratroopers and ground forces would have occupied Cuba.

Despite its lower cost and humane appeal, the U.S. Air Force didn't like germ warfare much. Pilots preferred bombs to bugs. Atomic, conventional, whatever. Spreading Q fever, the Cuban cocktail, or any other germ warfare agent would mean few bragging rights back at the base. Spraying germs silently down on villages and upturned faces and indifferent animals, and then flying away, didn't fit a pilot's sense of mission or complement his self-image. Most pilots saw themselves as tough guys with hearts intent on fighting Communism. They despised biological warfare. It was cowardly and coldhearted. It just didn't have the excitement and glory of blasts and flames and smoke. It produced phlegm instead of blood, mounds of Kleenex instead of body counts. And in the end, what did you have? A pissed-off enemy intent on revenge.

Yet, if the other side had bugs and poison gas and atomic bombs, could you really go without?

10

Clean Air Dreams, and Denials

As the 1960s opened, America had smog on the West Coast, radioactive fallout in the Southwest, and enough bugs, poison gas, and atomic bombs ready to launch into the air at an instant's notice to annihilate most everything that had evolved from the last known extinction event, the one 65 million years ago. But as the automakers had told the Southlanders in 1957, after conceding that smog did exist and might have something to do with cars—not to worry. Then in 1962, Rachel Carson published *Silent Spring*, London had one more of its weekends of respiratory hell, and fifteen hundred concerned individuals descended on Washington to attend a National Conference on Air Pollution, which President John F. Kennedy had announced early in the year. One result of the convergence of a Pulitzer Prize–winning book, which seeded the environmental movement, yet another killer smog across the Atlantic, and a liberal administration not intimidated by complaints about states rights being trampled was passage of the Clean Air Act of 1963.

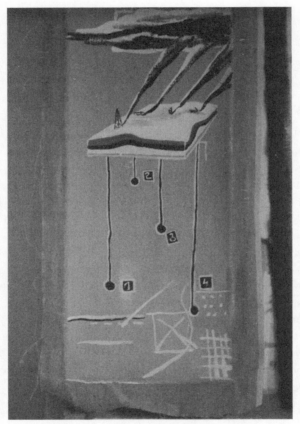

One of the ideas for removing excess carbon from the atmosphere in order to reverse global warming is called geological sequestering. It returns carbon dioxide, the planet's main greenhouse gas, back into the tunnels, seams, and pockets from whence it came as coal, oil, or natural gas.

THE KENNEDY ADMINISTRATION got behind the act because the administration was philosophically disposed to expand federal power. It justified its ambition with health statistics that claimed, with questionable scientific merit, that approximately 58 percent of Americans lived with moderate air pollution and 25 percent endured serious air pollution. The U.S. Public Health Service wanted little to do with the contentious arena of air pollution and took a dim view of the proposed act, which it would have to enforce. The conflict between the administration and the Public Health Service over the act became moot, however, when the Kennedy administration, fearing defeat of the whole package, cut the enforcement provisions altogether as a concession to get votes.

What resulted, interestingly enough, was an act with little muscle but a strong skeletal framework. It marked a major shift in federal philosophy toward regulation. The act put long-term control of America's air in the hands of the federal government. It made clean air, like war, the providence of the chief executive. For the short term, though, the act left air quality action up to the states and funded additional research and education, something it had been doing since 1955. At that time, the Air Pollution Control Act had been passed so that states could learn what air pollution was, where it came from, and what it did. All future air quality legislation in the country would build off the framework of the Clean Air Act of 1963.

As feared by the act's supporters, however, for the first couple years following its passage, most states, with the exception of California, dragged their feet over air quality. Few requests were made for assistance on problems that crossed state lines. Smog, acid rain, and the new airborne toxins such as DDT, which Carson assailed in *Silent Spring* as synthetic children of World War II, "sinister and little-recognized partners in changing the very nature of the world," were not on the radar screens of state health officials. Nor was radioactive waste, a by-product of the nuclear power industry, then in its formative stage. Radiation stayed under the control of the Atomic Energy Commission, a group with a hornet's nest of conflicting responsibilities. The commission regulated radioactive materials, storehoused U.S. war supplies, and was also a major booster of the nuclear power industry, which began building more than fifty nuclear reactors around the country in the mid-1960s.

In 1965, Congress did pass an act that allowed national emission standards for air pollutants to be set, but named no deadline for such a thing to happen. In 1966, however, President Lyndon B. Johnson said people had a right to breathe clean air in the Great Society he envisioned. In "Protecting Our National Heritage," a speech he gave to Congress, Johnson urged members to set auto emission standards rather than talk about it and to recognize the across-state realities of air pollution and its movements. When the Federal Air Quality Act of 1967 was passed, the car guys managed to get national emission standards dropped, but Johnson got most of what he wanted. First, America's air sheds would finally be identified because they were the regional fields of atmospheric change, the physical spaces in which pollution did harm. Second, air quality criteria and technologies would be clarified so that states could set specific emission standards based on them. Finally, the federal government was given power to seek injunctions to stop regional air pollution episodes if states failed to respond in a timely manner. All in all, the 1967 act was a federal coup d'état of air regulation policy.

In 1968, the Public Health Service was abolished. Air quality issues were briefly handled by a newly established Consumer Protection and Environmental Health Service. But then, in 1970, when Richard Nixon reorganized the government, a special Air Pollution Control Office, with a staff of three hundred, came into existence as a branch of the new Environmental Protection Agency (EPA). Air was the sole priority of the office, not a secondary and unwanted responsibility, as it had often seemed at the Public Health Service. In 1970 as well, Congress passed the Clean Air Act amendments of 1970, a visionary but unrealistic set of new rules to expedite the cleaning of America's air, which was now beset by a host of problems, from smog and acid rain to hazardous chemicals and radioactive waste.

This brief overview of the federal government's gradual move into air as a topic worthy of national concern suggests that the process was a timely, incremental response to a recognized problem. That's not quite the way it was. Air issues were often easy to ignore and challenging to debate. The halls of Congress echoed with both subdued and loud debates over the invisible, over the dimensions of the invisible, over whose job it was to control the invisible, the state's or the federal government's. How much was clean air going to cost? How dirty was it, anyway? Who could say how dirty was dirty enough to make a lot of people ill? As much as science and medicine moved the legislation forward, they were not as crucial as high idealism, symbolic gesture, and epic rhetoric—just as during any other progressive era. The real catalyst of change for air law had not changed, either. It was sudden alarm, anxiety, worry caused by a large tragedy, or some other precipitating event. That is, change needed a catalyst that gave the invisible heart, face, and soul, that humanized air a little and made it easier for politicians to persuade voters that here was something worthy of concern.

Kennedy may have wanted Congress to pass a radically new law, but no law at all would have been forthcoming without two important, humanizing catalysts. First was *Silent Spring,* a book that shocked both politicians and the public with its guilt-stirring vision of a great nation indifferent to the consequences of its acts and blithely allowing synthetic pesticides to poison the food chain. Then there were the deaths of thousands of innocent victims during a bad-air weekend in London right before a national conference on air pollution convened in Washington.

Not that Congress, after it passed the 1963 Clean Air Act, seemed to remember or care much about air pollution. "Nobody in Washington thought there was a problem with air back then," said Dick Valentinetti, who had just graduated from college with a degree in public health and taken a job in the trenches of the much-maligned Public Health Service as a junior air officer. Progressive ideas may have been in, but air quality concerns were not one of them. "Congress was supposed to be considering amendments to the '63 act signed by Johnson after Kennedy got shot," Valentinetti said. "But there was no urgency. Then all over Washington, you could smell the pall from this dump over in Maryland. The pall was over the entire city. It was open burning. There was a backfire at the dump and two kids got killed. Congress closed that dump *and* passed amendments to the air act because of the fire and two dead kids." In a tone of astonishment Valentinetti, whose subsequent career in air control took him first to California, and then to Vermont as the sole employee of the state's fledgling air control office, added, "That was an education to me. It was those things close to home that really moved legislators."

Two dead boys in Maryland, a pall of smoke floating over Washington—not a large tragedy on the national scale, but one close to home. It was a catalyst for change.

The catalysts for passage of the Clean Air Act amendments in 1970 were larger in scale, celebratory, and contentious: the first Earth Day, which brought millions of Americans into the streets and to rallies, and the formation of the Environmental Protection Agency, a brand-new regulatory darling about to enter a gauntlet few other federal agencies ever had to endure. Said Valentinetti of those idealistic times: "The environment was important then. I remember Vermont's first Green Up Day. Eighty thousand people got out there. The interstate was closed. One-quarter of the population gave up a Saturday to clean up the state. Today, you'd have mass protests and a lawsuit because you wanted to close the interstate."

The passage of the 1970 Clean Air Act amendments was an event. Air was suddenly important, not only in California but across the country. It had finally achieved the political status of water and sewage. All kinds of people got involved in shaping the new law, not just the typical cadre of politicians from smoky states and some subcommittee members with a scientific bent who actually knew some chemistry. Input came from bureaucrats, lobbyists, members of special-interest groups, environmentalists, health officials, and the public. The final bill built off its predecessors, off the frame of the original 1963 act. But the latest version strongly federalized the act. It made clean air a national priority and mapped out a means of achieving certain goals toward that end.

The act totaled seventy pages. It had a clear, simple framework. Health concerns were given priority over economic ones. Regulations focused on both stationary sources, such as smokestacks and vents, and on mobile sources, mostly cars and light trucks. Certain older stationary sources, such as coal-fired power plants, were treated differently than new ones because of the prohibitive costs of cleaning them up. For instance, existing power plants were grandfathered, or excused from having to install the expensive technologies demanded of new power plants. The act identified five so-called criteria air pollutants that needed to be significantly reduced: sulfur dioxide, carbon monoxide, nitrogen dioxide, ozone, and small particulates. Lead, a sixth criteria air pollutant, would be added to the list later in the decade. The EPA had to develop standards for each pollutant to protect public health. Every state had to clean up its emissions (no more funding for studies only) by either adopting the federal standards or setting stricter ones, which California did. Every state had to create its own implementation plan for meeting the standards, submit it to the EPA, and have it approved. The state implementation plans, called SIPs, were a compromise solution to the states' rights versus federal program debate.

Where the act got a little radical was with cars. Stringent national emission standards were set. By 1975, all cars sold in the United States had to reduce three tailpipe pollutants—hydrocarbons, nitrogen oxides, and carbon monoxide—by 90 percent. Defending the tough new rules, Democratic Senator Edmund Muskie from Maine, a liberal who happened to have his eye on the presidency at the time, said, "The first responsibility of Congress is not the making of technological or economic judgments—or even to be limited by what is or appears to be technologically feasible. Our responsibility is to establish what the public interest requires to protect the health of persons. This may mean that people and industries will be asked to do what seems to be impossible at the present time. But if health is to be protected, these challenges must be met."

Detroit didn't buy Muskie's line of reasoning. The car guys didn't like the details of the act, but they hated the deadlines. Deadlines changed everything.

Up until 1970, Detroit had been on a roll. Despite skirmishes with California over smog and with Washington over safety, Detroit had not veered from its vision of building bigger, gas-guzzling, stylized chariots with bigger engines that emitted more pollutants than ever per mile, and made bigger corporate profits. Since 1950, virtually everything about Detroit-built cars had grown: tires, horsepower, chrome, glass, interior space. To accommodate the cars, roads had been widened, and parking lots enlarged. Insurance rates rose because the cars cost more to repair. And, most important, the Interstate Highway System was being built. By the early 1990s, forty-three thousand miles of multilane highway would link all of the United States together.

Not that Detroit hadn't suffered some setbacks between 1950 and 1970. For instance, California had forced the automakers to develop and install their first piece of air pollution technology, the PCV valve, or positive crankcase ventilation valve, in the state's cars in 1962. A simple device, the PCV valve routed crankcase exhaust gases to the manifold, burning them again before they exited the muffler. The car companies fought the little plug as though it would bankrupt the industry. Then, in 1966, the National Traffic and Motor Vehicle Safety Act was passed by Congress. The act mandated seat belts, dual brakes, and padded dashes, again fought as expensive and of dubious value by the automakers (U.S. traffic fatalities declined by 39 percent over the next thirteen years).

Overall, however, Detroit did what it decided was best for Detroit, and what was best of all was building big, powerful sedans that brought in lot of profits. And, of course, emitted plenty of emissions as mileage-per-gallon averages dropped. Back in 1949, America's favorite luxury car, the Cadillac, had averaged 20 miles per gallon; by 1973, when the first oil crisis hit, Detroit's entire passenger-vehicle fleet averaged only 13.5 miles per gallon. Despite what was known since the late 1950s about the chemistry of smog, and its early appearances in many cities other than Los

Angeles, the federal government fiddled while Detroit was given carte blanche to start blanketing U.S. cities with smog. The bad situation was hastened along by the highway lobby, compliant car buyers, and the Interstate Highway System that was tying the country together with asphalt funded by taxes on cars, gasoline, and car parts. The giant auto and road oligopoly got what it wanted: more roads, fewer trains, and no trolleys in the streets to get in the way of cars. "The highway lobby had convinced government leaders that money spent on highways is a public investment, whereas that spent on public transportation is a costly subsidy," wrote transportation analyst Deborah Gordon. "Consequently, nearly all federal spending on land transportation went to highways."

Not that smog was caused just by cars. Power plants, smelters, and factories contributed about half its precursors, depending on the city and region. But the we're-not-going-to-do-it intransigence of the car guys, which was reminiscent of the tugboat captains in Chicago in the 1890s but with much bigger consequences, seemed to bleed over into much of the industrial sector. If the car makers could hang tough against California and then against federal regulators and their new watchdog agency, the EPA, why couldn't the energy, chemical, steel, and other big industries do the same? In the United States, the auto industry was the nation's largest private employer. It had tremendous power. It set trends. It made utilitarian machines at low prices that people loved for their design lavishness, power, and sense of freedom. But, as Dick Valentinetti said, "Detroit absolutely refused to change. The auto industry was the biggest dinosaur out there. The car guys didn't want to do anything unless they decided to do it."

When the 1970 air act was passed, Detroit still refused to acknowledge the atmospheric damages from cars. Closing ranks, Detroit's leaders insisted that their industry would take care of the problems it caused, yet could point to few examples of such responsive behavior.

Typical of the bean counters running Detroit during this age of style over substance, and of regulation over my dead body, was General Motors CEO Richard C. Gerstenberg, or "Old G. the bookkeeper," as Ed

Cray wrote in *Chrome Colossus*. Cray described Gerstenberg as "cautious, colorless, virtually unknown beyond the confines of the industry. He cooled tempers on the 14th floor and kept the focus on cost control and profits." Spending a lot of money on new technologies to clean exhausts was not high on the bookkeeper's priority list. In *The Decline and Fall of the American Auto Industry,* auto journalist Brock Yates wrote that Old G. "was obsessed by the financial structure of the car business and viewed the product mainly as an abstraction out of which profit or loss could be generated."

If we try to penetrate the minds of the auto chieftains during the late 1960s and early 1970s in order to better understand what they were thinking about, we can feel a little sympathy for them. Downtown Detroit, the site of their former glory, was fast becoming an industrial ruin, abandoned as modern factories were built on green-field sites in the suburbs. The feds were after the car guys on safety, conspiracy, and efficiency issues, as well as emissions. Once the Arab oil embargo, a backlash against America's pro-Israel stand in the Middle East war of 1973, pulled the plug on low-cost fuel and plunged the country into crisis, Detroit had no option but to respond as best as it could. And its culture, its mind-set, its last twenty years of business practices didn't make that easy. I was a student at the University of Michigan during the early days of that period and remember job hunting at the business school and feeling indifference toward the Big Three—Ford, General Motors, and Chrysler—as did many of my classmates looking for work. Nobody I knew, or knew of, wanted to join the auto biz. It was old, jaded, authoritarian, and out of step with the times. Given the circumstances and attitudes many Americans had toward Detroit, it's easier to see why the industry opted for a fortress mentality in regulatory battles, circling the chrome chariots and wondering how in the world to clean its overpowered engines and, after 1973, how to make the engines more efficient simultaneously because gas mileage also needed to improve.

No doubt, the car guys justifiably felt set upon and targeted for special treatment. After all, they were given mandates to force new technologies regardless of cost or difficulty, while coal-fired energy plants, chemical factories, and other polluters only had to install "technologically feasible" controls on their effluents and discharges. And many of the dirty old power plants got a free ride, one they would prolong right into the twenty-first century, and kept spewing emissions from smokestacks with no controls.

For air, the imbroglio created by Detroit's hands-off-industry attitude and Washington's do-it-or-else policies set in motion a Hatfield-and-McCoy-type feud, but not in some Appalachian backwater. This was a national battle of wills and values. The fight set bad precedents and ignored growing realities. As automobiles had become bigger and increased in number, smog was no longer just L.A.'s problem, despite allegations to the contrary by the auto industry. Smog was spreading to cities around the country and about to invade metropolitan areas around the world. In 1963, the Kennedy administration had used rather sketchy data to prove that 58 percent of Americans lived with air pollution, and 25 percent with bad air pollution. A decade later, monitoring tools had improved, along with health assessments of the hazards of overexposure to air pollutants. The sophisticated instruments and new techniques showed that air pollution was getting worse. By the late 1980s, when emission controls would have cut tailpipe pollutants by almost three-quarters of their pre-1970 levels, more than three hundred urban and rural areas would still fail to meet the federal clean air standard for ozone alone. And by then, as Deborah Gordon wrote, "more than 100 of our cities are choking on smog." The economic costs were huge. As were the mortality figures. In 1990, more than 120,000 people would die from air-pollution-related causes, according to the American Lung Association.

The oil crisis in 1973 was, ironically, a backhanded help for America's air. The rise in gas prices created a market for smaller, more efficient, and cleaner Japanese cars. The introduction of the 1975 models benefited air as well. Detroit showed that it could force a new technology, the catalytic converter, when it had to. Whether the technology was the best one for long-term emission reductions was the question.

Catalytic converters work by a sequence of reactions in which a catalyst such as platinum removes target molecules in the exhaust stream. General Motors selected and perfected the technology, "even though the National Academy of Sciences had declared it the least promising way in the long run of controlling exhaust emissions," wrote auto historian James J. Flink in *The Automobile Age.* Designed to work best with the big engines Detroit preferred and had made standard in its cars, the catalytic converter was not needed in most smaller imports, many of which complied with the emission standards without adding the pricey new technology to their exhaust systems.

In 1975, however, America's attitude toward cleaner air and the costs to achieve it was changing. And for the first time in over a decade, the car guys had a friendly face in the White House, that of the stumbler from Grand Rapids, a car guy through and through, the president by default, Gerald Ford.

Ford's Clean Air Fumble

While I had recognized that the economy was in trouble, I thought its illness temporary. I didn't think it would be wise for me as President to stand up and say, "Yes, we're in a recession." But after talking with Gerstenberg ...

—GERALD FORD, *A TIME TO HEAL*

Gerstenberg? Old G. the bookkeeper?

There you have it: Car guy ascended to the presidency talks to a GM bean counter extraordinaire because he's got trouble with the economy. But that was Gerald Ford, who goes down in the annals of the American presidency as an endearing, well-intended, likable conservative who seemed at times to have wandered off the set of a Woody Allen movie and into the White House to provide Americans comic relief during hard times.

When Ford took office in August 1974, after the resignation of Richard Nixon because of the Watergate scandal, the Arab oil embargo continued, unemployment was high, and inflation was rising. The times had changed unmercifully. Gone was the optimistic, upbeat, we-can-solve-it attitude of just a few years earlier. Long lines at service stations, price hikes at the pump, and rumors of a prolonged energy crisis put drivers in a "slightly hysterical" mood, wrote auto historian John Rae. The reversal in America's fortunes had also taken some of the pressure off Detroit to comply with the Clean Air Act. Right after the embargo began, the automakers complained that the emission standards they were being forced to meet by 1975 would push driving costs up even higher, which was true. Relenting, Congress had eased the emission deadlines back a couple years. Even so, General Motors went ahead and introduced the catalytic converter in its 1975 models, while President Ford, who was philosophically and ideologically allied with Detroit, appeared on national television, urging that the emission deadlines be pushed back five years, citing costs and energy security as reasons.

The mixed message—Ford urging delays and GM producing the requested technology—irked Congress. Ignoring Ford's request, it passed the Corporate Average Fuel Economy (CAFE) standards, which would go into effect in 1978. The CAFE standards said that all new cars in America had to average 27.5 miles per gallon by 1985, while also meeting the emission standards.

All the while, Ford smiled and waved as though everything was OK. But things were not OK. After Ford was in office only five months, an alarming 86 percent of Americans said in a Harris poll that they had no faith in the new president to steer the country out of its economic mess. So it was understandable why the economy was foremost in Gerald Ford's mind, with energy as a secondary issue, and environmental concerns way out there in the stratosphere. With America teetering on a recession, irritated about being reliant on foreign oil, paying higher prices for everything, and now populated with environmentalists who wouldn't let a classic conservative retreat to his favorite stance—hands off industry and let market forces rule—it was little wonder that the walk-on president had a tough time in office.

Not that people didn't chuckle at a president who was left-handed sitting down and right-handed standing up. And they laughed nervously when he took a header down the stairs of *Air Force One,* executed a decent roll, and came up standing (later Ford would attribute his agility to his having been an "activist" president, meaning he exercised and stayed in shape). But basically, Ford, once a star center for the University of Michigan Wolverines and now the accidental president, fumbled the ball. Then fumbled it again. He never quite got used to the fact that he was calling the plays, that he was America's designated interim quarterback as it ran up against an underestimated and feisty group of Arabs wearing OPEC jerseys who suddenly had the country not on a gridiron but over a barrel. And then for advice about what to do, he listened to guys instrumental in creating the situation to begin with: Old G. the bookkeeper; Henry Ford III, chairman of the board of Ford Motor Company; and Lee Iacocca, president of Ford.

Michigan-raised (he'd been born in Nebraska), Gerald Ford had breathed and lived the car biz most of his life. He'd grown up in Grand Rapids, working summers while in high school at the Ford Paint and Varnish Company, which was owned by his stepfather, breathing fumes, lead, and toxics, not thinking much about them. He bought his first car, a 1924 Ford coupe with a rumble seat, for seventy-five dollars in 1930, he wrote in his autobiography, telling the kind of anecdote that made him seem both likable and a bit foolish:

> The car ran beautifully during the football season (we were undefeated that year and won the state championship), but then cold weather set in. One December day, the temperature fell below zero and there was snow on the ground. Because I didn't know much about cars, I hadn't bothered to pour antifreeze into the radiator. I parked the car at school, attended varsity basketball practice and drove home for dinner that night. As I pulled into the

driveway, I noticed clouds of steam rising up from the engine. I lifted the hood, saw that the motor was a fiery red and decided—incredibly—that what I needed was something to keep the car warm all night. Some old blankets were lying in the garage. I laid them on top of the engine and went inside to eat. Just as we finished the family meal, we heard fire engine sirens loud and close. We looked out the windows and my poor car was in flames.

By 1975, the same could be said for his hapless presidency. He had granted Nixon a full and complete pardon for his role in Watergate, mistake number one. He had not regained the confidence of the American people that he could rescue the economy. Mistake number two. And he had set the Clean Air Act on a wobbly spin into an uncertain future, having misread its popularity and listened to the advice of the car guys. Mistake number three.

It is beyond the scope of this book to plumb the maze of politics, science, economics, and law that surround the history of the Clean Air Act in the wake of the Ford administration. It took Gibbon six volumes to tell the history of the rise and fall of Rome; I suspect it will take as many volumes for a scholar to put some life into the short rise and prolonged oscillation of America's federal Clean Air Act and its amendments of 1977 and 1990. The political decision of Congress in 1970 to micromanage the rules, with the EPA responsible for the fine-tuning, health studies, and enforcement, and with the president having the power to change the rules at his discretion, helped gradually transform the act into something its originators probably never envisioned. That is, into a huge, unwieldy package that after 1990 was often referred to as the Lawyers Relief Fund because litigation trailed it into the twenty-first century like hydrocarbons from a tailpipe.

What was important about decisions made during Ford's term in office was their retreat from the initial momentum behind cleaner air. The auto industry and economics quickly prevailed over health and environmental concerns. The mid-1970s appeared to codify—or maybe ossify—the regulatory dance around the nation's biggest natural resource, air. Subsequent dances between auto and energy guys and government regulators at the EPA continued in much the same rhythm, with many of the same steps, after Ford's brief stint as America's air-commander in chief, a position he never seemed to have taken very seriously. During his short term in office, Ford did bring a little goofy élan to a time when idealism ran head-on into the economic costs of change. And to a time that shined a bright light on the paradox that has battered regulatory efforts on behalf of clean air ever since: "society desires pristine air, but is not willing to sacrifice much of the standard of living to achieve it."

In the late 1960s, when the Clean Air Act had gathered supporters, the national mood had been ebullient, bold, ambitious. Cleaning the air had seemed feasible: All America had to do was go for it. Technology would comply. But when Jimmy Carter became president in 1977, confidence was in hiding, the public mood sour. Carter emphasized conservation as a way to reduce fossil fuel reliance and to lower emissions from both tailpipes and smokestacks. The CAFE standards established while Ford was in office came into effect under Carter's watch. But the changing times meant that again health concerns were held ransom by economics. When push came to shove, economic health took precedence over public health almost every time. Senator Muskie's arguments for health-based emission standards spoken so earnestly seven years earlier sounded ancient, from a bygone era, when gas was cheap, cars big and brassy, and meeting new emission standards hadn't seemed that big of a deal—except to Detroit.

During the late 1970s *trade-offs* became the mantra of regulatory decision making. Cost-benefit analyses compared the trade-offs. Ideally, the benefits of reductions in a particular air pollutant, say, ought to outweigh the costs to achieve them. One rub was that health and environmental savings were often much harder to quantify than the costs of the controls and technologies designed to theoretically produce them. How much was an inner-city kid's lungs worth, fifty thousand frogs floating belly-up in an acidic lake, a historic statue of a Civil War hero on a village green reduced to acned mush by corrosion from smelly, dank winds? Could a country in love with cars ever seriously deal with an accounting of their damages? Figures put out by the EPA to justify its decisions to ratchet up emission rules were often challenged by industry. In turn the EPA often called into question the cost-benefit figures of industry. For both sides in disputes, "good science" was crucial. But who had the best good science when both sides had experts lined up with impressive credentials, and they disagreed?

"You could get the science to back up your values," Valentinetti told me once, during a series of talks we had about the history of clean air legislation since the 1970s in the United States. "Good science is what's good for *you*." When disputes over clean air rules went to court, which they did more frequently in the 1980s, judges wanted to see all the good science, the cost-benefit analyses, the health implications of doing this versus doing that—then they often scratched their heads. Said Valentinetti, "You get so much data, hear so many interpretations, you get paralyzed. You get analysis paralysis."

All this while, what exactly did Americans think of their Clean Air Act and its amendments? It was hard to tell. The EPA wasn't interviewing the man on the street. Everyone wanted clean air, for sure. But was that going to cost jobs? Going to make it tough to run a second car? A basic disconnect to air, and to air pollution, often seemed hot-wired into the average American's existence. Ilya Prigogine's dictate that individual choices add up to collective action and shape the future was the type of thinking that many Americans did their best to avoid; it was too much like allowing psychic bacteria into your mind that could make you feel guilty behind the wheel or walking into your big house with the climate-controlled interior. Thinking about unbridled individualism was more comforting than thinking about individual responsibility. Few Americans connected their cars, driving habits, and housing preferences with air pollution. Many of them looked at conservation as an inconvenience, at

Jimmy Carter's support for alternative technologies as a weak president's concession to that inconvenience. After all, wasn't the term *alternative technologies* a synonym for battery-powered golf carts, and conservation a retreat from the American dream?

As Terry O'Rourke, an environmental prosecutor in Houston, told the car biologist K. T. Berger, most Americans wanted a car they could drive "right into their living room," even though that car was "literally and metaphorically the vehicle for the destruction of the earth. It is rare that metaphor and object are so much the same." When everything was said and done, O'Rourke added, keep this in mind: The relationship between a man and his automobile "was very intimate. Getting between him and his car and the gas pump was more serious than getting between him and his girlfriend or wife. What the car and gasoline meant for the American was freedom. It was the freedom to go wherever the fuck you wanted, whenever you wanted to. As long as you had your valid driver's license, your safety inspection sticker, and your insurance certificate in the glove box, you can go anywhere in the nation the size of a continent. It's amazing! That's freedom like nobody on earth has ever experienced anywhere."

That freedom got a breath of fresh air, if I may resort to an ironic metaphor, when Ronald Reagan took office in 1981, after defeating Carter in a landslide. Reagan promised to return America's focus to business, profits, and individual rights. It was time to forget the 1970s heavies: Vietnam, Watergate, environmental regulation. Conservation and alternative energy, although never warmly embraced, now went totally out the window, along with the notion of sacrifice for the public good. The more comforting, guilt-free view of the earth and its sky as domains put here for humans to exploit regained its vaulted position after a decade of anxiety and worry about the country's direction and values.

Barely in office a month, Reagan clamped a moratorium on all new regulations. He stalled the implementation of others in the pipeline, insisting his administration had to review them. During his first year in office, Reagan's environmental team tried to repeal, or relax, 115 provisions of the Clean Air Act. James Watt, a conservative Wyoming cowboy and the new secretary of the Department of the Interior, became Reagan's shoot-from-the-hip environmental spokesman. Watt lambasted the "ideological fanaticism" of environmentalists. He said battles over environmental issues were in truth fights between "liberals and Americans" and that America was becoming "a centralized, socialized society" because of excessive regulation.

Reagan himself, using the bully pulpit of the presidency, made some inane proclamations about clean air and the atmosphere. Before a group of steel and coal executives in Steubenville, Ohio, he resorted to his favorite role, the average guy taking a reading on the situation, and referred to Mount Saint Helens, which had erupted in Washington state in May 1980. "I have flown twice over Mount St. Helens out on our West Coast," Reagan said. "I'm not a scientist and I don't know the figures, but I just have a suspicion that that one little mountain out there in these past several months has probably released more sulfur dioxide into the atmosphere of the world than has been released in the last ten years of automobile driving or things of that kind that people are so concerned about." One of his science advisers could have told the president that the small volcano produced about two thousand tons of sulfur dioxide a day compared to the eighty-one thousand sent aloft daily by automobiles, in addition to their other emissions. The day following the speech, Reagan claimed "air pollution has been substantially controlled," and that the Clean Air Act had cost steelworkers jobs. Then he flew home to Southern California, but *Air Force One* couldn't land at the Hollywood-Burbank Airport. It was too obscured by smog-tinted mustard yellow by sulfur dioxide from factories and coal-fired power plants.

Eventually, the Reagan administration retreated from its broad-based assault on the Clean Air Act because neither Congress nor the public supported it. By the end of Reagan's first term, Watt was history. On the defense environmentally, the administration narrowed its air focus and tried to block intervention by the EPA into two areas of growing concern: acid rain and air toxics. Reagan's opposition to an air toxics program was undermined, however, by an atmospheric tragedy in India, one with the scale to grab global attention. In Bhopal, sixty thousand pounds of methyl isocyanate was released when a Union Carbide storage tank ruptured, killing twenty-five hundred people and leaving more than one hundred thousand with permanent disabilities, mostly respiratory problems. The scary story, which the media covered extensively, was reminiscent of a gas attack during World War I but with contemporary implications. Similar tanks, and similar poor safeguards, were commonplace everywhere in the industrialized world.

In response to this catalyst Congress approved America's first Toxics Release Inventory. Conducted in 1987, but only released in 1989, after Reagan left the White House for his ranch in San Clemente, the inventory listed a staggering 2.7 billion pounds of toxics—mercury, benzene, chlorine, manganese, dioxin, ammonia, and on and on—that had been put into the air over the country in 1987. That was roughly 11 pounds for every man, woman, and child. As for acid rain, all attempts to address it during Reagan's tenure came to naught.

In April 1987, showing just how badly things were going for clean air around the globe, *National Geographic* published a provocative piece titled, "Are We Poisoning Our Air?" The answer was an unequivocal yes. In the magazine's classic concerned-travelogue format, science writer Noel Grove went from a smoke-belching copper smelter being phased out in Arizona to the "valley of death," a complex of petrochemical plants and factories in Cubatão, Brazil, where the air sometimes got so bad, said one resident, "If you go outside, you will vomit." Living in Mexico City and breathing its glutinous air was the equivalent of smoking two packs of cigarettes a day, Grove reported. Forests in some countries he visited showed signs of mysterious decay. For instance, spruce and fir trees in the Black Forest of Germany were dying. Suspected causes included ozone from European cities and chemicals from industrial regions. Both sources of pollution weaken needles and harm organisms in the soil that keep roots healthy. In Athens, Grove found the Parthenon being nibbled away by sulfur oxides. Yet in many places that he visited, including Los Angeles and even Cubatão, Grove said that officials insisted that the air was cleaner than it had been, that air pollution was being reduced. But it sure didn't look like it.

"Globally new threats appear as fast as known pollutants are reduced," he wrote. The health problems from breathing many of them were now recognized as cumulative, with high concentrations often accumulating in body fat. More disturbing than what was known about air pollution, Grove added, was what was not known about it. And that was a lot, since "we are in the midst of a chemical revolution in which some 675,000 commercial compounds enter our environment each year." Some of the chemicals were known carcinogens; others suspect. Hundreds were toxic. Yet even in a country such as the United States, which had the most rigorous air controls in the world, only eight hazardous chemicals were routinely monitored by the EPA after seventeen years of responsibility, theoretically, for all of them.

Air Policy As Sound, Fury, and Frustration

We're going backwards!

—DICK VALENTINETTI, ON AMERICA'S CLEAN AIR POLICY, 2003

On my desk I have piles of folders containing information about the drama leading up to the next flurry of air regulation in America, the Clean Air Act amendments of 1990, which were signed by George Bush the Elder that November. There was no lack of infighting, good science, political maneuvering, lobbying, explanations of trade-offs, cost-benefit analyses, and so on. The act's renewal was "one of the longest—and hardest fought—legislative battles in recent congressional history," wrote Henry Waxman, a California Democratic congressman in the thick of it.

"Throughout the 1980s, thousands of hours were spent developing, debating, and blocking legislative proposals; hundreds of witnesses testified at hearings; and millions of dollars were spent on lobbying by interest groups. Eventually, the Speaker of the House and the Senate Majority Leader both had to personally participate in negotiations to resolve specific issues. The result of all this effort is a sweeping collection of programs that dwarfs previous environmental laws."

The act surely dwarfed its predecessors in size. It totaled 800 pages. By comparison, the amendments of 1977 had been a trim 177 pages, and the visionary 1970 act only 50 pages. The revised act further tightened tailpipe emissions. It introduced vapor-recovery nozzles, reformulated gas, and dashboard lights drivers didn't want to see blink on—they warned of emission system malfunctions detected by a modern car's "brain," an onboard computer that flashed a diagnostic code a mechanic could read. Toxics, 189 of them, were identified as hazardous substances that the EPA must monitor and control. Boats, tractors, and construction machinery now all had to have exhaust controls—as did wood-burning stoves. Acid rain finally got its own program, one with a new strategy, *emissions trading,* a market-driven measure that allowed emissions quotas assigned by the EPA to be swapped or sold in a program that the agency assured skeptics would work "as easily as a checking account."

Municipal incinerator emissions, stratospheric ozone loss, the transport of ground-level ozone across the Northeast air shed, new permit procedures—the list of things covered by rules went on and on. The costs, once everything was up and running, were going to be $25 billion annually, said the Bush administration, while critics put the figure closer to $50 billion a year. Alas, pointed out Kenneth Leung, an investment banker at Smith Barney, "Many clean-coal technologies—and other technical wizardry envisioned by the new clean air law, have yet to be developed." Yet technology remained America's clean air angel, at least in the eyes of the lawmakers pleased after the so-called lost decade of environmental progress under Reagan—to be finally wrapping this baby up.

Land use, zoning, and mass transit as additional keys to clean air were not covered in the 1990 act. They remained political taboos, even though sprawl lay at the heart of land-use patterns across America and spawned the commuters who drove those extra miles that added those additional pollutants. Zoning remained anathema in many regions of the country. Mass transit was still as popular with the average politician as having a bee fly into his ear, forcing him to drop his composure and bang his head against the table on a national television show.

The manufacturers of big-ticket items like scrubbers, which could add as much as $75 million to the construction costs of a 500-megawatt power plant, were pleased. As were environmental consultants. They suddenly were flooded with work advising affected parties on how to handle the revised law without breaking it. And the EPA was happy. It went into major hiring mode. The air control program added more than two hundred scientists, engineers, public policy experts, cost-benefit analysts, and writers, whose job it was to make everything comprehensible. The EPA's assistant administrator for air and radiation, William G. Rosenberg, made the overall purpose of the 1990 act sound imminently clear: "The goal of the act, the environmental objective, is to reduce pollution by 56 billion pounds a year. That is 224 pounds for every man, woman, and child."

A chorus of critics familiar with the long history of the act, and its previously inflated promises, were skeptical. Some said the whole package cost too much; others that industry now commanded too much power. Still others maintained that the EPA was turning into an unwieldy bureaucracy and the lawyers were going to get rich . . . or richer. Of those dubious of the act's future success, Craig N. Oren, an associate professor of law at Rutgers, summed up a flurry of criticism when he said, "the detail and complexity of the amendments may undermine their effectiveness. The amendments set forth a number of schedules that extend beyond the authorized appropriations, leaving the statute vulnerable to amendment if economic concerns come to the forefront."

If economic concerns come to the forefront? When had they not?

Oren also said rather prophetically, "the detail and intricacy of the amendments may overburden the regulatory process, create a self-reinforcing cycle of complexity, and create the opportunity for special interest groups to capture both Congress and the U.S. Environmental Protection Agency."

What sustained the act's momentum was a simple truth: It had achieved considerable successes that both Congress and industry could brag about. In 1990, the emissions from new cars were down 75 percent from their 1970 levels. Lead had been eliminated from gasoline. Scrubbers cleaned sulfur dioxide and heavy metals from many power plant smokestacks, although new source review continued to be ignored by many utilities whose upgraded older plants sent voluminous amounts of pollutants into the sky. Factories had baghouses to catch particulates and procedures in place to respond to emergencies. The EPA's "big six" air pollutants—nitrogen dioxide, ozone, sulfur dioxide, particulates, carbon monoxide, and lead—all had been substantially reduced, although ozone and particulates had proven very resistant to dropping to healthy levels. America appeared to be setting clean air standards for the rest of the industrialized, auto-loving, smokestack-impaled world, as it had for the past twenty years. Yet it had not set a pace that would return the atmospheric membrane of the planet to anything even closely resembling its pristine state, a state everywhere present only two hundred years ago. Attesting to that, and to a conclusion it was hard not to draw—that America was winning battles against dirty air but losing the war—was the fact that in 1990 more than three hundred cities and regions remained out of compliance with the ozone standard, that the costs from air pollution were rising, and that the easy fixes had mostly been exhausted—from here on smaller reductions in air pollutants were going to cost more per unit.

America's Pyrrhic victory over dirty air stemmed from well-understood causes. More automobiles were driven more miles each year by more people in a country with more industries and power plants—many as dirty as twenty years earlier—and there was no reversal of these trends in sight. Incremental technological fixes had indeed cut emissions substantially. But the total air pollution load remained intense, unevenly distributed, and unhealthy. The 1990 renewal of the air act tackled the same old problems in the same old ways, but with more tentacles. But no regulatory octopus could keep up with America's economy and its insatiable appetite for fossil fuel.

So, as the proverbial smoke cleared after the renewal of the air act, America had again put its hopes on the tech fix but more rigorously applied. Congress had refrained from meddling in lifestyle changes and had not pursued ways to decouple the nation from its carbon-based energy diet. At the same time, many regulated and unregulated pollutants continued to thrive and prosper regardless of the attempts made to eliminate them. In 1987, Dr. Irving J. Selikoff, a physician at Mount Sinai Hospital and the acknowledged dean of pollution health effects, had told *National Geographic*'s Noel Grove, "Air pollution is modern man's wolf at the door."

Nothing had changed in a few years to defang the wolf.

"If you look at the studies and the experiments with guinea pigs and mice," Selikoff said, speaking about the unknown long-term effects, "it looks like we should all get off the face of the earth. But we don't really know what many of the substances in the air do to people. It may take 50 years to know that. Can we wait for that long?"

It depended on the dose. It depended on the concentrations of the different pollutants and natural gases in the air whether they harmed people or not, whether they trapped sunlight, whether they created smog or acid rain or an asthma epidemic. In the early 1990s, epidemiologists and toxicologists were busy trying to learn which pollutants harmed people and how. In dozens of labs they sought conclusive proof of how specific air pollutants wended their ways from particular sources into a person's lungs and squirreled away molecularly, sometimes in membranous tissue, or went deeper into the body, into organs and fat, where many toxins tended to collect. The researchers studied how the invaders often set off a sequence of responses, some of which eliminated the intruders and others of which were overwhelmed by them. Medical evi-

dence continued to accumulate, for example, that high doses of cadmium weakened bones while too much fluoride thickened them, that
excess mercury could make you crazy, that too much nickel gave you
lung cancer, and that too much manganese screwed up your neurotransmitters and left you waving your arms around with Parkinson's disease.
A known, or an unknown, pollutant could be a nuisance, or it could be a
poison. It depended on the dose.

A healthy person could still hold her own against the daily onslaught
of new and old air pollutants now brewed into an atmospheric soup du
jour over dozens of smogalopolises around America, not to mention the
globe. But that could change, and there was no exact biological line separating the healthy from the vulnerable when multiple pollutants syner-
gized. The synergy of air pollutants complicated health matters
considerably. Acute exposures often had different impacts than chronic
ones. And children, especially in cities with lots of ozone-high days and
unhealthy concentrations of particulates, were emerging as a worrisome
segment of the vulnerable population.

Kids, being active and growing, breathe more air for their size than
adults. The consequences of this simple truth in a city such as Los Angeles, where the air remained the filthiest in America in the late 1980s
despite all the efforts to clean it (in 1989, the city violated one or more
federal clean air standards on 215 days), were cause for reflection. A study
by a pathologist at the University of Southern California, Russell Sherwin, found abnormal damage in the lungs of people between the ages of
fifteen and twenty-five who had lived in Los Angeles and died suddenly.
"These are pretty young people," Sherwin said, "running out of lung."

In 1991, Congress did try to curb America's thirst for gasoline a little.
It tucked money for public transit, bike and pedestrian paths, and other
initiatives to get people out of their cars into the provisions of a huge
program called the Intermodal Surface Transformation Efficiency Act.
ISTEA, pronounced "iced tea" by those in the field, was a mammoth
highway-repair project financed by the federal gasoline tax. But now that
America's interstate highway system was almost finished, with forty-

three thousand miles of multilane blacktop nearing completion, lawmakers felt they could skim a little off the tax loot and try to "balance" the transportation picture, as well as help cities achieve the goals of the 1990 renewal of the air act. ISTEA gave a boost to *new urbanism,* a design vision of cities with fewer cars, denser downtowns, and alternatives to getting around by the internal combustion engine. In other words, new urbanism was old urbanism; it tried to reestablish an older America, a place where some people walked, others took the train, and some folks chose to live downtown rather than in the suburbs.

ISTEA's green projects received almost $10 billion during the 1990s. This was a huge sum of money for clean air and anticongestion efforts, but not enough to weaken the grip of the highway lobby on the country's transportation system. Subway improvements were funded, and light rail systems built or improved in San Diego, Washington, San Francisco, and elsewhere. In 2003, even Houston, the fourth largest U.S. city, one in which the average driver spent fifty-eight hours stuck in traffic in 2000, approved a bond for a seventy-two-mile rail system. That followed a decade of haggling between groups that Rice University sociology professor Stephen L. Klineberg identified as representative of "old think versus new think." Federal largesse under ISTEA pitched in a big chunk of change for Houston's venture into light rail, although not as much as the car-loving city would have gotten from Washington for yet another highway of the same length.

The ISTEA money for anticongestion projects and cleaner air was the exception, however. Most efforts on behalf of clean air got pounded by opponents of regulation throughout the 1990s. In 1995, the Clean Air Time Out bill was introduced by Michigan Republican Fred Upton, but the bill failed to put the bloated act in a corner with a dunce's cap on it. That same year, the Contract with America, a conservative initiative, wanted to hack the Clean Air Act into bits and have it blow away, but also failed.

The act, all eight hundred pages of it, lumbered into the new century. A veritable colossus of interlocking rules, the act often seemed to require a legion of lawyers to prop it up. But the act ruled: It directed America's clean air policy.

In 2001, supporters could rightfully brag that the act had cut the six criteria air pollutants by a total of 48 percent from 1970 levels. That was a commendable feat, given that during the thirty-year period both energy use and miles driven had risen dramatically, and the gross domestic product had climbed 164 percent. Fewer cities had the unhealthy ozone levels of a decade earlier. One city, Denver, had even fully complied with the act, meeting a list of criteria for three years in a row. Yet the nationwide ozone problem continued, remarkably unchanged since 1990. The nefarious gas, which choked cities with smog down low and saved people's DNA up high, had spread to pristine rural places, including sections of sea coast and ranges of mountains, because of wind transport patterns. Unhealthy ozone levels now affected about 110 million Americans a year. One change worth noting, though, was the air quality in Los Angeles. It had finally improved quite a lot. Not enough to remove the city of angels from having the worst air in the country, but an improvement. Nationwide, acid rain had declined because of a drop in the total sulfur dioxide emissions. CFCs, which ate away stratospheric ozone and which will be discussed in the next section, had been successfully banned.

Still, America did not have clean air. It was a little cleaner than a decade ago, when the 1990 renewal had passed. But when everything was added up, progress to rid the air of pollutants had stalemated. It was literally, and figuratively, going nowhere. Most rule changes had become expensive, tedious, and time-consuming legal battles.

*

In late 2003, the last time I talked to air controller Dick Valentinetti, the legal battles were getting to him. So was the change in the political atmosphere created by the George W. Bush administration, a throwback to those of Ronald Reagan and Gerald Ford. "The present administration is trying to turn the momentum of thirty years around in a day," Valentinetti told me in his office in Waterbury, Vermont. "I'm just devastated.

We're going backwards. Just say no to climate change because it may upset the coal industry."

The administration's attack on cleaner air was a multilevel strategy: In the United States, the Bush administration was rolling back lawsuits against dirty coal-fired power plants, encouraging hard-nosed EPA enforcement officials to opt for early retirement, and pushing an energy policy harnessed to fossil fuels and disdainful of conservation and alternative energy; internationally, the administration had withdrawn from the Kyoto Accord on global warming and was backpedaling on the decades-old global agreement to fix the ozone hole. All these positions bothered Valentinetti, but he was most upset by the rule changes that had taken the pressure off fifty large, coal-fired power plants to reduce their emissions, emissions that had been drenching Vermont and the rest of the Northeast with acid rain for decades. The administration defended its move as cost-effective and job creating, as well as regulation reducing. What irked Valentinetti the most about the initiative was that after years of litigation and pressure by the Justice Department and the EPA, some of the utilities that owned the old plants had recently brokered deals to finally clean up their stacks, to finally comply with new source review. Now this. Basically, the rule changes threw years of work and money out the window. It was discouraging, he said, shaking his head, and the whole thing was headed to court. Legal battles were now the heart of clean air policy making. Lawsuits had become a popular tactic used by both sides in air battles during the Reagan years, he recalled, but now they were the modus operandi of change. No new regulations were really valid until tested in court, Valentinetti said, and everything seemed to end up there.

"These latest rule changes," he said, "I think they are clearly in violation of the air act, but that doesn't mean the administration can't pass them."

In truth, the Bush offensive was merely a resumption of the probusiness, antiregulatory stand taken by the inept auto guy, Gerald Ford, carried on zealously by Ronald Reagan, and sustained halfheartedly by George Bush the Elder, who oversaw the 1990 amendments to the Clean

Air Act. The antiregulatory fervor emanating from the new administration was just more intense and ideological. And more slick. "These guys are smart," Valentinetti admitted begrudgingly. "They're stealth ideologues. They give you half truths. Or aren't dealing with the issues at all. They deal in symbols. And now the energy industry is protected by an umbrella of security."

Of course, the president had the power to implement the rule changes that he wanted, added Harold Garabedian, Valentinetti's assistant director of the Air Quality Program in Vermont. The changes might be political payoffs to utilities that had made political contributions to the Bush campaign; they might be setbacks for clean air policy; they might echo the ecofriendly jingoism trumpeted by the president's new EPA administrator, Mike Leavitt, formerly the governor of Utah and a leader with a record of siding with industry in most environmental fights. But all these considerations were outside the sphere of regulation. They were in the political realm. "All government does regulations," Garabedian explained. "They're authorized to. The regulations provide the details. The administration can amend the rules, it has the authority. The only way to stop them is litigation. I expect that the challenge to the rule changes will prevail."

Lawsuits were already flying. A lot of money would be spent because the Midwest utilities the administration had let off the hook (the EPA had announced that lawsuits in court, and those pending, would be dropped) stood to save between $10 billion and $20 billion by avoiding the technology upgrades they should have done years ago to comply with new source review. Fourteen states, including New York, California, and Illinois, had filed a suit to block the Bush administration rule change. The states claimed that it was illegal and violated the intent and language of the Clean Air Act. That a lawsuit to get the federal government to uphold the clean air statute came from concerned states marked a sharp reversal of the philosophy behind the statute.

Displeased with the present state of affairs, Valentinetti grimaced and repeated, "There are just so god-damned many lawsuits out there now."

Turning clean air law into dust might not work, said Dick Baldwin, the former director of air quality for Ventura County in California. Baldwin said he thought the Bush agenda might flounder. "Reagan's anti-environmental tactics blew up in his face," recalled Baldwin. "The public hated them. Before George W. Bush, Reagan was the worst thing that's happened to air and the environment."

And the public in 2003—what did they think?

"The public's interest in air has tubed," said Barbara Page, information officer for the South Coast Air Quality District, which included the L.A. basin. "Environmental education used to be big," said Page, who had been in the air control business since the late 1980s. But not now.

"What about all the health problems associated with air pollution?" I asked her. "Don't people get the health message?"

"Health message?" she said, voice rising. "Nobody takes it seriously any longer."

What Americans took seriously in 2003 were unlikely terrorist attacks with biological weapons such as anthrax, ricin, or smallpox. Since September 11, 2001, and the attack on the World Trade Center, the country had been in a constant state of war against terrorism. Media attention and public concern about smog, acid rain, and the ozone hole had withered. And the public's perception of what constituted an air danger had changed, and not for the better.

On 9/11, when lower Manhattan groped through asbestos- and toxics-loaded air, the heat at ground zero had been so intense that one compound never seen before, 1,3-diphenyl propane, appeared as a unique trace of the tragedy. As did a respiratory stew that made hundreds of firefighters, city workers, and volunteers, especially those who did not wear respirators or wore faulty ones, ill from breathing it during the cleanup. The EPA's shoddy handling of the emergency (seven days after the towers went down, EPA administrator Christie Whitman declared the air in lower Manhattan "safe to breathe," possibly under pressure from the White House, despite little proof to substantiate the declaration), and subsequent deadly incidents involving anthrax spores got the public all wound up. Smog, acid rain, the ozone hole—all went into hiding as concerns of the public imagination. But since 9/11, find some white powder in an envelope in a center of power, such as the offices of the U.S. Congress, and the media and the public go ballistic. Yet the risk of the average person's breathing in anthrax spores, smallpox viruses, rickettsias, or ricin is virtually nil. The highly publicized episodes of air terror played with a truth already discussed but worth repeating: *All that fascinates us now seems to be what we can't take for granted*. We certainly can't take for granted that a terrorist is going to send us a great big dose of anthrax spores in a Valentine's Day card. But the wilder—the more out there—the danger, the greater attention it seemed to command in post-9/11 America. There was an inverse relationship between the likelihood of a danger and its fascination quotient. If the danger involved toxins that people knew little of, this stoked their fear quotient just as the vibrations of dinosaurs approaching must have stoked the fear quotient of mammals a hundred million years ago. If the air danger was new and silent and easily spread, it could capture attention. It was a catalyst. Not of change, but of fear.

Meanwhile, over U.S. cities the air was still making thousands of people sick, killing a good number, and eating away the masonry. And this air was anthrax free. The new era of terrorism, with its heightened focus on the particular event, on a reductionist take on specific incidents, could be put into sound bites, easily grasped, and disseminated as news while the general and pervasive air pollutants got pushed aside. Relatively, they seemed almost benign, part of the atmospheric status quo, part of the fabric of human existence. We inhaled smog, for instance, with almost as much indifference as we inhaled oxygen. For many Americans, something like smog had just become part of their daily air mix. They took it for granted.

Amid the new era of heightened fear of the not-likely-to-happen, rule changes affecting something like the smokestacks of power plants in the Midwest were not high on the worry list of most Americans. But they were all wound up over relatively inconsequential air dangers that few, if any of them, would ever have to face. But could think about a lot.

Anthrax, in particular, had captured public attention since its appearance in the offices of a supermarket tabloid in Florida in October 2001, where it killed a photo editor and brought in a team of fifty investigators in the wake of what appeared to be a bioterrorism incident. Fear of what might be in the wind displaced fear of what was in the wind. The last known case of anthrax poisoning in Florida had been twenty-eight years earlier, and throughout the United States only eighteen cases of pulmonary anthrax had been reported over the last hundred years. Anthrax bacillus has to be finely chopped into a powder if someone is to inhale the eight thousand to ten thousand spores necessary to cause the disease. But once in the bronchial tubes, the spores are hard to dislodge and lethal; the damp membrane makes a good petri dish. According to stories about the photo editor, he had gone to an emergency room feeling feverish and disoriented. He was vomiting and couldn't talk. In the aftermath of his death and the discovery of other anthrax-bearing envelopes in Washington, the public became very alarmed about anthrax. The bacterial spores were relatively easy to acquire, although tough to disperse. An anthrax expert said the bacterium was "a professional weapon." You didn't just open up a jar and sprinkle anthrax spores around like talcum powder—it was "not for the amateur." The germs were said to possess a high "degree of lethality," an attribute the U.S. Army liked to ascribe to its better-killing machinery; the spores could take out 80 percent of those infected if the victims had not been vaccinated or couldn't get antibiotics into their bloodstreams before the first symptoms appeared.

Persistent declarations by health officials that anthrax was not contagious did not quell the sense of fear that pervaded the air in Florida the first two weeks of October 2001. As American planes bombed the Taliban and missiles rained down in Afghanistan, frightened Floridians lined up to buy ciprofloxacin, an antibiotic for anthrax. People wore face masks, talked about spores flying through the air—invisible particles that could enter their lungs and set up shop and kill them. Cooler heads advised the anxious that if spreading anthrax spores was as easy as many people said, then the kind of people who hated the United States would have done it years ago. Still, schools were closed when students came into classrooms with baby powder on their sneakers, and wild rumors could send people flying out of a building faster than someone shouting the word *hurricane*. It was the scare factor of germ warfare, the terror of the invisible finding someone quickly and silently and anchoring inside, deep in a person's biological vulnerability. All that the powder had to do, people learned by reading the scientific descriptions of its pathology, was skirt the lung's defenses, flow down the bronchial tubes, navigate the ever smaller airways, and insert itself amid the alveoli. There, scavenger cells, the macrophages, had the last chance to haul the intruder germs away.

In the wake of 9/11 and the germ-warfare scares immediately following, there also came detailed reports of Saddam Hussein's use of poison gas to kill Kurds in northern Iraq in the late 1990s. Helicopters had dropped a supposed mix of mustard gas and two nerve agents, sarin and VX. Because the gases were heavy, they slithered into cellars, where Kurds were hiding, thinking the choppers were going to drop bombs or fire on them.

The power of such air terrors to relegate conventional air pollutants to a kind of inferior status as dangers bothered me. The germs and gas were headline grabbers. They riveted people's attention onto airborne stuff that could harm or kill them, and do it quickly and silently. But the attention also obscured the real atmospheric problems. This paradox—attention on harmful air substances but on the wrong ones for the wrong reasons—might be addressed with a necrology of air, I thought, a kind of death-by-dirty-air display. It would keep the reality of what was really bad in the air right in front of people's faces somehow. In this way, the public would not be blinded by the highly improbable, though also highly entertaining, air villains that they read about in papers, heard about on television, and now powerfully associated with the war on terrorism. If we devised a powerful necrology of air in the United States, maybe other nations would follow our lead. We could even get high-tech about it, although a necrology of air, no matter how you presented it, would be awfully hard to make very entertaining.

What the necrology would make painfully clear was that the shadowy figure of air pollution—not germ warfare or poison gas—killed people by the hundreds of thousands every year and had been doing so for
well over a century. The most reliable numbers I found were mostly from
the 1990s, probably because it takes a long time to count so many dead
people, but these numbers would have to do.

A report from the World Bank in 1992 estimated that the deaths from
air pollution ranged between 300,000 and 700,000 annually. In 1996, the
prestigious Harvard School of Public Health put the yearly total at a
more precise 568,000. In 1997, the World Health Organization lowered
the annual figure, claiming it was closer to 400,000.

The necrology might also enlist the more expansive statistics com-
piled by the environmental historian J. R. McNeill. He wrote that "the
health consequences of air pollution in the twentieth century were gar-
gantuan, although hard to measure precisely. I reckon that air pollution
killed about 20 million to 30 million people from 1950 to 1997.... All told,
a 'guesstimate' for air pollution's twentieth century toll would be 25 million to 40 million, roughly the same as the combined casualties of World
Wars I and II, and similar to the global death toll from the 1918–1919
influenza pandemic, the twentieth century's worse encounter with infectious disease." In the United States alone, the necrology would make
abundantly clear, probably 2 to 3 percent of all deaths since the late 1970s
could be traced to polluted air. That meant between 30,000 and 50,000
deaths each year, which was more than deaths from gun wounds, but
fewer than from tobacco products. These last figures from a country that
prided itself in the success of its clean air policies.

The necrology would leave few doubts about what substances in the
air killed all these people. Not anthrax or ricin or sarin or VX, though all
were certifiable killers and not to be ignored. But they weren't the real
culprits. Not in the United States, and not around the rest of the world.

In that world in 2003, the United States might have been leading the
technological pursuit of sophisticated monitoring systems, might have
been spending more money on cleaning its air, and undoubtedly had the

biggest air control law. But it had retreated from the air issues of international scope, while at home it remained guided by a thirty-year-old law
that had evolved into a monstrously swollen package, a broken dream
seeded in now distant and idealistic times, around which lawyers circled
with glee while regulators like Valentinetti wrung their hands in mounting frustration.

In the final analysis, the sound and fury of U.S. air policy signified
something, but not success. In the early twenty-first century, industry still
ruled, probably more than it had in the previous fifty years. Symbolic
political action remained popular, especially gowned in cute phrases,
such as the Clear Skies Initiative, another Bush administration euphe-
mism for deregulation. Good science could be manipulated to spur
action or to grind it to a halt. And air terrors now threatened to obscure
the real health culprits in the sky.

Ultimately, the approach the United States had created, and spent
hundreds of billions of dollars on, to clean air was too simple, too influenced by industry and fear of the wrong things, and in denial of the big
picture because of its reductionist bias.

Air needed a fix. And American air policy needed a fix. Not just
because the world's biggest air polluter wanted healthier cities and air
sheds. But because the world looked to America for leadership on ways
to deal with the thickening haze and the global threats. Back in the 1980s,
during the Reagan administration, ironically, the country had lived up to
high expectations by responding to the biggest atmospheric threat yet:
upper-level ozone loss. America demonstrated that when kicked in the
butt hard enough, and alarmed by a potential catastrophe even a school-
child could understand, the nation could react with urgency, intelligence,
and muscle. All it took was a hole in the sky the size of Mars through
which came radiation that could fry DNA. The international cooperation
that galvanized to close the hole suggested that there might even be
some hope, eventually, to do something about the Big Worry, global climate change.

Ozone: The Atmosphere's
Dr. Jekyll and Mr. Hyde

That the greatest effects come from the smallest causes has
become patently clear.

—C. G. JUNG

Ozone is duplicitous. It's "the central species in the stratosphere's chemistry," Paul Crutzen and Thomas Graedel tell us. "In the bloodstream it
can act like a sledgehammer in the brain," says Fairfield University professor and toxicologist David Brown. Ozone high overhead protects human
life; ozone near the ground makes smog. And though we're irritated by
the concentrations of ozone in our cities and along our beaches in the
summer, we're pleased with high concentrations in the stratosphere. Up
in the stratosphere, around fifty thousand feet, a layer of ozone acts like a
colander, like Gore-Tex. That is, it lets in what we desire and keeps out
what we do not want. The difference between Gore-Tex, what we call a
miracle fiber, and stratospheric ozone, what we call a trace gas, is that the
gas really is a miracle. We could get along fine without the fiber.

Over a century ago, the fact that ozone is a quintessential paradox,
life-giving and life-damaging, a colander and a shield, a miracle and a
health threat, was not at all understood. Ozone was only identified in
1845. A German scientist, Christian Schoenbein, detected trace amounts
with starched paper treated with iodine. Other scientists verified the presence of the gas around the world.

Health-conscious Victorians decided ozone was good for them and
sought it out on mountaintops and by the seashore, far from their smoky
cities. Pale blue in color, distinctly metallic in smell (the word *ozone* comes
from the Greek *ozein,* which means "smell"), the gas could be detected
during thundershowers because lightning produced it. Ozone created by
crackling veins of light excited normally reserved Victorians to no end.

The English lyricist William S. Gilbert, who teamed up with the composer Arthur Sullivan to produce such popular musical comedies as *H.M.S. Pinafore*, *The Pirates of Penzance*, and *The Mikado*, even wrote an odd little poem about ozone and its peculiarities in 1865. The last stanzas reads

> But if on Ben Nevis's top you stop
> You will find of this gas there's a crop—but drop
> To the regions below,
> And experiments show
> Not a trace of this useful ozone is known,
> Not a trace of this useful ozone!
> It's because I'm an ignorant chap, mayhap,
> And I dare say I merit a slap or a rap,
> But it's never, you see,
> Where it's wanted to be,
> So I call it Policeman Ozone—it's known
> By my friends as Policeman Ozone!

Today we're not quite as eager to breathe a lot of ozone. Victorians might have thought it harmless (they even pumped it into churches, hospitals, and their early subways), but we know that too much of the gas irritates the deep throat and can permanently scar children's lungs. We know that ozone has killed thousands of ponderosa pines in California and black cherry trees in the Carolinas. Of course, high overhead, the gas protects everything on the earth, from green plants to plankton to reptile eggs, from getting fried by shortwave ultraviolet radiation. Stratospheric ozone lowers the odds of melanomas cropping up on the skin of beach-loving photophiles who neglect to smear on enough sunblock.

Ultimately, despite the fact that the automotive lifestyle creates too much ozone every day in dozens of megalopolises around the globe, ozone is humankind's ally. Natural processes don't dump it in our cities. Cars and combustion do.

Because of its paradoxical nature, ozone fascinates us more than most atmospheric components. We can't take it for granted. In a way, ozone is sort of a spokesperson (or should it be spokesgas?) for air. It demands our attention.

Think about it. When a breeze blows, or you take another deep breath, or your alveoli are enjoying another blood-rich afternoon of keeping you alive—you find nothing fascinating there. But have a hurricane wind tear the roof off your house, or a chicken bone snag in your trachea, or the tar from too many years of smoking give you emphysema, and you're paying attention. Air, like everything else in a noisy world, needs its spokespeople, its spokesgases. And ozone is a good one because of its complexity, its devious Jekyll and Hyde personality. Not to mention that it's been here for almost three billion years, constantly involved in a balancing act of creation and destruction that has ushered life into existence, has given it shelter, and can also let life go if whacked too far out of balance.

By the mid-1980s, that happened: Ozone was out of balance, both unnaturally high on the ground and unnaturally low in the stratosphere. Sunlight and emissions were creating too much of it in cities while human-made chemicals, mostly CFCs, threatened to destroy most of it overhead. We could live with ozone in our cities, but not without it to protect our DNA from the sun.

A theory that CFCs threatened upper-level ozone had appeared in the early 1970s, but was not proven for another decade. The realization that CFCs threatened to eliminate the ozone shield, which not only allowed people to get suntans but also helped warm and stabilize the atmosphere, resulted in something almost as rare and as breathtaking as ozone itself—international cooperation to halt, and then reverse, the damage.

The story of ozone between 1845, when it was discovered and soon embraced as a health elixir, and 1985, when the first international meeting convened to stop its loss in the stratosphere, is both sobering and encouraging. Sobering because of the scale of the threat uncovered and encouraging because of the demonstration of the political will to do something about the global threat.

In 1876, during the height of the ozone health craze, Albert Levy, a meteorologist at the Paris Observatory, began measuring ground-level ozone. Then, of course, that was the only ozone scientists were aware of. Levy religiously monitored ozone levels in suburban Paris for thirty-four years, establishing an important baseline for the gas and one helpful to scientists today for the purposes of comparisons. During Levy's lifetime, the ozone concentrations he measured were nothing to worry about; they were generally low, about 10 parts per billion of air. He detected them with an advanced version of Schoenbein's ozonometer. It was a flask through which ambient air entered via a tube, then was filtered through an arsenate-iodide solution, and was passed through a gas meter, the movement assisted by a pump. Today, by comparison, ozone levels in most major cities average two to five times the concentrations Levy detected over a century ago in Paris. Levels on really smoggy days can be ten to fifteen times as high, or between 100 and 150 parts per billion (the highest level ever recorded was 680 parts per billion in Los Angeles one day in July 1955).

Toward the end of the nineteenth century, ozone was found to absorb short ultraviolet rays from the sun, rays of less than 290 nanometers, and to exist in greater concentrations at higher altitudes than close to the ground. In the 1920s, after the stratosphere was discovered, Oxford scientist George Dobson devised an elegantly simple device, which relied on a prism to scan the light spectrum at different times each day and at different angles. The instrument allowed Dobson to calculate the total ozone in a column of air from the ground into space. In 1930, Sydney Chapman advanced a theory to explain how sunlight striking molecular oxygen in the stratosphere created ozone. By midcentury, ozone monitoring of the stratosphere had become a global enterprise, with monitoring stations on every continent. The South Pole stations found that ozone levels there oscillated seasonally, increasing by 35 percent each spring and then returning to normal, or to what was thought to be normal. But in the late 1970s, the ozone levels over the pole mysteriously dropped. By the early 1980s, they were about 35 percent lower than what they had been in the 1960s. When Joseph Farman, head of the British Antarctic Survey Team, first took the low readings at the Halley Bay research station, he had such a hard time accepting them that he sent off to England for new dobsons, as the ozone-measuring instruments were called. Once they arrived, freshly calibrated, Farman took ozone readings over the next twelve months at widely dispersed locations in Antarctica to double-check those from Halley Bay. The new readings corroborated the originals and lent credence to the decade-old theory of ozone depletion. The findings bothered Farman, a not-easily-alarmed kind of guy who had been collecting data at the South Pole since 1957.

The first hint of an atmospheric emergency in progress had come about indirectly. Back in 1971 James Lovelock, the atmospheric chemist and creator of the Gaia hypothesis, had used an instrument he built, an electron capture detector, to show that CFCs were virtually everywhere in the air around England. CFCs, the generic name for a wide spectrum of compounds manufactured since World War II, had become standard ingredients in all sorts of coolants and sprays. But Lovelock found the compounds ubiquitous in air, from the ground level up to where passenger jets flew daily by the hundreds.

An independent scientist of the old school, without an institution or an industrial sponsor, Lovelock applied for funding to put an electron capture detector aboard the *Shackleton,* a research vessel traveling from England to Antarctica. He wanted to see if CFCs had migrated to the middle of the ocean and to points further south. But he was turned down. One of the scientific referees said he could think of little less worth doing than what Lovelock proposed. Undeterred, Lovelock paid to have an electron capture detector onboard. Fifty air samples taken en route to the South Pole detected CFCs. Lovelock concluded that wind patterns carried the chemicals all the way to Antarctica, but he also decided that CFCs did little harm and were probably not hazardous to the environment.

Thomas Midgley, the inventor of CFCs, certainly would have agreed. A sort of roving chemical genius at General Motors, Midgley had discovered that tetraethyl lead reduced the knock in engines. His discovery resulted in the development of ethyl leaded gasoline, and the more powerful, high-compression, high-octane engines it made feasible, in the 1920s. In 1930, Midgley unveiled the first CFC, Freon, a refrigerant that wouldn't catch fire or hurt people if they breathed it. To demonstrate Freon's safety to hundreds of his fellow chemists attending the annual meeting of the American Chemical Society, Midgley inhaled a lungful and blew out a candle flame. Given the subsequent records of leaded gasoline and CFCs to do atmospheric and bodily harm, with this display, reminiscent of Priestley's inhaling an air five or six times more pure than common air, Thomas Midgley set in motion a rather dark legacy. As the environmental historian J. R. McNeill put it, Midgley "had more impact on the atmosphere than any other single organism in earth history." Given the bubblers' record for putting oxygen into the atmosphere at the beginning of life on earth, I'm not sure if that is correct. But Midgley's impact was way up there. He became an air villain long after his untimely, accidental death (a polio victim, Midgley designed a special contraption to help him get out of bed, only to become tangled up the ropes and cables and to hang himself accidentally above his own mattress in 1944), and only after CFCs became a huge commercial success.

CFCs gained their popularity after World War II. Used in refrigerators, air conditioners, aerosol cans, and elsewhere, not only were CFCs and related gases nontoxic and nonflammable, but they did not break down in sunlight (ironically, because of the protection of the upper-level ozone layer they began to destroy). Nor were they soluble, except in rare conditions. This meant that CFCs could not be stripped from air by the hydroxyl radical, a major atmospheric scavenger. A fragment of water, the hydroxyl radical (chemical symbol, OH^-) "wants" to be water. So, in air, hydroxyl bonds with hydrogen atoms, which happen to be what most pollutants ride around on. By incorporating things such as acids and particulates into water, hydroxyl radicals clean the air, dropping the pollutants to the earth as rain. CFCs, though, being insoluble, did not take part in this basic rinse cycle of air. Instead, presumably benign CFCs just kept moving around in the winds. And they kept rising. They kept going up toward the clear, dry, and cloudless stratosphere. But once there, they were not benign. In the stratosphere, with about 99 percent of the atmosphere's molecules below, shortwave radiation got a stronger bead on CFCs, which began to break down. They formed chlorine atoms and residual fragments. It was the chlorine that wreaked havoc. The chlorine atoms and ozone reacted together; they formed a free radical, chlorine oxide, and it set off a chain reaction. A single chlorine atom from a compound that was benign below could destroy as many as one hundred thousand ozone molecules from the stratosphere.

The two scientists who pieced together this theory of ozone depletion in the early 1970s were F. Sherwood Rowland, a chemistry professor at the University of California in Irvine, and one of his graduate students, Mario Molina. "I was just looking for interesting things to understand," Rowland said later, explaining how Lovelock's work, which Rowland first heard about at a lecture in 1972, tweaked his scientific curiosity. Rowland found himself wondering about the fate of compounds like CFCs sent aloft. Some, of course, caused smog. Others, captured by the hydroxyl radical, came down as rain. But what about those that stayed in the air, like CFCs, as Lovelock had validated, yet spread around the entire planet in a few decades? What became of them? Intrigued by the question, Rowland put Molina to work on figuring out what ultimately did become of CFCs. And on a corollary: Should people worry about what became of CFCs?

After three months of intense research, "not to worry" was not the answer Molina gave Rowland. A practical idealist, Molina had hoped to use chemistry to help society rather than to harm it. But the student had not thought that an opportunity to do this would present itself so quickly. Molina concluded that CFCs were slowly rising to the stratosphere, being broken down by sunlight, and then setting off an ozone-destroying chain reaction. Work done previously by Paul Crutzen, which linked ozone destruction to concentrations of nitrogen oxides, helped buttress the theory. In 1974, Molina and Rowland published their findings in the prominent British scientific magazine *Nature*. Instead of being hailed as saviors of the planet, however, the two chemists found their well-constructed theory greeted by resounding silence. Maybe it was shell shock from Vietnam, withdrawal after Watergate, or myopia caused by the oil crisis, but politicians in particular seemed unconcerned with the article's implications. As for their fellow scientists, well, they didn't exactly applaud their colleagues, either. But that was the nature of science, especially science funded largely by industry hostile to regulation. And if there was ever a theory that carried regulation in its wake, it was this one. For if the ozone was disappearing, then without action in the not-too-distant future, so would human beings.

Rowland and Molina made a personal crusade out of publicizing their theory. And their efforts worked. Soon politicians were talking about ozone depletion, as was the media. A sure sign of their success were the rebuttals from CFC manufacturers, attacks from groups opposed to regulation, and the trumpeting of alternative theories devised by other scientists who put the blame of ozone loss on natural phenomena rather than synthesized chemicals.

In 1976, though, the National Academy of Sciences got behind the theory. Then, in 1977, the amendments to the U.S. Clean Air Act included a provision to regulate any substance "reasonably anticipated to affect the stratosphere." The following year, the United States banned CFCs in aerosols. But it wasn't until Joseph Farman's ozone data from Antarctica, data collected not by costly high-tech satellites overhead but by a dobson, a sixty-year-old ground-based device that cost only a few thousand dollars, was subsequently validated by satellite images that an emergency global conference was called by the United Nations.

Held in 1985, the Vienna Convention on the Protection of the Ozone Layer attracted twenty nations, most of which agreed that the ozone layer needed international protection. But how? And when? The language that emerged from the convention brought to mind that found in the weak smoke ordinances written in the 1800s: "appropriate measures . . . protect human health and the environment . . . human activities which modify or are likely to modify." The agreement didn't ban CFCs or, for that matter, even mention them except in the fine print. It did support additional research and more cross-border cooperation among nations to resolve the ozone problem. But it was far from an aggressive response. Still, it was a first. Nations had agreed in principle to do something about an atmospheric problem probably threatening all of them.

Then damning evidence continued to pour in. In August 1986, the National Aeronautics and Space Administration (NASA) announced that its satellite data since 1978 had identified an ozone hole; the reason the data had been discounted was that it had been so far out of the anticipated range that it was rejected as erroneous. In the autumn of 1986, an ozone expedition went to Antarctica. Its members used sounding balloons and ground observations to validate Farman's measurements, confirming the existence of "a hole in our atmosphere that you could see from Mars," as Rowland put it. In short order, the competing theories explaining ozone loss were laid to rest, and even more importantly, DuPont, the world's largest manufacturer of CFCs, announced that it would find alternatives to use in aerosols, refrigerants, and wherever else the ozone-depleting compounds were used, and do it within five years.

In 1987, twenty-four nations ratified the Montreal Protocol. They included the United States and Japan, the world's biggest producers of CFCs. The protocol set rules for gradually eliminating the production of ozone-depleting substances worldwide. It granted developing countries more time to comply, and contained provisions to tighten, or weaken, the rules if future information warranted. Subsequently, first in 1990, and again in 1992, the rules were tightened.

Not that everyone was completely in accord with the ozone-loss theory, or the Montreal Protocol and the later tougher amendments. When Sherwood Rowland, Mario Molina, and Paul Crutzen shared a Nobel prize for chemistry in Stockholm in 1995 for their work on the theory, Tom DeLay, minority whip in the U.S. Congress, complained that the award came from "an extremist environmental country." An outspoken opponent of environmental regulation, and the former owner of a pesticide company in Texas, DeLay, with other congressional representatives, called for hearings on the ozone hole scare and its dubious science. DeLay said he wanted to see a direct link between the chemicals and ozone loss. He wanted to "make sure the so-called UV radiation that's supposed to make people drop like flies is actually making people drop like flies."

"It's ignorance," Molina said of DeLay's remarks, "real ignorance."

DeLay's side lost this environmental battle, although Arizona sided with him. Its legislature declared the Montreal Protocol invalid inside state lines—a curious position to take in a sunny state, given that stratospheric ozone losses of only 1 percent were estimated to increase basal cell carcinomas by 4 percent and squamous cell carcinomas by 6 percent. An antiregulation zealot, DeLay ignored the health predictions from ozone loss, either for ideological reasons or because he had inhaled too many of his own pesticides and his judgment had been impaired. The link between many other air pollutants and cancer remained only tenuous, but not between upper-level ozone loss and cancer. The connection was well documented, practically irrefutable. After all, it was ozone that had allowed life to crawl out of the oceans and evolve into animals to begin with. Evolution put darker-skinned people near the equator, and their lighter-skinned cousins in the northern latitudes, because ozone concentrations overhead varied. Above the equator, the concentrations were thinner, a result of synergistic weather and natural systems, and in the temperate latitudes, they were thicker. Because the human body could be harmed by high exposures to solar radiation, additional melanin, a skin pigment, became the biological protection from the greater bombardment of shortwave ultraviolet radiation in the earth's tropical regions. During Reagan's presidency, there had been a brief and absurd stand taken by Donald Hodel, Watt's successor as secretary of the interior, over ozone loss. Rather than support regulation of the CFC industry, Hodel suggested that Americans buy sunglasses and wide-brimmed hats and stay in the shade. A decade later, Tom DeLay, the former bug man from Texas, wanted to see people laid out like flies before he'd let go of his skepticism that the whole ozone scare was anything but an environmental hoax.

In 1996, despite pockets of resistance, a worldwide ban on ozone-destroying chemicals went into effect. Substitutes had been developed to replace Freon, solvents, aerosol sprays, and related gases grouped as halocarbons, including some pesticides probably close to DeLay's heart, such as methyl bromide. Still, in 2003, outlaw CFC manufacturers continued to supply a black market with banned compounds. Global enforcement of the ban was a challenge. As were parties that had signed the Montreal Protocol but that were now backsliding. Foremost among them was the United States, which, in 2003, waffled about its commitment to keep ozone-depleting substances out of production. The Bush administration wanted to relax the rules and pushed for an exemption for methyl bromide, the pesticide that ate ozone with bromine atoms rather than chlorine. U.S. officials attending a regularly scheduled forum to discuss the Montreal Protocol argued that the ozone layer was healing, a contention few atmospheric scientists agreed with, and that the pesticide was vital for the economic health of American farmers growing strawberries, tomatoes, and other crops, primarily in Florida and California. The American request was denied, though mutterings were then heard from Congress about approving methyl bromide production increases, anyhow. America's shift from a leader fighting ozone loss to a nation anxious to break the rules gave support to antiregulation groups. These groups used U.S. waffling to continue claiming that the science justifying the ban was bogus, that ozone loss was just hype, and that doomsayers had scared politicians into making bad decisions on bad science, thereby rationalizing government intervention in the first place.

With the ban in place, ozone loss had slowed considerably, although it had not completely stopped. In 2003, the latest projections showed that ozone would begin to accumulate before 2010, then gradually return to a balanced state around 2050, if the ban remained in place.

The ongoing story remained a positive chapter in recent atmospheric history. It also galvanized attention on air pollution more than smog or acid rain had because ozone loss in the stratosphere impacted the world, not just urban centers and regions downwind of coal-fired power plants. Ozone loss demonstrated the scale of the response needed to deal with a global air alarm, along with the time frame it could take to get the earth's atmospheric house back in order—if the tenants weren't too busy bickering over who owned which room and where the furniture should go. It also suggested that eventually the international community might find a way to face the Big Worry, global climate change.

Such a Big Worry

About climate change, George says: "You still get these people who say, 'Do you really think it's happening?' and I'm like, 'What is it you don't understand?'"

—DARCY FREY, "GEORGE DIVOKY'S PLANET"

Ornithologist George Divoky spends three months a year on a barrier island off the northern coast of Alaska and is one of a growing legion trying to warn people about what has become "an inescapable truth: our world is melting away."

Divoky first went to Cooper Island, three hundred miles north of the Arctic Circle, in the mid-1970s. He wanted to study a rare colony of black guillemots. Pigeon-sized, ducklike birds with white shoulder patches and red feet, the black guillemots were nesting in fifty-five-gallon drums and old ammunition boxes abandoned on the uninhabited slab of sand by the U.S. Navy in the 1950s. Returning year after year to his remote outpost buffeted by ice and sea, Divoky plotted an increasing number of the birds there each June: 70 in 1978, 220 in 1981, 600 in 1990. But he also noted that the birds returned earlier each year because the snow was melting quicker and was staying melted longer.

In the 1990s, Divoky's attention shifted to the changing weather. Global warming was becoming the next *big* problem, and scientists like

Divoky are drawn to big problems. By then he had put in fifteen years of tough, isolated field work, and he had a lot of unique temperature and bird behavior baselines from which to hypothesize. Besides, he liked coming to Cooper Island. He camped out and lived rough, almost like his subjects. With his tent, binoculars, cached food, and notebooks, Divoky had built his own little research microcosm on a little island that was an outpost for global climate change. That was because temperature shifts were playing out much faster at the earth's physical extremes, in places such as northern Alaska and the coasts of Antarctica, than anywhere else. On Cooper Island Divoky's black guillemots were analogues to the yellow canaries used in coal mines during the nineteenth century to detect dangerous levels of explosive gases that the miners were oblivious to. But with one big difference: The guillemots found the changes in the Arctic to their liking; the canaries had died.

Much of the world may not share the guillemots' appreciation for the warming trend that made Cooper Island more hospitable to mating earlier each year. But the outpost and the birds got a lot of attention. Darcy Frey wrote about them in "George Divoky's World," in the *New York Times Sunday Magazine* in January 2002: "Whether the glaciers of Greenland will continue to melt and the southern oceans rise up to flood Bangladesh, whether Cape Cod will erode to a sand spit and the American prairie dry out like the Mojave, whether thunderstorms will one day reach Antarctica and sparrows the North Pole—whether all the disasters predicted by climatologists and their computer models eventually come to pass, one piece of the puzzle is already in place: the earth's climate will change first—and change most substantially—in the Arctic, that enigmatic expanse of snow and ice, of ancient peoples and unspeakably hostile temperatures that spans the top of the world."

Top-of-the-world changes were in the news a lot. For instance, the north coast of Ellesmere Island in Nunavut, a solid shelf of ice for three thousand years, had lost 90 percent of its mass in the twentieth century. Arctic permafrost was letting heavy trucks break through during months when it used to be rock hard. The thumb-shaped Kenai peninsula in

Alaska had become a magnet for bugs drawn there by warm temperatures. Kenai's thick blanket of spruce had been annihilated by bark beetles, which showed up in the early 1990s and killed more than two million
acres of trees. To envision two million acres, think of a rectangle. Go
thirty miles along the base, take a right turn, go one hundred miles along
the side, take a right turn. Go thirty miles along the top, take a right turn.
And return to your starting place. Now fill the space with dead spruce.
That's a portrait of bark beetles cannibalizing a forest, leaving skeletons
as far as the eye can see, ravaging ecosystems dependent on the spruce,
all because of a few degrees' change in the weather.

Behind these changes in the Arctic, as well as more subtle ones elsewhere (butterflies migrating farther north each spring, maple sugar trees
dying off in Vermont, atolls in Micronesia slipping underwater), was the
out-of-whack carbon cycle. What atmospheric scientists call the natural
carbon cycle has been here a long time, since the atmosphere stabilized
more than 300 million years ago, and is pretty easy to understand. Basically,
animals give off carbon as waste from cell metabolism fired by oxygen, and
as decay, once they die and break down. Plants take in carbon as part of
carbon dioxide from the air to drive their growth and functions, then give
off oxygen as a waste gas. If I haven't made that real clear by now, well,
there it is: Oxygen in, carbon out, for animals; and carbon in, oxygen out,
for plants. In the oceans, the carbon cycle is similar. Marine plants and creatures play the same balancing game fundamental to their ongoing existence. Our oceans also act as carbon sinks into which carbon can diffuse
from air into water, and vice versa. Scientists tell us that 100 billion tons of
carbon diffuse each year back and forth between the oceans and the sky.

Man's role upsetting this grand balance, though still resisted because
its acceptance would mean a fundamental rethinking of the way civilization works, began with the little atmospheric disruptions of man, escalated with the special case, the twentieth century, and now lurches into
the twenty-first century pretty much out of control—but with the big
money still betting on breakthrough technologies to insure a future
much like the present. Despite technology's record of failing to have

cleaned the air of pollutants (carbon dioxide is not classified as a pollutant, because it occurs naturally and is fundamental to life) and evidence of a steady increase in greenhouse gases, betting on technology still makes sense to those looking at the problem in a reductionist mode. But if you look at the whole earth, at the big piles of evidence that say, hey dude, the carbon cycle is out of whack, it's past a tipping point, you are not going to reverse the warming of the planet now, then the future is sobering. And what's really nightmarish, of course, is that we don't know how this is going to play out, or just where, or exactly when. But the change is in motion, the game is under way. All because of increasing concentrations of gases that seem such a paltry thing, a veritable sneeze in a hurricane, a fist shaken at the huge face of a laughing sky god, a kick in the pants of *deus absconditus*.

As George Divoky said, echoing a growing chorus of scientists from multiple disciplines around the world, "What is it that you don't understand?"

At the outset of the new millennium, there was 30 percent more carbon dioxide in the air than there had been two centuries ago. And the rate of change of a gas crucial to the biosphere and climate was about one hundred times faster than its average rate of change in the nineteenth century. Earth hadn't seen, or had to adapt to, such high levels for probably twenty million years. Twenty million years ago, the Andes and the Himalayas were just rising from the planet's crust, ocean currents supported a great surge in plankton, and land bridges appeared, connecting Siberia to Alaska. The fossil record tells us there were camels, bears, dogs, foxes, beavers, ground squirrels, even horses around then, . . . though the forebears of human beings had a few more million years to wait before they started swinging from trees and loping across the savannas.

With the descendants numbering more than six billion, and our lifestyles a brand new earth game, there were multiple factors at work to ensure that the carbon concentrations kept trending upward. The first factor was the world's increasing appetite for fossil fuels, an appetite with only one blip during the twentieth century: the Great Depression. A second was the jump in the other major greenhouse gases: methane, nitrogen oxides, and CFCs. Methane, like carbon dioxide, has a natural balance in the atmosphere, but since 1800, its concentrations have doubled. This increase is due mainly to landfills and cow manure. Decaying garbage gives off the gas as do the flatulence and waste of cows, animals that more and more crowd the world's feedlots as protein-rich diets spread in popularity. Nitrogen oxides, which are a result of all combustion, are produced in larger quantities by high-temperature combustion, as found in automobile engines. Because of the global ban on CFCs, these compounds are declining as a major greenhouse gas, although their atmospheric lifetime (CFCs can stay aloft for a century, as can carbon dioxide, whereas methane remains in the air only about eight or ten years) means CFCs will be heating the atmosphere throughout the twenty-first century, along with the others.

During the twentieth century, the average global temperature rose one degree Fahrenheit. That doesn't sound like much. But the impacts, because of the size and natural systems of the earth, have varied greatly from region to region. As noted, the Arctic has warmed the most. It's on average about five degrees Fahrenheit warmer than a century ago, with seas rising, tundra thawing, and icebergs calving. By 2050, if the world continues on its present path, temperatures are projected to rise as much as three more degrees Fahrenheit, and by 2100, by as much as ten degrees. Computer models, which give us graphic displays of the many variables and their relationships, get more dystopian after that: sea levels up thirty feet, the world's coastlines reshaped, its great ports partially submerged, ocean currents altered, floods, droughts, the whole shebang. Something right out of the Bible. It's little wonder that many politicians and industrialists, as well as people in the street, hearing such doomsday visions of despair, violence, and ruin all over the damn place, stick their heads in the sand whenever they hear the term *global warming*. They want more data, more good science, less alarmism because these scenarios they keep hearing say that the world as they know it is coming to an end. And the change sounds as though it's right on the earth's doorstep.

The stuff going on was pretty fascinating, and a bit scary.

One report I read that really shocked me said that Chalcaltaya, the world's highest ski area, located on a glacier above La Paz, was disappearing. Since my trip there in the early 1980s, the glacier, despite being at 17,500 feet, had lost half its mass. Predictions claimed all the ice would be gone in another ten or fifteen years. If I wanted to go back, I'd better hurry.

All the glaciers in the tropics seemed to be disappearing, many at alarming speeds. They were melting much faster than their likenesses farther north. The meltdown in Bolivia was symptomatic of the lack of attention being given this phenomenon, which, in Bolivia's case, was more complicated than ice simply reverting to its liquid state or skiers needing to go elsewhere for a few turns. A high, dry place, Bolivia would most likely face a freshwater shortage once its glaciers melted. They supply the lakes and reservoirs. Yet Bolivia lacked both the resources and the skills to adapt to sudden changes brought on by climate change. Government officials in La Paz acknowledged that there were no plans in place to deal with probable impacts, no studies in progress, no estimated timetables on the table.

Meanwhile, as Bolivians watched their glaciers receding, a glacial accident in the Caucasus Mountains of Georgia put a human face—in this case, a famous human face—on global warming as something to worry about. In September 2003, after snapping apart at sixteen thousand feet, an aircraft-carrier-sized block of ice plowed ten miles down a gorge, killing thirty-year-old Sergei Bodrov Jr., who had just wrapped the second day of a film shoot. The hot young movie star, often called Russia's James Dean and known less for his acting skills than for his persona as the new Russian young man, was crushed. Also killed were many of the film crew and more than a hundred villagers in nearby Karmadon, which the runaway glacier took out as though it weren't even there. Appalled rescue teams faced five hundred feet of dripping ice.

Since the major player here is carbon, the question is, can more of it be removed from the atmosphere to reverse the warming trend? The simple answer is, sure. Just replant the world with trees. The complicated answer is exactly that: complicated.

Nothing science has discovered yet to do the job is as good at carbon removal as a forest. And growing forests are best, just as they were during the massive carbon banquet of the Carboniferous period. Growing trees are hungry. They eat carbon, digest it molecularly with the aid of sunlight, and store the carbon in pulp, leaves, and bark. A young forest is like a bunch of teenagers attacking a refrigerator. And the trees aren't fussy. They'll eat almost any old kind of carbon dioxide, whether it came from your lungs or your utility's power plant or a pet dog in the yard.

Of course, a planet with six billion people, and growing, and an insatiable hunger for wood and its fossil fuel cousins is not going to quickly reblanket itself with young forests that solve the carbon balance problem. Nor are the rain forests of the Amazon and the great northern forest that arcs across Canada and Alaska going to take in much extra carbon. They are mature forests. Their carbon dioxide intakes are balanced by their outputs. So we have to be more creative. Or adapt. Or both.

Most of the so-called revolutionary technologies out there hoping to do the job of rebalancing the atmosphere's carbon and slow global warming involve capturing the carbon in the sky and storing it, sequestering it, in the earth. For a while, something called carbon fertilization looked very promising. Carbon fertilization is basically a jiving up of mature forests with high concentrations of carbon dioxide, in expectation that the trees will get hungry again, want to grow, and then use greater amounts of carbon dioxide than they emit. Experiments to date, though, suggest the hope may be misplaced. Mature trees do seem to throw off their carbon slumber when exposed to higher concentrations of carbon dioxide. But they only go into short growth spurts, then settle back into just being mature, and don't take in extra carbon. Another process, called geological sequestering, repacks the world's most voluminous greenhouse gas back where it came from. That is, inside the tunnels, seams, and pockets left after fossil fuels have been mined or pumped out. Still in its infancy and being tested most rigorously in the North Sea by the Norwegians (people in one of those radical environmental countries), geological sequestering works to a degree. It has problems to be solved, however, including leaks of carbon dioxide out of holding zones and into freshwater sources, where it contaminates drinking water.

Then there are the oceans. For billions of years they have been absorbing carbon dioxide from the air. Couldn't the world load them up with additional carbon, now that we desperately need help? No. Same problem as with the mature forest, actually. The oceans seem to be enjoying a carbon balance. They are not as hungry for carbon as they used to be. One ingenious attempt, dubbed the Geritol solution to global warming because Geritol ads boast of the product's iron, involved spreading iron sulfate on the water. The theory was that more iron would boost the appetite of green algae that live in shallow depths for carbon dioxide, thus cleansing the air. Didn't work well, though. Decay from the Geritol-enhanced algae gave off as much extra carbon dioxide as the marine plants had taken in, canceling the possible benefits.

Although our oceans presently absorb more carbon dioxide than they give off, that balance may be upset as water temperatures rise. Since carbon dioxide is less soluble in warm water than in cold, warmer ocean water is a less efficient global sink for carbon dioxide. One oceanographer at Princeton predicts that by the end of this century the warming oceans will be 10 percent less receptive to carbon dioxide than they are now—not a very promising prospect. This type of positive reinforcement loop set off by unpredictable complications threatens all the predictions about global climate change. Global warming could very well have its own built-in accelerators, and then again other changes may dampen the speed of temperature change. What seems most unlikely at this point, though, is that the warming trend will reverse. There is simply too much riding on the carbon load in the atmosphere for that to occur.

Carbon fertilization, geological sequestering, the Geritol solution—these and most other so-called revolutionary solutions to counter the carbon are old-think. They are technology-based fixes with little to no inconvenience to the way people in the developed world live now. And how Brazil, China, India, Indonesia, and other countries appear to think their people ought to live as well: using lots of energy with no inconvenience. The subtext is, let technology fix it. But even the most inventive of the new technologies can't hold a candle to a young forest. None of them, even at their most optimistic, promise a solution commensurate with the scale of the problem. A scale the world has never faced before, or hasn't faced since man has been here, and surely does not show much eagerness to face now.

The bottom line is that both the global powerhouses and powerhouse wannabes are avoiding the hard truth. Carbon loads in the atmosphere are out of whack because of the conditions bred by the industrial structure embraced by the world, by the reliance on fossil fuels to energize that structure, and by the gasoline-powered automobile as the machine of choice and the symbol of the structure most widely loved and longed for by the majority. The conditions are characteristics of our world, a world that men and women have shaped, and can reshape, given the will. First, of course, we must recognize what the choices have done before we can change them. Recognition of what the world has done to the membrane most important to its birth, growth, and future existence is not high on the to-do list of leaders around the globe at the outset of the twenty-first century.

Trashing the Kyoto Accord, which the United States did in 2001, was of little help. Ratified in 1997 by 120 nations that agreed the industrialized countries would cut their greenhouse gas emissions 5 percent below 1990 levels by the year 2012, the agreement was badly flawed. It ignored such things as China, giving that developing country a free ticket to contribute to global warming for a while. But at least the Kyoto Accord was *something*. When George Bush the Younger declared in 2001 that the accord was unfair to the United States, he was right in a sense. The United States was going to have to lead by reducing its greenhouse gas emissions more than any other country. On the other hand, the United States produced 25 percent of the world's greenhouse gases, with 4 percent of the population, who had been consuming one-quarter of the planet's fossil fuels for half a century. Was that fair to Bolivia? To Micronesia, which might disappear by 2050 because of rising seas? Shaping global-warming policy did demand fairness, but fairness could only result from leadership and compromise and a historical perspective.

The United States during the Bush administration retreated from both demands. Not that the administration worked in a vacuum. It was helped along by conservative think tanks such as the Competitive Enterprise Institute, a see-no-climate-change, hear-no-global-warming, think-no-regulation group that filed a lawsuit to scrub computer modeling descriptions of global warming clean from congressional reports. But then Congress also treated global warming with the kind of receptivity shown to a courier declaring he'd come with a whole lot of anthrax—now, where should he put it? Back in 1997, immediately following the Kyoto Accord, which the Clinton administration supported, irate lawmakers slashed funds from energy efficiency programs as a token of their disgust, then tried to stop the EPA from making climate control a public debate. Bills were introduced to halt information seminars on global warming and to prevent spending on the "contemplation" of the accord's ever being ratified by the United States.

In June 2001, when the Bush administration said the United States would not ratify the Kyoto Accord, sending it into a protracted death spiral, the energy plan being hatched by the White House praised the same old black egg: fossil fuel exploitation. In the United States, market forces would drive down emissions. In the United States, voluntary emission controls by industry and breakthrough technologies would do what Kyoto could not do. That was the message. As was this: The kindness of corporations was going to make the world's atmosphere a cleaner place without the assistance of any international regulatory body America did not control.

Many longtime observers of the global-warming story agreed that the Kyoto Accord was badly flawed. But it was a beginning. It signaled a choreographed pulling of heads from the sand. By rejecting the accord completely, the world's biggest producer of what needed to be controlled kept its head in the sand and its back to the international community. Why? To shift the issue into the hands of corporate chieftains, who could deal with global warming as one more externality on the balance sheet? The rub, of course, was so old, so known, so tender that it wanted to avoid regulatory touch. The rub was that the auto, oil, and other industrial sectors of the American economy did not work, had not worked, and would not work as the central force driving clean air policy. Profit-making corporations did not, by their very nature, serve public health. They served investor health. They did not have a line on the balance sheet for the public good. They had a line on the balance sheet for the private good. Yet clean air is about as public a concern as it is possible to imagine in a world stained by industrial pollutants and shadowed by the prospect of weather changes that reshape civilization. Meanwhile, most of American industry cheered. But why? What were these executives at the top thinking about? That they'd dodged the carbon bullet again, for a couple of years, maybe ten or twenty? Was there gloating in the boardrooms because industry had turned its back on probably the biggest dilemma facing the world, not only in the new century, but ever?

Of course, other eight-hundred-pound gorillas stalked the ever-larger global-warming story. Although, it occurs to me, even as I write those words, that to use such an awful metaphor of a creature so close to us genetically, yet pushed into the maw of our time's slow-motion extinction, won't do. Scratch the eight-hundred-pound gorillas. Replace the metaphor with names of specific countries: India with a billion-plus people, China with 1.3 billion, and Brazil with about 175 million. Those three will do for starters since they were all putting as much carbon into the sky as they could, even if they couldn't quite keep up with the United States.

China, though, was making India and Brazil pale as carbon producers in the first decade of the twenty-first century. It wasn't just the larger number of Chinese or the coal resources at the country's command, with their voluminous carbon emissions. It was the Chinese growth phe-

CLEAN AIR DREAMS, AND DENIALS

nomenon. The coming-out-of-nowhere pace of it. The avalanche of cars. With auto workers in China willing to make as little as fifty cents an hour, the nation was fast becoming a global player in the auto biz. In 2003, the Chinese built 3.8 million cars and light trucks, an increase of 70 percent in three years. The output placed China a million vehicles a year behind Germany, though far behind Japan, which made around 10 million vehicles in 2003, and the United States, which made 12 million. China added 1.8 million cars to its own fleet in 2003, bringing the total of cars in the already traffic-clogged nation to 10 million. That number was projected to double every three or four years indefinitely.

The automobile revolution in China happened quickly. Thirty years ago, most Chinese walked or biked. Trips meant the train. A typical house had a lightbulb or two, a wok, a rice cooker. Today, like their American and Western European counterparts, the majority of Chinese families own a couple of color televisions. Their woks and rice cookers plug into electric sockets. Four million families own the 10 million cars. Luxury-house construction is booming. As, of course, are the emissions that George Bush pointed out, correctly, as not capped by the Kyoto Accord.

Some scientists and futurists said that the United States, if it didn't want to follow or cooperate, ought to lead out of this mess. It ought to jump-start a Manhattan Project about what to do with all the carbon. Or pull together an Apollo project, but *not* to go to the moon. The nation ought to roll back the future now being painted by an ever-growing cross section of experts wielding sophisticated and persuasive arrays of everything from three-dimensional models to data about black guillemots returning to Alaska to lay eggs earlier every year. Martin I. Hoffert, a physics professor at New York University, said he thought multiple Apollo-like projects might do the trick. "Maybe six or seven of them operating simultaneously," he suggested. "We should be prepared to invest several hundred billion dollars in the next 10 to 15 years." The goal for spending all the dough? To narrow the industrial and political world's fractured attention spans and profit-making schemes into more intense beams focused on how to slow down the rising temperatures by doing something about the carbon, and on how to get the equivalent energy from alternative sources that did not yet exist.

Just so that you don't think that it would suffice to put a few hundred of the world's brains in a room with a bunch of computers, keep in mind that the world we inhabit presently gets 85 percent of its energy from those forests laid down in drawers like corpses. That is, fossil fuels. Along with 12 percent from nuclear power plants and hydroelectric dams. And a measly 2 percent from alternative sources, mainly windmills and solar collectors. To keep the global temperature rise by 2050 to only 3.6 degrees Fahrenheit, the alternative sources that emit no greenhouse gases would have to increase their energy output between four and fourteen times what it was in 2003. In a classic bit of understatement on the subject, *New York Times* environmental reporter Kenneth Chang wrote, "No existing technology appears capable of filling that void."

Even Bush administration secretary of energy, Spencer Abraham, agreed that some new, state-of-the-art launching pads would be required: "That means creating the kinds of technologies that do not simply refine current energy systems, but actually transform the way we produce and consume energy."

The launch pad for transformation still looms somewhere on a warming, hazy horizon. There is no Apollo rocket energized by solar power upright on the pad. And the clock is ticking.

Soon, of course, the United States, would have to act. It could either breathe new life into, or suffocate, the next version of the Kyoto Accord, or its equivalent. The nation had led the world into making the atmosphere a soupy mess, in asphyxiating its great cities, and in changing the world's climate The modern industrialized world looked to America for leadership, inventiveness, and philosophy. Brazil, China, India, Indonesia, the European nations, Russia, Japan, and all the rest, didn't want to admit they were killing their own futures. Instead, following America's lead, it was a lot easier to shift the chairs around on the *Titanic*.

But the atmosphere (and the oceans and the glaciers) didn't give a damn about the *Titanic*. It was history. When you came right down to it, air didn't give a damn about anything. So humankind had to get its act together for this one. The sky wasn't going to pluck *Homo sapiens* up out of the impending extinction disaster.

Someday, I thought, what we might see is a Carbon Inspectorate for the planet. A kind of Globe Land Security force. Its inspectors will be nice, tactful. But they won't screw around. They'll carry sensors that ferret out the smallest carbon leaks and will have the authority to take appropriate action to stop any emission without anything as old-fashioned as a court injunction. These folks will be the equivalent of superheroes. Their predecessors, admired in videos made especially for schoolchildren, will be the overworked drudges of the historic smoke-inspection departments of American cities during the nineteenth century; members of England's Bad Odor Task Force of 1873, a.k.a. the Royal Committee for Noxious Vapors; and the unlucky bastards who went into Chernobyl immediately after the meltdown to clean up the radioactive mess and died for it. The Carbon Inspectorate, nicknamed the De-Carb, will turn to future legal mechanisms (whatever they may be) to elicit humanity's cooperation. It will be one very authoritarian time, probably not democratic, but effective at keeping the window of life open for the species. Carbon-emission violators will be guilty until proven innocent, their rights squelched because the perils of procrastination will have overcome the old protections of outmoded legal institutions and laws. After a carbon maker proves he or she is innocent, the person will be back in business. Not before. It will be a ruthless world, one that seems, I say with some heartache, inevitable . . . one that is already, like global warming, showing its future dimensions. Dimensions more than likely to create a much different, much more controlled, and not necessarily cleaner world, than what we have now. But a world nevertheless.

The Black Triangle

There are places in the world whose names have become synonymous with damage: the Aral Sea, Bhopal, Hiroshima, Verdun, Pompeii.
Historically, the damages wiping out a city like Pompeii came from
lava flows or shifting tectonic plates. Once nations fielded large
armies and refined weapons of destruction, the world had its Gettys-
burgs and its Verduns. The twentieth century saw its share of both
natural and military disasters, but it also introduced a new category
of destruction: the environmental wreck. A place such as the Aral Sea,
the fourth largest lake in the world, died from strangulation almost
like a murder victim in a remote closet. Los Angeles, once a Shangri-La
surrounded by mountains and facing the sunny Pacific, gagged in the
effluents of its adopted automotive lifestyle. Steamy Cubatão, home
of Brazil's uncontrolled chemical industry, became the "Valley of
Death," a place where you sometimes wanted to vomit as soon as
you stepped outside. A stretch of the balmy Gulf of Mexico southeast
of Houston became known as "Cancer Alley" after refineries and
chemical plants were packed together mile after mile. And a formerly
lovely basin of lakes and fields between the mountains of Saxony and
the fertile hills of the Czech Garden lost its towns and health to coal,
garnering it the sobriquet "the Black Triangle."

Boys playing soccer near one of the eight coal-fired power plants in the Black Triangle of the Czech Republic.

THE BLACK TRIANGLE in North Bohemia is not really a triangle at all. It's a rectangle roughly forty miles long and ten miles wide, with a hazy penumbra that extends into adjacent mountain ranges and hills and even into southeastern Germany and southwestern Poland. From the air the flat rectangular basin looks like a golf course from hell, with divots miles wide and very deep. Through the divots crawl gigantic insectlike excavators that suggest a postnuclear Armageddon. They chew with steel teeth into seams of brown coal that is one-third water, one-third ash, and one-third fuel. Perched in the steel framework of each excavator like a ganglion is a small structure, inside of which sits the operator. He is a fundamental neuron in this large-scale recycling of carbon from the pit mine to the nearby power plant and thence to the atmosphere, from whence it came.

Blažena Hušková grew up in the Black Triangle and its penumbra and has spent most of her adult life trying to restore the wreckage. "I am an heir to loss," she told me one gray morning in April 2002, as we drove from the Jizersky Mountains, where she was the director and sole employee of a conservation organization that replants trees and educates people about pollution, toward Most, the physical heart of the Black Triangle, to visit her father, Tony. "The places I played as a small child," Blažena said, "they exist no more."

One such place was the village of Kopisty. Her grandmother had lived in Kopisty, and Tony was born there. Toward the end of World War II, Tony had led the grandmother and other family members from their home to nearby coal mines for protection during bombing raids. Today, Kopisty was gone and her grandmother rarely spoke of the village anymore, Blažena said, though recently she had broken her silence.

"She was walking me through the streets by hand with her," Blažena said. "It was like I lived there. So I went to Kopisty after, because I couldn't believe it was gone. I cried because all there was, was a hole."

We passed through Ústí nad Labem, an industrial town that was pounded at the end of World War II by Allied bombs and where angry Czechs lynched and tortured hundreds of their Nazi oppressors, tossing their bodies into the Elbe. Now Ústí was in the news because it attracted Romanies and refugees from the Balkans and the former Soviet Union looking for work of any kind, along with German men who came at night, shopping for sex. A major highway links Ústí to the German border, only ten miles away as the crow flies. The roadside was often busy with prostitutes, some working solo, others in groups waving with the enthusiasm of high school kids in rural America trying to cajole you to pull in and have your car washed.

On Ústí's outskirts, you can see dozens of tall *paneláky,* or "rabbit hutches," eight- to ten-story prefabricated complexes built by the Soviets to hold the locals displaced from the towns that were razed to get at the coal beneath their streets. Tall and monolithic, ugly and impressive, some of the *paneláky* were quilts of color from laundry hung to dry on balconies; others bled stains down their fronts, suggesting brown tears. Between 1938, when the Nazis had occupied the area and tried to refine gasoline from the coal, and 1989, when their successors, the Soviets, had left behind decades of black snow and the golf course from hell, around one hundred villages had been razed, leaving dust, ash, and a panoply of heavy metals, toxins, and gases where communities had thrived for centuries.

Just past Ústí, the first pit mine came into view off to the left of the highway. Shafts of sunlight had broken through the clouds, and light danced in the dust. A faint haze, a nimbus, trickled off to the sides of the pit. I could make out the skeletal frame of an excavator parked on the rim. The distance to the rim of the pit was too great for me to tell if the excavator was removing overburden, the soil above the coal. In the dust and light, I could barely make out the operator's shack perched in the steel framework. Inside it, there might be an operator, an electrician, and a mechanic eating lunch, sausages and sauerkraut, or dumplings with beef, most likely—unless the mechanic or electrician was fixing something. Or maybe, I thought, the excavator was simply not working, just perched there and waiting.

Run by an operator, an electrician, and a mechanic, a coal excavator resembles a 300-foot long insect with a virtually insatiable hunger for dark seams of fossil fuel buried in the earth.

Blažena paid the machine no heed. She was forty-two, had a round, youthful face, dark brown eyes, thick dry lips—they looked unused to lipstick. Her auburn hair was cut short, tucked behind her ears. She looked lost in thought.

In Litvínov, a small city not far from Most, Blažena turned north and headed up a narrow, winding road into the Ore Mountains.

Tony

> I am not so much optimistic; I am one of those working to do what is possible.
>
> —TONY ŠINÁGL

We entered Meziboří, a village carved out of the midriff of the mountains. After a steep climb, Blažena parked by the modest house of her father, Tony Šinágl (pronounced "sha-naw'-gul").

Earlier, Blažena had told me, "My father, there are a lot of bad things that happened during his whole life where he lives, and he wants to show the good things."

I knew he'd been raised in Kopisty, now a hole. That as a boy there, he'd dodged bombs dropped by the Allies. I had met enough older Czech men to know that the transition from Communism to capitalism had destroyed their self-confidence and self-esteem, that they hadn't been able to make the shift without losing their bearings. So I was a bit startled to meet a man in his midsixties who radiated good cheer, had no lines on his face, only a touch of gray in his sideburns, and brown eyes that occasionally even twinkled. Warm and welcoming, Tony ushered Blažena and me upstairs, where I was immediately introduced to the grandmother, then ninety-one. She was just in from her daily four-kilometer walk, for which she relied on two canes for balance and wore gloves to keep her hands warm and shielded from ultraviolet rays. As bright-eyed as her son, she had a remarkably strong handshake.

During lunch—soup and bread, dumplings, meat and gravy—Tony told me about his present job as a consultant for the oil giants Agip, Conoco, and Shell. An expert on the pipeline delivery system that brought oil from Russia and Western Europe to refineries in the Czech Republic, he traveled frequently to Brussels for meetings and had an office in Kralupy, a city just north of Prague, where there was a refinery. Then, in greater detail, he recounted his youth in the shadows of the German-built refinery in the royal town of Most, most of which had been dynamited for coal when Blažena was twelve.

Born in 1935, Tony had grown up in Kopisty, then only a mile or so from Most. He remembered the German refinery, which had turned brown coal into gasoline in a dirty process, taking shape to one side of Most, and the nearby coal mines that supplied the refinery with its rough fuel. During much of World War II, the refinery was beyond the range of British and American bombers, he said. But in mid-1944, they began coming from Italy and France. The children he knew understood the whistle of a bomb as a cautionary sound, and a noise like thick Sunday newspapers slapping down the stairs "as a dangerous sound because the bombs was coming straight down towards you." If you got caught out in the open, an unwritten rule was to run as fast as possible to a crater because it was believed that a bomb never fell in the same place twice. The waves of planes came in two- to three-minute intervals, Tony said, so there was time to run between the craters toward the tunnels leading to the mines and some protection. He had only gotten caught in the open once, so he, the grandmother, and others had to crouch in a bomb crater. "The pieces of iron were flying. And stones and dust. I remember all this when I smell dust from demolition. And from blood—it smells sweet."

In the 1950s, Tony had gone to the Technical High School in Most, then to Plzeň Technical University to study atomic reactors and steam turbines. After graduation he returned to the refinery in Most, now called Stalin's Factory. It still made gasoline from coal in a complicated, expensive process that relied on catalysts—nickel, cobalt, and uranium—which came from the Ore Mountains. A bright young engineer, Tony managed the construction of a power station built at the refinery to provide additional electricity once the decision was made to convert the operation from gasification of coal to conventional refining, with oil piped from Siberia replacing the local dirty coal. Tony was paid only a little more than the laborers, but he had a big black Tatra sedan and a driver, along with considerable responsibility. All this ended in 1968, with the Prague Spring rebellion against the Communist occupation of Czechoslovakia. Recalling those days, Tony said there had been a gradual but relentless loss of freedom, together with a hardening of the Soviet attitude, ever since they had taken over in the 1950s. "By the early sixties, it was not possible to do what you could do in the fifties," he said. "After 1968, it was two times not possible to do something. I was history with the Communists, and I was glad. It was crazy in that time. Half the Communist people were playing Communist for the job, the children. I was put on the list to be transported to Siberia with my family."

"Were any of the families ever sent there?"

He shook his head no.

The oil refinery was renamed the Chemical Plant of Czech-Soviet Friendship. It kept growing, liberally supplied with oil from more than five thousand miles away via one of the mammoth construction projects on which the Russians so prided themselves. By the early 1970s, the refinery employed twelve thousand. Tony was a clerk. He stayed inconspicuous, kept his mouth shut. He remained in Meziboří, a village on the slopes of the Ore Mountains, where he had moved when he had been an up-and-coming young engineer. At 1,500 feet, Meziboří was usually above the inversion layer that often sealed the basin below and that lidded the Black Triangle and its dense air pollution. The haze ebbed and flowed around Litvínov, which was higher than Most, which was frequently lost in the murk. Some mornings, from Meziboří, Tony said, he could see only a dark cloud spread below Litvínov. Other mornings, Litvínov had disappeared. "You drove into the cloud and it was terrible: stinking and smoking. The snow was sometimes black."

Although the Germans had introduced pit mining to feed the refinery they had built to oil Hitler's war machine, it was the Russians who ramped mining up to a village-eating, landscape-ruining, air-polluting monstrosity, one the Soviets boasted served the working class and its fight against the capitalists. Beginning in the 1970s, new pit mines eliminated dozens of towns and villages on all sides of Most. Steel mills and chemical factories were built adjacent to the coalfields. New power stations went up, also fired by coal. As the need for coal intensified, it became obvious that Most was doomed. The royal city, eight hundred years old, with a population of around thirty thousand, was perched on a column of coal that the Soviets intended to cut away. In 1974, the city was dynamited: row after row of two- and three-story flats, an opera house, theaters, public buildings, and churches—all except the Assumption of the Virgin Mary, a sixteenth-century Gothic cathedral the Soviets moved about a mile on railroad tracks with a hydraulic system being tested by army engineers—dropped like a herd of elephants buckling at the knees, collapsing, clay bodies crumbling into shards. The coal that Soviet geologists claimed lay under Most was not there, however, and what coal there was, was dug up in several years.

Meanwhile, the emissions from the Chemical Plant of Soviet-Czech Friendship, the new chemical factories, the power plants, the *paneláky*, the public buildings, and the busy pit mines generated air pollution of record-setting blends and concentrations, since there were virtually no controls at all. The pollutants tended to ride the prevailing winds eastward, with the Ore Mountains to the north and the Czech Garden to the south, and head in the direction of the Jizersky Mountains, about fifty miles away. There, they drenched the forests in acid rain and soot. Only fifteen miles from Most, the once great spa town of Teplice, formerly a haven for the composers Listz, Wagner, and Beethoven, was atmospherically mauled, its natural springs polluted, its towers and steeples gloved with the dust of lignite and the stink of sulfur. Even Casanova's bones, interred in nearby Duchcov, where the old satyr had lived the humiliating last years of his life as a librarian while dictating *Histoire de ma vie,* a chronology of conquests as rapacious as those then ravaging the land surrounding the village, became another casualty of the insatiable need for coal. Casanova's cemetery, like dozens of others, got churned up and spit out by the excavators and forgotten. The acid rain that fell in the Jizersky Mountains did not go unnoticed by Blažena, then living in them with her mother, who had divorced Tony. Trees were being killed at certain altitudes, she recalled, with a clean sweep. Hectares of weakened trees were often hastily cut on the orders of Soviet foresters or bureaucrats before the trees died and were completely worthless. Dead trees by the thousands sometimes stood for a few seasons, an eerie spectacle. With their gnarled trunks and blackened limbs without leaves or needles, when the trees toppled, they exposed roots which, like those of trees in the Black Forest of Germany, had lost their ability to provide nutrients because of the acidic soil. Some Czechs, hiding their dismay, explained away the losses with a popular local legend. It said that Krakonos, a hoary old bachelor in the Great Mountains, the Krkonoše, continually lusted after Kacenka, the beautiful young goddess of the smaller mountains, the Jizersky Mountains, and of another range farther east called the Orlicky Mountains. His advances continually rebuffed, Krakonos became angry and blew hard. The wind from the Great Mountains then rattled windowpanes, kept people inside, and knocked over trees. It was a comforting story for bleak times.

It was during these times, in 1978, just before Blažena began college at Charles University in Prague, that Tony managed to get her into the Chemical Plant of Czech-Soviet Friendship for a visit that proved a seminal event in the emerging environmentalist's life. The air stank that day, Blažena remembered. But she sort of liked it. Tony had divorced her mother when Blažena was ten, so she saw him infrequently. The smell of the plant, a blend of ammonia, urea, polyethylene, synthetic rubber, and other products refined there, with the smoke from the stacks, reminded her of what her dad used to smell like when he'd come home from work when she was a girl. She could hear the hiss of open valves, the rumble of trucks, the trains grinding along steel tracks. Her nostrils twitched and

The Chemical Plant of Czech-Soviet Friendship was renamed Chemopetrol after the fall of Communism in 1989.

suddenly it bothered her that her father worked here amid figures that moved like apparitions. No birds flew. No trees grew. Something alien, a core of hatred for what this place was doing to the natural world, knotted in her suddenly. "It was overwhelming," she recalled twenty-five years later. "It was a different story being close to the pipelines and the towers. From three hundred yards, they look almost human in scale. Closer, I felt swallowed. I was eaten by all the construction." She climbed one of the cooling towers behind her father. "We were inside the stomach but had a view of the rest of it." The rest of it was holding tanks, field flares, smoke, ash, dust, the squeal and clunk of railroad cars, metal rubbing metal. She heard fluids hissing under pressure, combustion taking place somewhere out of sight at devilishly high temperatures.

"You imagine you can fight against it?" Tony asked, sensing his daughter's discomfort.

"There must be different options."

For him, no. For her, then, maybe. He had a second family to take care of and few options to mull over late at night, unless being packed off to Siberia managed to force its way into his thoughts.

An Energy Sink, An Atmospheric Toilet

Always the hope remains.
—BLAŽENA HUŠKOVÁ

That evening, I stayed at Koldům, a blockish Communist-era hotel in Litvínov said to be Corbusier inspired. The epitome of blasé functionalism, kind of a *panelák* for travelers with all the warmth of a bunker, elevators the size of upright coffins, and windows that wouldn't open, it felt like a tomb.

I lay on my bed and thought about some of the things Blažena and Tony had told me. What must it be like to have had your childhood haunts poisoned or ruined or totally eliminated and replaced by dusty air? As a pampered American who could still return to the village of his birth and see the house in which he had played, run, and taken breath after breath of clean air, I could barely get my imagination around these visions, never mind my heart. And I thought about the father and daughter bonding of Blažena and Tony. They were somewhat like oil and water, or acid rain and roots. She so earnest and concerned, he so chipper and positive. I was curious how they would respond to Chemopetrol tomorrow.

I recalled what a photographer I knew in Prague had said about the refinery a few miles down the road, the place formerly known as the Chemical Plant for Soviet-Czech Friendship (where did they come up with those names?), and before that as Stalin's Factory, and now as Chemopetrol. Like Tony Šinágl, Ibra Ibrahimovič had worked at the refinery during its heyday under Communism. He hated it. "I want to live until that plant is destroyed," he said. He knew the place better than Tony did. For the last fifteen years Ibra had been documenting the refinery in all seasons, and from all angles: with cows in the foreground, from the wing of an airplane, from inside what Blažena referred to as its stomach. In 1986, right out of high school (he graduated from the same technical school in Most that Tony had attended), Ibra had gone to work at the plant and began taking forbidden shots. The snow *was* black then, he confirmed, but it was a nice contrast to footprints, which were white. He served two years of compulsory military service in 1987 and 1988, then returned to the refinery with his awareness of it, and of the region where he had spent most of his life, altered. He knew then, as he put it, "North Bohemia was a place not good for life."

After the fall of Communism, Ibra documented impoverished Gypsy families in the Black Triangle, campaigned unsuccessfully with other young activists to save the town of Libkovice from destruction in the mid-1990s, and witnessed a gradual decline in environmental concern throughout the Czech Republic. "The human memory is really short;

that is the problem," he told me over coffee in Prague, his wife translating. "Czechs think the environment has been improved. There is a gigantic unawareness about the globe. Under the Communists, awareness of environmental problems was large but action small because of the danger. Now awareness is small and action is smaller."

Blažena agreed. Earlier that day, she told me that environmental activism had been one of the main forces behind the Velvet Revolution of 1989, which overthrew the Communists, but after the fall of the Iron Curtain, the activism had quietly lost its momentum. In the 1980s, she explained, because Czechs had so much leisure time under Communism, activists of all stripes had been able to organize meetings, compare notes, and make plans. The leisure time the Communists made so much about had come back to haunt the regime. For a growing number of Czechs, raising their voices about air quality became a vent for the outrage they felt about a long list of deprivations, insults, and indignities, which the country had endured for the last fifty years, beginning with the occupation of the Germans. Rallying around air pollution, they learned, was relatively safe. "We weren't criticizing officials," Blažena said, "or the police." Activists held large rallies about air—air that people could taste and almost eat, not only in North Bohemia but in Ostrava, a steel town; Prague, where coal smoke blackened everything in the winter; and every other industrial and mining region in Czechoslovakia. Dirty air was a shadowy villain that Czechs could openly complain about "without alarm bells ringing in police headquarters."

Now, at the beginning of the twenty-first century, that had all changed, Blažena lamented. The old environmental fire of the 1980s was gone. She had felt people's fervor cooling during the 1990s, displaced by concerns about jobs, privatization schemes, cell phones, flats, cars, and glitz from the West. "Today," she said, "in the Czech Republic, everybody just makes money. Makes money, makes money, makes money. You are a good guy when you work eighteen hours a day and have two cars and two houses and three girlfriends."

And you know, she added, sometimes she wondered why she wasn't making more money herself. Businesspeople asked her why she was working for a tiny conservation group in North Bohemia. "I guess they think if she is good, she'd not be working for that organization for that little money. But it's my solution. I am not going to solve the problems of the world. But I try. Always the hope remains."

*

The next morning, Blažena and Tony picked me up in front of Koldům. We headed down toward the Chemopetrol complex beneath a gray but relatively clear sky. Cooling towers soon appeared, horizontal pipes, railroad cars filled with chemicals, then a health clinic named Paracelsus.

It made me smile.

I rolled my window down a little. Blažena said she couldn't smell anything unusual. Tony was too busy looking toward the power station that he had managed in the 1960s, when he was a big shot under the good graces of the Communists, to talk about odors.

"Most dangerous place!" Tony suddenly announced.

He was pointing toward large, round storage tanks painted white. They contained ethylene and propylene, refined hydrocarbons kept under high pressure.

"Drive by faster," Blažena urged, only half in jest.

Tony didn't. It was one of the tit-for-tat little exchanges they liked to have, latent conflicts defused by silence, maybe a slight smile, a look, a look away. Going by the tanks slowly, Tony appraised the pipes, his chin up, his lips pursed. They were his specialty, his post-Communist area of expertise. Without pipelines, a refinery, after all, was nothing.

As we continued through the four miles of ramshackle industrialism, I counted the cooling tours, watched thin plumes of mist drifting from the lips of one. The structures had a certain sensuous beauty, a benign utility. They made me think of Temelin, the controversial Moravian nuclear power plant that had been built by the Soviets on the Chernobyl design and that was meant to be the biggest such plant in the world. Coming up on Temelin's cooling towers on the flat Moravian plain, I had been struck by their scale. Monoliths jutting out of the flatness, they made me think of structures left behind by some cold and hard civilization. Yet they radiated an undeniable beauty. At the time, I had no idea that a village had been obliterated for them to be there, nor that the village was the birthplace of Blažena's grandmother and Tony's mother, the one of the four-kilometer daily hike, of the later Kopisty displacement. Little did I know then that Temelin, now operating at about one-quarter capacity because of its inherent dangers and hated by the Austrians because it was located by the border, was another scar of loss for Blažena's family. But these cooling towers at Chemopetrol seemed almost babyish compared to those at Temelin. Yet what had been razed here, I wondered now, what had disappeared so that they could be built?

I also counted four smokestacks. None were emitting visible smoke. Yesterday, around his kitchen table, Tony had assured me that Chemopetrol and the other power plants in the Black Triangle were much cleaner than they had been under the Communists. Since 1989, new technologies had been added, plants and factories retrofitted. The new controls included machines that pulverized coal into dust so that it burned better, electric filters that removed particulates from smoke, calcium-milk slurries in stack scrubbers that captured even tinier particles. Much of the final stack waste ended up going into gypsum wallboard. All in all, the coal-fired power plants in the Black Triangle were now cleaner than many of their counterparts in the United States—and would remain so with the clean air rule changes and Clear Skies Initiative of the Bush administration unless courts overturned them.

After Chemopetrol we headed up a winding, very steep lane to Hněvín, a small thirteenth-century castle on the top of a thumblike projection. On a promontory by the edge of the castle wall, we had a sweeping view of Most and of much of the Black Triangle. There was a huge pit mine directly below, the Assumption of the Virgin Mary cathedral facing it. To the south rose the low hills of the Czech Garden. To the west columns of smoke lifted from the tall stacks of tandem power plants in Kadaň. Directly below in another direction, so tight to this volcanic

thumb that it had been spared demolition in 1974, was a historic neighborhood of old Most.

Tony pointed out his old technical high school, now a museum. He
directed my attention to the house in which Blažena was born. He aimed
a finger toward Kopisty's former location, where he had once zigzagged
through craters.

Blažena said little. It was her dad's thing, this brief flirtation with nos-
talgia. With the Most that was, with the Most that had disappeared when
she was only a girl.

For a moment I felt a sharp disappointment. I glanced at the *paneláky*
around Most, at the cathedral, at the pit before it, at a race track several
miles in the distance. None of them looked quite ugly enough. Even the
new Most, a shabby conglomeration of pseudomodern buildings and
broad streets, wasn't some smoke-enshrouded hellhole with people dark
from ash and coughing their lungs out as they stalked along grimy
streets. I could actually see some of them quite clearly from Hněvín. The
air right then was clean, didn't stink, even let through a pale bluish sky.
Most was failing dismally at living up to its advanced billing as an atmospheric hellhole out of Dante. Though maybe, I realized, as Blažena and
Tony wandered off, talking, their broad backs the same size, their heads
forward, their shoulders almost touching, Most, the real Most, was gone.
Obliterated for coal. Into dust and faint memory had gone three squares,
the opera, the train station, the library, and all the rest, except the tiled
roofs immediately below. There was no Most, only an absence. Most was
extinct. It existed only as the memory of coal might exist, if you thought
about such a thing, as a very ancient forest that had lain quiet in the dust
for eons and then burned.

That afternoon, we visited another castle, the 244-room Jezeří, tucked
into the Ore Mountains ten miles west of Most. What distinguished
Jezeří, besides its scale and horrendous view—the castle overlooks the
largest pit in the Black Triangle, the Czechoslovak Army Mine, once the
site of the largest lake in the country, called Komořany—was its
grandiose history. Beethoven and Goethe both spent considerable time in
this border region in the late-1700s and early 1800s and stayed here.

Beethoven composed parts of several symphonies, including the *Eroica*, at Jezeří under the patronage of Franz Maximilian Libcowicz.

When Tony, Blažena, and I first arrived at the castle's gate, a no-nonsense Czech woman all in black and wearing knee-high leather boots emerged from a small stone dwelling, blonde hair loose and a couple of dogs leaping around her legs.

"She is the spirit," Blažena whispered into my ear.

She was Hanna Krejová, project manager for Jezeří's protracted restoration. Hanna K. wore a gold cross around her neck and had an alluring gap between her front teeth. With Blažena interpreting, Hanna K. told me she had visited the castle as a child, watched the pit below eat her village on the shore of the lake, and approach within a half mile of the castle wall. She had gotten the job as the assistant to the director of the restoration and then took over his position in the mid-1990s. "It is my destiny," she said with a weary laugh.

Built in the early 1700s on a site castled ever since the Renaissance, the castle was named Jezeří, which meant "castle above the lake," she explained. The baroque pile had been sited overlooking Komořany and surrounded by a thick beech forest. In addition to the 244 rooms, Jezeří had an oval-shaped theater in which Franz Maximilian Libcowicz listened to his favored composers perform.

Hanna K. showed us tinted photos from the late nineteenth century with people strolling in the lower terraces, the grand castle perched above. The Depression had been tough on a place with such grandeur and an insatiable demand for upkeep. During World War II, to camouflage the castle and keep it from being bombed, the Germans had painted the walls green. They incarcerated prisoners in the dungeons and rooms. The Communists just let the place go to hell. By the 1970s, Jezeří was abandoned, its treasures gone, its rooms and salons vandalized, its historical importance dismissed. Hanna K. said, "The Communists wanted the public to forget about it completely. It was much more important to mine coal."

Then, in the early 1980s, a minor miracle occurred. Expansion of the Czechoslovak Army Mine stopped half a mile from the walls. The pit

had consumed the lower terraces and threatened the integrity of the castle's foundation. But suddenly the authorities granted Jezeří a reprieve.
Acknowledging its historical significance after thirty years of denial, the
Communists drew up a plan of reconstruction. Work began in 1988, with
funding and a twenty-year plan. A year later came the Velvet Revolution.
Jezeří was returned to its owner, the Libcowicz family, in 1992. Not want-
ing another derelict castle the family donated it to the state. Work on the
restoration continued, but erratically. Prague got the lion's share of state
money for historic preservation because that was where the tourists
went. Presently, said Hanna K., the challenges she faced included getting
adequate funding from the Institute for Historical Monuments in Prague,
dealing with chauvinistic Czech workmen, and debating preservation
experts, some of whom wanted the castle to look old and weathered
rather than rebuilt as an early-nineteenth-century showplace, complete
with period furnishings and reenactments, which was her vision. Waving
a free hand at a forty-foot wall of a wing that looked to me as though it
had not been touched since the Thirty Years' War, Hanna K. said all the
work didn't bother her. She took a long drag on a Petra, exhaled, and said
defiantly, "This place has a special spirit."

A young guide gave Tony, Blažena, and me a tour. Soon we found
ourselves in the north tower game room, an antler chandelier overhead
and bristly boar-skin rugs on the parquet. Tony and Blažena went down
the spiral stone stairs, but I lingered awhile. I leaned out a narrow window for an elevated look at the Czechoslovak Army Mine, a kind of pale
cocoon inside of which I could discern nothing. I did, though, imagine
what it might have been like for Beethoven to have leaned out the same
window and to have seen water and trees and, if it had been spring, flow-
ering vineyards and orchards below in the castle terraces not far from the
shores of Komořany. Beethoven certainly had not heard the squeal of
excavator wheels suggesting schoolchildren's shrieking, since the composer was deaf, or the faint groans of the coal-eating insects chewing into
the basin's lodes. Nor was he able to see, as I did gradually, a coal-separat-
ing plant far in the hazy distance beyond the pit, and even farther away,

what I slowly realized were Chemopetrol's cooling towers, ghoulish and faint, with Hněvín a blurry thumb above them.

Dirty Air and the Human Spirit

North Bohemia is perceived as a colony for coal and electricity; it is ruled by an energy mafia.

—JAN PINOŠ

In the Black Triangle coal mining had gone into a slump after the Velvet Revolution. A loud outcry over the devastation, the shortened life spans, and the filthy air in North Bohemia moved the new government of Czechoslovakia to issue decrees promising no more pit mines.

In 1990, the freshly minted democracy quickly got in step with the oil-consuming world. Its leaders hastily agreed that coal was the black sheep of energy resources. Coal's dust and smoke, its toxics and sulfur, all tainted the fuel's popularity. Added to that was the role coal played in upsetting the carbon cycle. So when the government in Prague openly criticized the nation's mining history and the coal miners in the Black Tri-angle, the leaders were in step with the times. Only recently considered the aristocrats of the proletariat, the miners had now suddenly lost respect, privileges, and jobs and found themselves vilified for having ruined the environment and poisoned the air. These coal workers thought their days were numbered.

But pit mining in the Black Triangle was just entering a brief hiatus.

The twentieth century had opened with the United States leading the world in coal production. In the 1930s, Russia surpassed America's out-put. And in the 1980s, China became the number one coal-producing nation. China got most of its energy from a vast belt of coal that arced almost three thousand miles across its northern border and was four hundred miles wide in some places. But about 20 percent of it, or roughly 200 million tons in a typical year in the 1990s, went up in wasted smoke. Underground fires similar to those simmering beneath the vacated streets of Centralia, Pennsylvania, and burning in India, and consuming Burning Mountain in Australia for the last two thousand years, blazed across the nation's forehead like a glowing tiara. A hot tiara. Or that's how it looked in photos of China taken from satellites. Temperatures in some of the hundreds of underground fires topped one thousand degrees Fahrenheit. Miners just worked around them. They dug into seams of coal to feed China's ravenous energy needs, which by the year 2000 were only surpassed by America's.

Compared to two carbon-burning giants such as the United States and China, the Czech Republic and its Black Triangle were puny players in the globe's carbon-burning sweepstakes. Still, the Czech Republic was not an inconsequential player. For its relatively small scale, the Black Triangle lofted hundreds of thousands of tons of carbon skyward to help warm the planet. It had adopted the modern extraction methods popular in America, Australia, Russia, Poland, and other places where coal lay close to the surface. Like other places scarred with open-cast mines, the triangle had beheaded hills, scalped forests, drained lakes, and razed towns to get at the fuel. But that kind of devastation was finished in the Black Triangle, or so it had been decreed by the Czech government in 1990, during the short-lived euphoria following the Velvet Revolution.

Then came Libkovice.

Between 1991 and 1994, the village of Libkovice, about ten miles northeast of Most, became a post–Communist-era sacrifice for coal. In an attempt to stop the demolition, a Czech environmentalist named Jan Pinǒs and another Czech activist, Petr Pakosta, worked with Greenpeace. They organized protests and publicized the plight of the village, spotlighting the hasty retreat of politicians and officials from the recent decrees promising that this kind of devastation had ended. Despite intense media coverage and legal action, Libkovice was razed. Pinǒs lost a couple of teeth during scuffles with police, and Ibra Ibrahimovič captured the whole story on film. Said Pinǒs of the bitter experience: "I took one lesson away from the fight. You can't campaign for a piece of land if you don't live there."

Pinŏs said that ten years later, few of those who did live in the Black Triangle had stepped up to stop the exploitation. "Today," he told me, "North Bohemia is perceived as a colony for coal and electricity; it is ruled by an energy mafia." He described the mafia as a loose coalition of mine owners, politicians, businesspeople, and developers who had reestablished the region in the public's mind as the North Bohemia Coal Basin, a name synonymous with exploitation, pollution, and a lack of spirit to shape its own destiny. In other words, a place to be colonized and exploited for national energy needs.

Petr Pakosta, Pinŏs's former ally during the Libkovice fight and a battle-scarred environmentalist from numerous other conflicts, agreed. Pakosta was vice mayor of Saint Kathryn of the Mountains, a village tucked high in the Ore Mountains about twenty miles from Most and bordering Germany. With his gold-capped teeth, white hair, and grizzly appearance, he struck me as a Czech legend in the making. After Blažena

The Počerady power station in 1993.

introduced me to him, we wandered around the backyard of Pakosta's small farm with its motley array of critters: llamas that came bounding at his call, carp in a big pond, monstrously huge pigs, friendly horses, free-range chickens, and assorted cats and dogs. Later, in a village pub, Pakosta made it clear he wanted to blow a strong wind toward the airless bureaucracy in Prague, whose attitudes, he felt, were further poisoning the people and landscape of North Bohemia. And whose indifference to the environment was now a disgrace.

When I asked about air quality here in the mountains north of the pit mines, Pakosta said emissions were down. But with the increase in mining and the reliance on coal-fired power plants, even upgraded plants, for energy, sulfur emissions and heavy metals continued damaging forests and giving people respiratory diseases. And now, oil-related emissions from factories, automobiles, trucks, and diesel-burning equipment compounded the situation. "In the mid-1990s, we had London-type fog in the winter," he said, gold teeth flashing, "and Los Angeles–type smog in the summer. Today the pollution is less visible, less smelly. So the government puts the issue on the edge of its interest. Air looks better, the numbers are better. The problem is not a visual thing any longer."

That sounded familiar. As in the West decades earlier, invisible air pollution had displaced the visible, and one consequence was people's losing interest in air problems because they could no longer see them. A false sense of security that everything was fine atmospherically . . . or at least OK, had been a consequence.

Another change in the Black Triangle that convinced people that things were getting better were the so-called reborn landscapes. That was the word a brochure employed to describe the renovation of the divots and vast pits of the region. The curious publication was given to me by Tony's stepdaughter, a quiet and felicitous elementary school teacher in Meziboří. Published by the city of Most, the pamphlet described the rebirth of landscapes in the mid-1990s, when bulldozers, earth movers, and other machinery had filled in holes with slag and topped it up with soil to create artificial hills, plains, and even small lakes. The results proved that "mining is a temporary land exploitation," that the new landscape, in some aspects, "was even more attractive" than the old. The publication included flowery testimonials. There were black-and-white photos of smoldering pits and garish excavators from the 1960s and 1970s juxtaposed with recent colored shots of a vineyard, an artificial lake, a racetrack with jockeys galloping on thoroughbreds with *paneláky* in the background silhouetted by an Idaho-blue sky, examples of the rebirth of the land. No mention was made, or photos printed, of the hundred villages that were no more. Nor was there mention of Komořany. There were no interviews with the tens of thousands of the displaced. One of them, Karel Krejčí, had become an unsung hero in a book illustrated with photos taken by Ibra Ibrahimovič documenting the destruction of Libkovice. Krejčí had been a coal miner for thirty-seven years, but refused to leave his house for coal to be mined beneath it. He'd finally been hauled away by the police, with the media on hand to report the event. He was placed in a *panelák* in Most, joining thousands of other displaced Czechs whose gardens, homes, pubs, businesses, schools, churches, and public buildings had been taken for coal.

The gradual revival of coal power in the Black Triangle was being repeated around the world and especially in developing countries blessed, or cursed, with the planet's still most abundant fossil fuel. Countries were digging away with the same zealousness found in Wales and England during the industrial revolution, but with excavators rather than picks, shovels, or small steam engines. Mountains continued to be beheaded, lakes drained, towns razed, just as they had been in the formerly alluring part of North Bohemia. Now a golf course from hell, but with some reborn greens and reshaped fairways atop tons of slag, the region was not seeing the end of pit making and air pollution, despite promises and rhetoric. Coal mining in the Black Triangle was making a strong comeback. In a country often perceived as a little more enlightened and progressive than most, the environment was under a sustained and relentless attack in subtle and pervasive ways that were less visible than during the Communist era. But at least then, the damages had been criticized and became a rallying point for a revolution. "Today the Czech government is sometimes said to be less responsive to enviro problems than the Communists were in the eighties," said John Crane, an American expat who had taught Czech history at the International School of Prague for fifteen years. "Then the environment was so bad it couldn't be denied. The environment was something over which you could dissent. Under the Communists, coal miners were the aristocrats of the proletariat. After the Velvet Revolution, they were the enemy." Made the poster boys of environmental ruin and the scapegoats for an era, many miners had lost not only their jobs but their identities and their self-respect, while government made its symbolic, but ultimately pointless, decrees about banishing pit mining, decrees that it never meant to keep.

The comeback of coal mining in the Black Triangle, for all its negative impacts, did have some appeal. First, of course, were the jobs it provided a destitute region. Second, compared to mining in places such as Ukraine and China, where explosions, collapsing tunnels, and poison gas remained everyday dangers, miners here worried more about dust and heavy-equipment accidents.

The comeback of the pits to the already scarred landscape she loved did not please Blažena Hušková, even if the miners here were a little safer than their contemporaries digging away farther east. The renewed emphasis on coal undermined conservation and weakened the pressure to seek alternatives to the fuel most closely associated with global warming. "If you bring up anything about climate change in the Czech Republic," she told me, "you are an environmental terrorist." If you brought up a recent survey, which claimed that half the country's electricity load could be eliminated by a state program to tighten up homes, flats, and office buildings with better windows and insulation, you were stifling growth. If you worked for the Foundation for the Rescue and Restoration of the Jizersky Mountains, located in a log cabin in Liberec, along with low-paid employees of other small nongovernmental organizations (NGOs), you were wasting your future in a losing battle. Still, Blažena battled on. She seemed motivated most powerfully by some commitment she had made to herself years earlier to be relentless in her defense of nature, and to be an enemy of nature's exploitation if it harmed the human spirit, which derived nourishment from the natural world. In the Black Triangle, it seemed that the new era had arrived, yet the past had never been thrown off. The shadowy legacy of exploitation, manipulation, and opportunism epitomized by the Nazis, then the Communists, loomed over the region like some awful albatross, keeping its luck bad. The place had been down so long that anything looked good: more coal mining, more prostitution, more exploitation by people who were natural predators. And predators found the pickings here easy because, as Blažena put it so clearly, "In the Black Triangle, the local population is easy to manipulate; they are happy if they have enough pork and beer."

The most blatant example of manipulation, and one with bad implications for air over the Black Triangle, though good for the job market, was NEMAK.

*

In 2001, NEMAK, a Mexico-based multinational, announced its intentions to build an aluminum-engine factory on one of the last greenfield sites in the Black Triangle. The site was seventy-five acres of good meadowland beside the fields and farm buildings of Jan Rajter. Rajter (pronounced "writer") took issue with the idea of a factory that would totally change the rural environment, be located on good farmland, and emit dioxin, a poison often compared to the widely banned DDT. He expected that dioxin would probably ruin his crops and contaminate the milk from his cows as it worked its way up the very short food chain.

Over the course of three years, Rajter emerged as a hero in barn boots. Probably the pivotal incident that turned the sixty-two-year-old farmer into an avowed enemy of NEMAK and those pushing it down his throat occurred in early 2002. Having heard that a farmer was making a big stink about the plant being built next to his fields, the Czech prime minister, Miloš Zeman, called Rajter "an environmental terrorist," and labeled his family "a ticking time bomb" during a national television broadcast. Hearing this, Rajter couldn't sleep at night. And for a man with his disposition, that said a lot.

Rajter had been a farmer since the 1950s. In Silesia, as a young man, he had refused to join an agricultural collective and been promptly thrown in jail by the Communists for insubordination. He was eventually relocated to Most, in the heart of the Black Triangle. He had no flat, and he worked for the railroad. He began farming on the sly, raising cows and running a small dairy in the no-man's-land around the Stalin Factory. When the factory made gasoline from the local brown coal during the early years of the pit mines. In the 1970s and 1980s, Rajter maintained probably one of the strangest farms in the universe. "A farm in a dump," he called it. His pastures and meadows were the marginal fields surrounding the sprawling Chemical Plant of Soviet-Czech Friendship. Once the Iron Curtain fell and the land was privatized, he bought a ruined farm in the same vicinity. Then, in 1992, Rajter had the chance to acquire a derelict farm in a village about ten miles from Most, but with some of the last good land left in the basin. He bought the land, house, and buildings with his oldest son and

In the 1980s and early 1990s, Jan Rajter ran a small dairy farm—he called it "a farm in a dump"—in the marginal fields surrounding the Chemopetrol refinery.

poured more than $100,000 into it over the next few years. By 2001, when NEMAK suddenly appeared out of thin air, announcing it was about to build an aluminum-engine factory amid a patchwork of fields Rajter owned and rented, it seemed like an ill-fated wind had sought him out just when he was finally making a decent living on what he had shaped into the most successful farm in the Black Triangle.

Relying on front men and behind-closed-door deals, NEMAK slipped into Rajter's village by eliciting the cooperation of regional rather than village officials and by securing approvals that the farmer was not privy to. Co-owned by Ford Motor Company, NEMAK was also flush with cash and had influential connections with politicians in Prague, including the prime minister.

Rajter tried to react. But it had all happened so quickly. He normally rose before sunrise and was in bed before sunset. Taking on a foreign corporation with spokesmen, deep pockets, and signed papers was not something he could do by himself or even with his sons who ran the farm with him. Help appeared in the form of the Environmental Law Service. An NGO dedicated to legal challenges of government decisions that it feels contradict the law, the service claimed that the NEMAK deal was illegal, that impacts on third parties had been ignored, that the dioxin assessment provided by the company was wrong, and that the siting of a factory in prime meadowland violated the Czech Republic's Farmland Protection Act.

In both the regional courts of North Bohemia and the Constitutional Court, the highest in the land, the challenges were either denied or dismissed. Mostly they managed to get the farmer and his sons labeled as a terrorist cell for opposing a factory that threatened to make their cows' milk toxic.

The Rajters were not opposed to the plant. They just didn't want it anywhere near their farm. But then it became apparent that NEMAK intended to be only an anchor factory of an industrial park dedicated to automobile parts production. What would come next? A paint factory with fumes? Why locate the complex on the last decent farmland, when so many brownfield sites were nearby? Most, Ústí nad Labem, Teplice, and other nearby cities had industrial sites aching for tenants, for renewal. So-called brownfield sites, the alternatives had rail connections and utility infrastructures. They would require upgrades, but not new roads, water, sewage, and power, as NEMAK did out in the countryside on its lush, green acreage.

"It's terrible," Petr Pakosta said shortly after a groundbreaking ceremony was held on the site. "Here, we have a *kind* of protection for good farmland. But the minister of the environment can give land an exemption."

And he had.

When the final challenges filed by the Environmental Law Service went nowhere, one of the attorneys fighting with the Rajters to stop NEMAK, a young law school student named Pavel Franc, told me the courts were intimidated: "Our courts and judges are not very brave." Looking for action on a higher level, Franc said he was shifting the issues up to the European Union level, hoping for a more favorable ruling, although EU courts might not have any jurisdiction in the Black Triangle.

A NEMAK spokesman I talked to insisted the company was a good corporate citizen and played by the rules. NEMAK's European chief, Jorge Rada, said that the industrial park initiated by his company would be "an industrial zone forever." Siting the zone on greenfields in the Black Triangle was defended by NEMAK as a choice it made because it liked virgin land. It hated brownfields. Brownfields had to be cleaned up. Ford Motor Company, which advertised itself as America's green carmaker and had a $140 million stake in NEMAK, claimed that it had no influence over NEMAK's decisions and the company had done everything by the book.

The Black Triangle had not been the company's first choice for its new factory. Initially, NEMAK tried to locate the plant in Plzeň, home of Pilsener Urquell beer, in Western Bohemia. Despite the promise of fourteen hundred jobs, the city vetoed the plant in a public referendum. Activists in Plzeň made a big deal out of the dioxin levels common around aluminum-making operations, even with state-of-the-art pollution controls, and the public had decided it didn't want to breathe a carcinogen for which there was no acceptable minimum healthy dose.

Dioxin, which is often compared in its toxicity to the pesticide DDT, is not a single compound. It's a generic name for hundreds of related compounds that are by-products of intense combustion. Dioxins are usually associated with municipal incineration and certain manufacturing processes, such as aluminum and copper smelting. A few large municipal incinerators in the American Midwest were hit hard by adverse publicity about their dioxin emissions in 2000 following a study by the North American Commission for Environmental Cooperation (NACEC). The NACEC report, based on a sophisticated air transport model originated at the U.S. National Oceanographic and Atmospheric Administration and written by a research team headed by Barry Commoner, said dioxin emissions from the Midwest, as well as from Mexico and Canada, were traveling thousands of miles to the far north and harming the food chain of the Inuit in Nunavut. Basically, plumes of dioxins rose high in the atmosphere from the incinerators' tall stacks, caught a ride on trade winds, then came down on the tundra, where they were absorbed by algae or lichens. The foods were eaten by seals, walrus, caribou, and elk, and in short order the dioxins ended up in the mothers' milk of the Inuit because they ate the animal fat, which is where the carcinogens tend to concentrate. Inuit leader Sheila Watt-Cloutier is part of a group working for a global ban on dioxins, PCBs (polychorinated biphenyls), and other airborne carcinogens that all seemed to migrate to the once-pristine Arctic. She decried the injustice of garbage-burning plants thousands of miles away contaminating the Inuit's natural food and threatening to put an end to hunting and fishing, replacing it with fast foods and expensive staples flown in from the south. "Now when we put our babies to our breasts," said Watt-Cloutier, "we feed them a noxious chemical cocktail that foreshadows neurological disorders, cancers, kidney failure, reproductive dysfunctions, and more."

Despite NEMAK's claim that its dioxin emissions would not be harmful, the plant's opponents in Plzeň made the point that such a claim was a lie. Less than a hundred miles farther east, though, the appeal of jobs in the already exploited, atmospherically hammered, and notoriously apathetic Black Triangle trumped the fear of dioxin. Either a rep from NEMAK, or maybe someone associated with the energy mafia, decided this region would be more amenable and selected a likely 75-acre factory site on a nice, flat, green place where a 500-acre industrial park could easily expand around it.

Once NEMAK had the deeds, the approvals, and the support of powerful politicians, not to mention the Ministry of the Environment, which relaxed the Farmland Protection Act, it was virtually impossible to uproot the company without judicial intervention. The analogy of an unwanted particle entering the lung and getting situated despite the efforts of the macrophage to get rid of it may be a good one here. NEMAK was unwanted by the atmosphere, but embraced by local power brokers and those looking for work. In all likelihood, given the circumstances, Jan Rajter's efforts to dislodge NEMAK from the precious soil by his farm, though heroic, were doomed from the outset. The best that could be hoped was that the negative publicity and the contested processes would influence subsequent similar projects elsewhere in the Czech Republic. And that was no small feat, given that twenty-five prime greenfield sites around the countryside had been earmarked by industrial groups for NEMAK-like offenses. All would harm the in-place rail system, neglect brownfield sites in cities, and ape the American paradigm of greenfields first, to hell with cleaning up those old messes. Besides, new roads would go into these new sites, along with all kinds of systems, off of which profits would spin. Most auto workers would go back and forth to work in cars, adding to the positive-feedback loop of placing industry where it should not go. From what I could learn, the greenfield sites were being pitched to auto, electronic, and other manufacturing multinationals with promises of tax breaks, infrastructure support, and other amenities standard for sustaining this loop around the globe. As for the virtually universal complaint that companies voice when brownfield sites were suggested as alternatives, there was a pilot project going on in

Ostrava, a steel-mill town near the Polish border. A 150-acre site, formerly a steel mill, was being cleaned up for reuse based on a British model. If successful, the reclaimed land might soften the stigma associated with "this awful American name, brownfield site," as the manager of the project said, emphasizing the negative connotations of the term.

NEMAK began construction in the summer of 2002, optimistic that legal decisions would go its way. In 2003, as the plant neared completion, the European Bank for Reconstruction and Development, a financial arm of the EU, canceled its involvement with NEMAK. The efforts of Pavel Franc and the Environmental Law Service to put the facts on the table were instrumental in persuading the bank to withdraw its offer of a loan for the NEMAK plant. The bank cited possible illegalities, a questionable dioxin-emissions assessment by NEMAK, and allegations of political and economic misconduct. But NEMAK had other sources of cash. Machinery was moved in, roads widened, employee interviews begun.

Next door, on the Rajter farm, photographer Ibra Ibrahimovič documented six months in the life of the family, a time during which the patriarch continued his dogged attempts to get some justice. The Environmental Law Service funded Ibrahimovič in the hopes that art might prove more powerful than law.

As things turned out, it was a smart choice.

In late 2003, Ibra won the Czech Press Photographer of the Year Award for his documentary on the Rajters' stand against NEMAK. The photos were strong, the background compelling. Just ten years ago, on his first visit to the Rajter farm with his camera gear, Ibrahimovič had been met by Jan Rajter, who had been carrying a gun. Rajter told him to get out of there before he got shot. Even now, the family remained suspicious of the lens, said the photographer who shot hundreds of rolls of film of all the members of the family while NEMAK bulldozed away the meadow next door. "They knew my pictures were important for the case," Ibrahimovič said. "But they hated it. Every time I came, it got worse."

An early exhibition of select photos premiered in the Czech Senate in

November 2003. At the reception, Pavel Franc said the Environmental Law Service hoped the large, black-and-white photos adjacent to the senate's chambers would stir the senators into thinking about what was going on with the nation's landscape as new development opportunities poured in.

Looking rather haggard and tired, Jan Rajter chatted with supporters. Ibrahimovič said a few words. A brief film clip shot by a television crew played on a TV in a corner. In the clip, Rajter claimed that a farm "was the family jewels," and that "the Lord blessed me with a wife so patient; without her help I doubt I'd have made it." The farmer concluded by saying, "I regret I'm so old now, and losing my ideals and strength."

A couple weeks later, at the more formal and media-thronged award ceremonies for the Czech Press Award in the Prague Town Hall on Old Town Square, Jan Rajter attended only as an image. A shot of him, head back, laughing, and a young boy named Vašek leaning in the window of his car and angling his eyes at his elderly friend, whom he occasionally helped out around the farm, appeared on a large screen while Ibrahimovič received the award from the country's new president, Václav

Vašek, a local boy who helped out on the farm, liked to show Jan Rajter that he could still laugh.

Klaus. Klaus gave a brief speech to the standing-room-only crowd. Then he gave Ibrahimovič a trophy and a check. As they shook hands, he whispered into the photographer's ear that he really liked that winning photo, the one on the big screen, apparently oblivious that a year ago he'd labeled Jan Rajter an environmental terrorist and a threat to the nation. And that Ibrahimovič's photos may have opened a can of worms about government malfeasance.

For his part, Ibrahimovič simply nodded and kept shaking the president's hand, refusing to let it go. His fellow photographers, around twenty of them, clustered in a scrum, shooting like crazy. They knew the absurdity of the situation here in Kafka's hometown. And they made the most of it.

"All of my friends had told me to keep shaking Klaus's hand," Ibra said, "so they could get a good shot of me and him."

*

Tony Šinágl used to hunt on the fields of Jan Rajter's farm, near what was now NEMAK and the future industrial park. When I asked Tony about the factory, he seemed unaffected by the fate of the farm. A lost farm was part of progress. Technology was the future. Food would come to the Czech Republic by various pipelines, just as oil did. Not that Tony didn't feel some sympathy for a man roughly his own age and who had endured many hardships. But jobs at a plant in a region with high unemployment were more important than a couple hundred cows, seemed to be Tony's conclusion, although he never said that directly. A realist and a survivor, he knew the odds one faced standing up to authority in this part of the world, and the odds had not improved tremendously with the disappearance of the Soviets. "I am not so much optimistic," he told me in a summary of his philosophy. "I am one of those working to do what is possible."

For that attitude, Blažena had little patience. But she understood where it was coming from.

"My father is not against nature," she said. "He absolutely believes that by using technological solutions, you can solve anything. With machines and the right finances, you can save the earth. But there is a stumbling block. Man can construct a spaceship, but not a frog. Not a snail. Not even a leaf of grass."

As my guides to the Black Triangle, both father and daughter had been wonderfully giving, had been only themselves. Blažena was a relentless crusader for cleaner air. Tony wanted it as well, but did not campaign. Father and daughter had different visions of how their decisions would affect the future. Tony led by a kind of passive enthusiasm for life itself, for its renewal that had lifted up his life at the age of fifty and given it new purpose, new zest. He often struck me as a younger man stuck in an older man's body. Driving along the winding roads of the Ore Mountains, stopping to take a long, appreciative look at Saxony, even passing eager young prostitutes who waved like schoolgirls from the trollop-thick roadside leading to Chomutov, he always emitted a low-grade excitement about life. And for good reasons: The air was cleaner now, the Communists were gone, he had a loving family, and he had a daughter with high ideals who made him proud but whose goals were a bit unrealistic. He accepted the environment's drop on the concern list of the average Czech. It had its years of fame, had helped free the country from the Soviets. Now corporations and a pro-business government ran things. For that, the environmentalist should be pleased, was his attitude. That's just the way it was. Did you think you could fight it?

Blažena was Blažena. Strong positions taken when we are young never abandon us. Blažena's decision to fight it, to oppose the black snow, the dying beech, the acidic soil of the Black Triangle remained her anchor. She had a husband, two children, a house, a mortgage, and a car. She flew (though she didn't like the energy-use equations, or the jet-engine exhaust profiles, especially of trans-Atlantic flights); she used the Internet (though she much preferred face-to-face encounters and the subtleties of human connection). She even took vacations and enjoyed them. But she worried. She worried about the air, about the forests, about the world her kids would inherit. Yet she retained her hope. She could still laugh, even at those running her country, some of whom she wished were oiling the tracks of the excavators in the hazy pits rather than making bad decisions in Prague.

The last time we met, in the Café Imperial in Prague, I asked her how the forests in the Jizersky Mountains were doing.

Her lips were dry and cracked. "Nobody knows."

Since 1989, she said, ash, sulfur dioxide, and heavy metal concentrations had all declined. But nitrogen oxide levels had increased, and no one knew for sure what was happening to the soil. "It's a kind of synergy. The nitrogen compounds seem to harm the trees, but how is not well understood. I don't know what is going on in the soil. And global warming—nobody thinks about it. Nobody makes an issue of it. What will rising temperatures do to the forest? Nobody important is thinking about it here. When you say 'forest ecosystem' to the foresters they still don't know what you mean." Her brown eyes gleamed a moment, sparked with friction. "Timber, timber, timber—that is what they see. The big animals, the deer, are still there. So they say everything is OK."

And the energy mafia, what about them?

She poked a fork at her carrots and peas. She didn't think the energy mafia cared much about the forests. Not now. Not one hundred years from now. They cared about NEMAK, about power. Assisting them was a new regional government structure, one that allowed each region in the country to set its own environmental, energy, and economic agendas. Conservation and an increase in alternative energy, biomass and solar power primarily, were supposed to be on every region's plate as ways to generate energy. Unfortunately, she said, the newly formed region that contained the Black Triangle was now headquartered in Ústí nad Labem, the triangle's northeasternmost point. Its administrators and politicians were historically attached to coal as the lifeline of their territory. As for the national government, the new president, Václav Klaus, lacked an environmental ethos or much sympathy for anything but money.

"'Tell me the price of trees, then we can talk,'" she barked, leaning over her plate. "That is the attitude of Václav Klaus. 'Tell me the price of a cubic meter of air, then we can talk.'"

Surprisingly, she did approve of the president's doubts about entry into the European Union, even if it was a done deal.

"Most people believe that a united Europe is the only way to survive. I do not know if that is true. We are famous for our potatoes. Once we join the EU, we will be able to go into a store and buy potatoes grown in Spain, peeled in Great Britain, frozen in Sweden, and shipped to the Czech Republic. This is what the EU is built on."

Before hurrying off to catch a bus to Liberec, she told me that she had gone for a walk in the forest recently with her grandmother, who was now ninety-two. The memory made her smile. "She may be ninety-two, but she feels more than I do. She is more aware. I think it is because she grew up hearing birds singing, hearing wind in the trees."

I said that I wanted to tell her that Tony wasn't as sold on technology as the savior of the future as she thought he was.

"He told you that?" Her eyes looked astonished.

"He told me he doesn't believe technology will solve all our problems. He said he believes in the practical."

She seemed to chew on that. Then she slowly smiled. She said with a laugh, "Maybe next time you see my father, he will say that he believes in God!"

12

One Clean Breath

When I began this book, I thought I might try to find the place with the cleanest air in the world and go there and breathe a little of it. I thought such a goal might take me to Micronesia or to Nunavut or to Cape Grim. Or maybe to the middle of the Sahara or to some barren stretch of Outer Mongolia. One of these places must have pretty clean air, I thought. To find out, I intended to haul along a slick new piece of technology, a portable air quality monitor made by Pax Analytics. Suitcase-sized, fitted with various sensors for different air pollutants, and computer compatible, the unit would sniff and under-stand the contents of the wind like a good hunting dog. With the monitor as a tool, I'd gather anecdotes about enjoying such a rare thing as one clean breath in a very polluted world.

Tasmania's Cape Grim, off which the ocean is littered with sunken ships, has one of the earth's more remote atmospheric monitoring stations. It regularly detects traces of man-made chemicals sweeping out of the southern oceans and Antarctica.

Wow, was I ever dreaming.

Between the idea and the reality, as T. S. Eliot wrote, falls the shadow. The coral-rimmed, emerald-green bays of Micronesia, as I learned, can be tinged with smoke from biomass fires in Borneo and New Guinea, now parts of Indonesia undergoing intense development. Nunavutian Inuit mothers don't like to breastfeed their babies these days, because dioxins and PCBs are sent aloft by medical and municipal incinerators in the American Midwest, ride the trade winds, and fall to the tundra and ice. There, in the Arctic, the chemicals are ingested by the caribou, whales, and seals, and then are concentrated in Inuit people, who love the carcinogen-laced fat of the creatures they've hunted and eaten for as long as they can remember. Cape Grim, a meteorological outpost on the northwest shoulder of Tasmania, enjoys cleaner air than either Micronesia or Nunavut, but traces of human-made combustion and industry, having blown thousands of miles across the oceans, are detected there as well.

With some disappointment, I abandoned my idea that there was clean air out there in the natural world and I was going to find it and breathe it. The idea was naive. It was an excuse for adventure travel couched as a scientific journey. Clean air was a myth, not a reality. No offbeat trip with a suitcase-sized monitor was going to change that.

Still, the *idea* of clean air intoxicated me. But what exactly was clean air, anyway? Wasn't *clean* the wrong word to describe our air, the air in this "flask without walls"? Air contains thousands of trace elements, both natural and artificial, in addition to the long-term standards: nitrogen, oxygen, carbon dioxide, and the so-called noble gases—helium, neon, krypton, radon, and xenon—none of which reacts with other gases or compounds very well. Atmospheric chemists tell us that we normally breathe about 170 of the thousands of trace elements, which are not distributed as evenly as the basic gases, by just walking around outside the house. On such a short stroll, we also inhale some odd stuff, such as fragments from dead stars, fungal spores packed with DNA, and radio waves sent through space from volcanoes erupting on Io, a Jovian moon. "That seemingly empty cube you're looking through is a howling blizzard of bizarre matter and even more exotic radiation," said the air-aware social theorist David Bodanis of Oxford University.

So, one clean breath in the world we inhabit? Forget about it. Even if I went up to the stratosphere somehow, anywhere in that windless, clear, tranquil zone of the earth's membrane, where ozone has been fighting for its future against CFCs and, thanks to the Montreal Protocol, holding its own, I'd find about forty trace elements just floating around.

What gradually became crystal clear was that regardless of where I might have gone—to the far north, the far south, the far off—the air I breathed would not have been really *clean,* not according to my dictionary's definition of clean as "free from dirt or pollution; free from contamination or disease."

While I was still dithering over whether I could somehow go on this misguided pursuit of one clean breath, Michael Oppenheimer, an atmospheric scientist at the Environmental Defense Fund, kindly replied to an e-mail I sent, asking him about possible trips into other dimensions. At the time, I was hazily considering going back in time, like Jules Verne in *The Time Machine,* or maybe back to the future, like Michael J. Fox in the movie of the same name, to experience one clean breath—imaginatively, of course.

Oppenheimer advised: "Drill a hole and go deep below the surface [of the earth] to find air trapped in earlier times. I've seen reports from Eastern Europe of people going hundreds of feet down into old coal mines to breathe air that is cleaner than at the surface."

Well, that was interesting. I couldn't drill the hole, but while I was in the Czech Republic visiting the Black Triangle, I did manage a trip south to the Moravian karst near Brno. The karst was once a limestone plateau but is now a series of canyons, tunnels, and caves because of the sculptural power of rainwater. For millions of years, rain laden with small amounts of carbon dioxide has shaped the limestone, dissolving it, washing cracks into canyons, and finding outlets that snake miles beneath the ground. The Moravian karst has become a tourist attraction, with a miniature railroad and shallow boats that ply a subterranean creek. The limestone itself was laid down as lake bottom during the Devonian period. But the spectacular results of millions of years of erosion was only discovered in the early twentieth century by avid speleologists venturing into an oddly magical world. Cool and very damp—the humidity is near 99.5 percent and the average temperature around forty-eight degrees Fahrenheit—the caves suggested an underground cathedral spun out of calcium, with slimy statuary and webs dangling from walls and with ecclesiastical drapes in creamy angel-wing white. The air in my nasal passages had an unusual density, a palpable and soothing weight. Not that the air in the caves was clean, I was informed later by Jay Leiter, the expert on all things respiratory. The caves and their expansive spaces were not the equivalent of a subterranean spa offering relief to the cilia, bronchial tubes, and alveoli of asthma sufferers, Leiter told me.

"We went down into these caves in southern Moravia with a doctor," he said, recalling a visit of his own to the karst. "He was a very old gentleman who, I must admit, seemed on his last legs. Apparently, they had once taken schoolchildren down below for five- or six-hour spells to treat asthma. The kids probably felt better, because of the moisture, and the strange surroundings. But the long-term prognosis for improving their symptoms with cave spas isn't very promising."

After the karst failed to pan out as a site for one clean breath, my choices narrowed. Going back in time retained its appeal. Especially going back to the 1870s to visit the meteorologist Albert Levy at the Paris Observatory, when he began taking his measurements of ground-level ozone and establishing a benchmark so helpful to scientists today. I'd ask Levy if he had had an inkling of how important his disciplined attentiveness might be to future scientists of air. I was curious to see how accurate the description written by the novelist Henry James, a frequent visitor to Paris toward the end of the nineteenth century, had been. James called Paris a "bright Babylon," in contrast to the darker Babylon across the

English Channel. The air in Paris in those days contained its share of coal smoke, although it lacked the pea-soup quality found in London and Pittsburgh. In the early twenty-first century, of course, Paris had joined the world's smogalopolises. During the summer months, the city often had ozone levels so high that health officials urged Parisians to stay home, to carpool, or to take the bus.

While I was thinking about the differences in the air of modern Paris and the air back in Levy's day, I chanced upon a short review of a French duo called Air. In the summer of 2001, Air had moved from Versailles, where ozone wasn't bad, to downtown Paris, where it was terrible, the reviewer informed readers. Why had they moved?

"We are very freedom," said Nicholas Godin who, with Jean-Benoit Dunckel, was Air, two mop-heads suggestive of the early Beatles playing lounge music. "We wanted to be more in touch with urban feeling," Godin went on. Did he mean watery eyes, lung burn, lots of coughing—all reactions to too much ozone? No. The upside of Air's move into the smoggy heart of Paris had been, Godin said, "Sex with a lot of girls." Wasn't everyone out of breath too quickly for that? I had the urge to e-mail the duo and tell them to reconsider their name, to think about changing it to Bad Air, given their decision to move into smoggy Paris. I wanted to tell Air that the air in late-nineteenth-century Paris had surely been better for prolonged trysts that required heavy breathing than the air they had inhaled throughout the summer of 2001 in the hope of more sex. But I never found their Web site.

Just staying around my home in northern Vermont, if I so desired, I could have joined Social Oxygen, a networking group. Social Oxygen was for individuals sick of bars, childless, and busy. It promised members a minimum of eight activities a month with a little heavy breathing. That sounded intriguing, although I was uncertain how much it had to do with one clean breath.

One delightful spring day near my home I saw a male woodcock using clean country air to seek a mate, and his mastery of the medium took my breath away. Using his syrinx, the stumpy, long-billed, ardent little game bird let out a series of notes, seconds apart. They sounded a little like the word "paint" repeated again and again. Then the bird spiraled upward, almost out of sight. Suddenly, he plunged earthward, calling plaintively— *paint, paint, paint.* He leveled off with all the skill of a pilot at the stick of an F-104, a jet with stubby wings and the maneuverability of the woodcock, landed on some moss, peeked around, then began the ritual all over again.

All those pheromones, those estrous exhalations, those elaborate, airy dances of various winged Casanovas such as the woodcock and the pheasant, move better through clean air—though, admittedly, they seem to work pretty effectively in dirty air as well, as long as it's not too dirty. Animals that rely heavily on air to mate manage to do it in dusty climates, to the leeward of smoky volcanoes, and even in urban parks thick with smog. The male ibex breathes in a female's scent the same as you might sniff at the cork of a bottle of wine. The object of ibex desire cooperates by aiming her rump toward a prospect's curved horns, a good sign for him. If she doesn't bound away, notes Irma Almados, an ibex ecologist, "he'll persist with chasing and mounting."

I understand that the albatross, the omen of doom in Coleridge's "Rime of the Ancient Mariner," initiates a rapturous dance, yellow beak raised, black-and-white wings stretched. If turned on, a female will fan the air between herself and the male with her wings, pushing billions of atoms over her pending mate's feathers and through the nostrils of his beak. More oxygen will find its way to his beating heart than finds its way to the heart of a man or another male mammal because of the bird's unique physiology.

Birds, ibex, people—we all desire a mate with the qualities of air. Air is tough and resilient, yet also vulnerable and fragile. It's flexible, within limits. It cleans up most of its messes. It can awe us with sudden displays of power, exude the blue perfume of ozone in a lightning shower, ease us to sleep with an almost silken touch.

Led Zeppelin's Robert Plant said that one of the greatest things "about being a pompous, jumped-up rock god" was that "there's plenty of air around you." Though no rock god, I still feel my body breathing longingly at times. Not as often as I'd like. I'd like to get more *prana* in my breath, more soul. When I'm breathing shallowly, I'm hardly living at all. I'm depriving myself of vital spirit.

We long, of course, for big, plentiful, life-affirming breaths. Breaths redolent of myth, legend, and love. We long for a thrilling molecular miasma. It descends through the tree of our being to the furnaces of our cells, a reminder of just how alive we are because we are on fire. Such breaths make us feel luminous, suffused with a mystical aura, so conscious of our energy that we glow. Breathing like this we are radiant. It is such breathing that makes it possible to see life anew, fresh, and as promising as we never saw it before. Breathe deeply of the air that thrills you for it is life that you pull in. *Prana*. Spirit. Oxygen.

But then, just as suddenly, we can lose it. Sometimes, as Loren Eiseley wrote, "life is lost in the years behind us. Or sometimes it is a thing of air, a kind of vaporous distortion above a heap of rubble."

Whether life is luminous or a heap of rubble, it is good to pay attention to the breath, even if we do live in a time where pursuit of one clean breath is a futile mockery of intent. So pay a little more attention to the daily twenty-three thousand, I often tell myself. Appreciate the air, this mystical and scientific elixir, whatever its condition. We may live in a time when, as one wit put it, "it is no longer Cleopatra's nose that determines history, so to speak, but the germs within her nose." We may live in a time when many people live by the words of the defiant rapper Tupac, who sang, "Give me a reason to be the last motherfucker breathin'." But every one of us would do well to remember the advice of the Chinese sage Huang Po:

> You must take the most strenuous efforts
> Throughout this life, you can never be sure of
> Living long enough to take another breath.

And then to sing air's praises to the sky.

Afterword

My Short Stay at
the Buddhist Breath Center

EN ROUTE TO the Center for Buddhist Studies in Barre, Massachusetts, I stopped and reread a passage from Herman Hess's *Siddhartha*. Hesse wrote that Siddhartha, as a holy man wandering through India with an alms bowl, fasted, grew thin, dreamed strange dreams, eyed wealthy businessmen with contempt, and discovered that all life "stank of lies. . . . The world tasted bitter. Life was pain." At the center of the pain was Siddhartha's breath, which he "learned to save. . . . He learned, while breathing in, to quiet his heartbeat, learned to lessen his heart beats, until there were few and hardly any more."

Once I arrived at the Buddhist center, instead of jumping right into meditating, I first settled down in the quiet library. I was surrounded by books about Zen and breathing and the Buddha. I spent several hours with Veter Tilman's *The Ideas and Meditative Practices of Early Buddhism*.

Tilman made the Buddha's bio read a little like that of Jesus. Both were sacred men who taught disciples their beliefs. Buddha presaged Jesus by five hundred years and outlived him chronologically, eighty years to thirty-three. Buddha was conceived, though, by wealthy parents; he was not the result of an immaculate conception in a manger. The Buddha got married and had a son. He didn't become a wandering seeker near the Ganges in northern India until he was almost thirty. It was over the next five years that he achieved his goal of enlightenment, which meant a release from the eternal cycle of rebirth and the suffering of life, by following the Eightfold Path to nirvana. The route to nirvana, the Buddha discovered, was through his own breath.

On my second day at the center, when I interview Mu Soeng, the director, I ask him if he can keep his mind on his breath all the time when he meditates. And why the breath is the focus of meditation, rather than, say, the beating of the heart, or the blinking of the eyes, or the twitching of a finger.

Mu Soeng parries my first question, telling me that he was once a Zen monk, so he meditates in the Zen tradition, which differs from the Buddhist tradition emphasized here in Barre. As for the focus on the breath, Mu Soeng says that long before the Buddha, yogis across India knew that if they controlled the breath, they controlled the psychic system. They learned to regulate the levels of oxygen reaching their brains. This practice heightened ascetic visions and refined mind control. "If you effectively control the breath," he explains, "you are controlling life. And you can transcend your limitations as a human being."

Calm and centered, Mu Soeng slowly lifts a finger to his upper lip and taps it lightly. He says that the Buddha focused on the air moving across the upper lip. The Buddha concentrated on the breath rather than retreating inward, deep into the labyrinth of the mind and shutting himself off from both the joys and the sufferings of the real world. By keeping his mind focused on the physical sensation of air entering and leaving his body, the Buddha gained what he called open concentration. That is, he sharpened his awareness of all of life going on around him. Instead of escaping into the mind, as the other yogis did, the Buddha calmed his mind, then allowed it to investigate its thoughts as they arose, not grasping any single thought but looking at each and letting it pass away. In this manner the Buddha learned that his thoughts followed a repetitive sequence. Each thought arose, stabilized, decayed, and passed away. "Just like matter," says

Mu Soeng, who not only is a former monk but also has written extensively about the linkages between meditation and quantum physics.

If you follow your breath, he continues, "you establish a wholesome relationship with desire, with fear, with doubt." You do not get hung up on the breath, you do not become so calm that you are catatonic. You do not investigate your own mind so intently that you chase after its thoughts with your logic. A refined relationship with your thoughts comes, he reassured me, as your practice grows and matures, as it deepens. But, he cautioned, you also discover that not only feelings of anxiety, grief, and pain rise, stabilize, decay, and dissolve. Those of joy and happiness follow the sequence as well.

As I leave the office, Mu Soeng thrusts a copy of A. L. Basham's *The Wonder That Was India* into my hands.

I go immediately to the meditation building, leave the book in my room beneath the main hall, then go up the gleaming wood stairs and into the hall and sit.

Ignore everything but your breath, I tell myself. Don't think about the motes of dust in the air, about carbon dioxide, about oxygen. Just stay with the breath. Rid your laundry chute of a mind of its dirty thoughts, its wild abstractions, its clutching desires.

But I have a tough time concentrating on my breath, on my upper lip. The air expanding my lungs seems to be irritating a spot in my upper back. It feels pinched. And my breathing is rough. Why am I breathing so roughly, so anxiously? Is it because earlier, in the library, I had the urge to kill two flies? Killing the flies would have been a Buddhist faux pas. I had this weird fantasy, a reincarnation thing, I think it must have been. I was a frog crawling out of the primordial ooze onto the tartan plaid algae of the Cambrian explosion and zapped a fly with my tongue. Stuck it in the air. And my skin breathed as I chewed. Strange. Now I keep trying to follow my breath to a count of eight but keep hearing flies and thinking about frogs and get up and go back down to my room. I open *The Wonder That Was India*.

Basham wrote that in the *Rg Veda,* India's oldest sacred text, there are descriptions of the first encounters of "a class of holy men different from the brahmans, the 'silent ones,' who wear the wind as a girdle, and who, drunk with their own silence, rise on the wind, and fly in the paths of the demigods and birds." These were the early yogis that Mu Soeng mentioned. They were drunk on silence and went deep into their own minds during trancelike meditations. They lured the future Buddha into their midst once he abandoned his princely life to seek a spiritual one. In 533 B.C., he was attracted by their style, which was less self-lacerating. These holy men had forsaken worldly goods and cares. They begged for food and slept out in the open, but didn't lie on spikes for hours, wrap themselves in thorns, or hang upside down from trees like the extreme ascetics. The moderates mostly sat. They focused on the breath. Their ultimate goal was ecstasy, immeasurable joy, bliss—all attained by riding inward on a current of air into the deepest recesses of the body and the mind.

The pre-Buddha yogis were religious aberrations of the Aryan civilization. A wild, turbulent people who lived in bands in the vast steppes of Central Asia, the Aryans first surged into northern India around 2000 B.C. They were an odd group, I thought, to have spawned yogis who spent their days begging and following air going in and out of their nostrils. Aryan warriors drove chariots drawn by pairs of chestnut-shaded horses—it was the appearance of the chariots that terrified the peoples of northern India when the Aryans first swept in. The newcomers loved poetry, song, and strong drink. They danced and gambled and honored multiple gods. They attended showy sacrifices conducted by their high priests in honor of prosperity and victory in war. It was the very wildness and the excesses of the Aryan civilization that created an ascetic backlash, said Basham. The showmanship and display, in particular, were intended to pave the way to salvation for the Brahmans, the elite. The poverty, insecurity, and hunger of ordinary people were ignored. So when the first silent ones—half-naked yogis wandering through villages with alms bowls in outstretched hands, sleeping on the ground in public parks, performing physical and mental tests of endurance—appeared amid the Aryans, they caught people's attention. Living on virtually nothing, a yogi's lot was much closer to that of the common folk than it was to the lifestyle of a priest or Brahman. And though the wilder yogis did their thing on nails and swung upside down from trees, the moderates who sat in the lotus position and meditated were the ones who attracted the pampered and spoiled son of a prince, once he decided to renounce the world of privilege.

Yet, as Mu Soeng said, "Buddha plugged into their system, but it didn't satisfy him. His great discovery was the awareness of the breath."

I leave the book on the table and go out into the sunlight. I sit on a bench, thinking about yogis, the breath, and air and what's in it that I can't see or feel. When I return to the meditation hall, I sit on the same pillow that I had left earlier on the buffed hardwood floor. I concentrate on air. I feel it entering and exiting past the tip of my nose. I can't feel it brushing across my upper lip. Counting each inhalation and exhalation as one breath, I can't get above four before my mind goes off on a tear, wandering into the past, where it loves to find some aging chestnut to reexamine. I keep bringing my attention, like a restless puppy, back to the tip of my nose. I start counting again from one.

"Breathing in, I know I am breathing in. Breathing out, I know I am breathing out," the Buddha wrote in the *Anapanasati Sutra,* his five-minute sermon on the breath. He first clarified the sutra during a talk to about a thousand monks in a park in Savatti about 2,500 years ago. The *Anapanasati Sutra* is likened to Jesus' Sermon on the Mount, which included the Lord's Prayer, and to Mohammed's Last Sermon. Sitting cross-legged in a grove of trees, the Buddha told the monks how to breathe with full awareness, how to do it right. He made it sound simple, just a series of sixteen exercises anchored around air. The first four helped you be in your body in order to look at it, to create ease and harmony, to unite body and mind. They focused on long breaths and short breaths, on awareness of the whole body's taking in the breath and being calmed by it.

"Practicing awareness in this way, we see that our breathing affects our mind, and our mind affects our breathing," wrote the Vietnamese monk Thich Nhat Hanh. "Our mind and breath become one."

Or, viewed another way, I think, mind and air become one. Not absolutely one, in the quantitative sense, as the cells in our brain are made of ingredients other than oxygen, nitrogen, carbon dioxide, noble gases, and those thousands of trace elements floating about. But one in the sense that my physical essence is like that of air, of breath. As a house is mostly wood, nails, and glass assembled in a unique yet functional way, so are you and I mostly oxygen, hydrogen, and carbon united in a mystifying biological complexity. A complexity that entices the mind with the promise of calming and relaxing the body with air inhaled again and again and again.

Even I, with my restless puppy mind and knotted back, sense this. But not for long. I suddenly want to leap past the first breathing exercises the Buddha told his assembled monks at twilight in Savatti so long ago. I want to start the second set described in the *Anapanasati Sutra,* the set that begins, "Breathing in, I feel joyful. Breathing out, I feel joyful."

First, maybe I should be able to count to eight breaths without getting distracted, so I'm mindful of the Eightfold Path fundamental to Buddhist adepts.

But I'm not an adept. I'm a writer meditating on the breath for a mixed bag of reasons. I'm often as indifferent and unaware of air as the next guy. Yet I'm a little afraid of what the world I live in is doing to air. Fear is not something to meditate on. Nor worry. Let them rise, level off, and pass away. It's air that you're interested in, I tell myself. And mindfulness of air.

I inhale and exhale, feel the sweep of molecules pass the cilia in my nose, and see Mu Soeng's finger touch his upper lip, let it pass away, and briefly enter something close to a peaceful state of being.

The Rapture of Air

Always be a beginner.

—SHUNRYU SUZUKI, *ZEN MIND, BEGINNER'S MIND*

On my third day at the Center for Buddhist Studies, I find joy in the words of Shunryu Suzuki in *Zen Mind, Beginner's Mind*. My puppy mind is OK, Suzuki assures me. My mind that can't count past four breaths without racing off somewhere is fine. My doubts and desires, showers of thoughts pummeling my consciousness like meteors blasting into a molten earth during the Hadean Eon—all fine. All beginner's mind, which is Zen mind, says Suzuki.

I really like this guy Suzuki. He was a pacifist in Japan during World War II. A pacifist among the samurai. He came to America in 1958, aged fifty-three. The atmosphere in California in 1958 was tense with the Cold War, with ongoing nuclear bomb tests in Nevada and the Pacific that dwarfed the explosions that wiped out Hiroshima and Nagasaki. But Suzuki did not dwell on the negative. He discovered that he loved America. In California in particular, he found that Americans had beginner's mind, Zen mind, with its liberating freedom, its good questions, its openness. Suzuki founded the first Buddhist monastery in America in California, a place focused on the breath. He was rich beyond the understanding of many Americans. He was a *roshi,* one who had attained the pure freedom available to all humans. The translator of his books, Trudy Dixon, wrote that Suzuki "exists freely in the fullness of his whole being. His consciousness arises spontaneously and naturally from the actual circumstances of the present."

I'm relieved that the *roshi* says it's OK if I never achieve intense enough concentration to see a white light, the so-called *nimitta.* The *nimitta* signals a new level of mindfulness. Many yogis get hung up on it, Suzuki says. Try as they might to stay with the breath until it becomes luminous, becomes suffused with the *nimitta,* they can't do it. They stress out, white light or no white light, clutching for something that sounds like a ghost to me, or maybe an angel or a sylph.

I brush my teeth to clear away the taste of honey from too many cups of tea, then go again to the sitting hall. The floor shines. It's again empty. I get a pad and pillow from the closet, sit, and immediately get hung up on distractions, though I do see spirals of purple a couple of times on the back of my eyelids at about breath three. I notice I'm taking short breaths.

"Breathing in a short breath, I know I am breathing in a short breath. Breathing out a short breath, I know I am breathing out a short breath." That is the second exercise in the *Anapanasati Sutra*.

Does that mean I'm on exercise two? It can't be. I lack any awareness of my whole body. I feel nervous, sense neither happiness nor joy except for one weird instant when my neck makes this wicked crack and I think (mistakenly, as it turns out) that the stiffness in my back has somehow shot out of my body. The next instant, the idea of following my breath for minutes, and hours, day after day, in a half lotus on a *zazu,* the name of the pillow beneath my aching knees, seems so beyond me, so in another realm, that I want to cry. Discernment, calm, focus—they are so beyond my grasp that I fight back the urge to get up from the pillow and flee.

I keep breathing. I follow the air in and out of my nostrils. One possibility does arise, stabilizes, drifts off, but then returns. I can't get it out of my head!

Beginner's mind is OK. That comes up again. It's true: Beginner's mind is me.

And the world can clean the air. The air I feel going in my nose, going out.

It's all air.

There it is: air and breath. In and out.

What am I? Always return to that enigma, Suzuki said.

Is what I call "I" just "a swinging door which moves when we inhale and when we exhale," as he also said?

Is what I call "I" a biochemical synthesis of a half dozen elements recycled since the Big Bang and now jittery with electricity?

Am I going to be reincarnated as a frog because of bad karma, because I had that thought about killing the flies in the library?

One thing I'm sure of: I'm not enlightened. If I were, I wouldn't *think* about air; I'd just breathe it. I'm not holding my breath in anticipation of reaching nirvana. That doesn't mean I don't want to improve. I can conserve more, maybe plant some trees, kick a few congressional folks in the ass, egg some SUVs. No, no, forget the eggs. Well, maybe a Humvee. Probably get shot, though, for egging a fucking Humvee. Especially if it's a governor's car.

Feeling irritated and perplexed, breathing through my mouth, I unfold my stiff knees. I return the *zazu* to the closet and stand there looking at all the other *zazus* so peaceful and relaxed with one another. I go outside and walk around the beautiful grounds. I contemplate the impermanence of the stones, listen to bird song, stare at ferns. It's fall. Colored leaves crunch beneath my soles.

I check my watch. I need to return to the hall to meet a guy who has been mindful of his breath for five years, ever since his son hauled him along to his first meditation session. Earlier, the receptionist told me I might like to meet the guy. She arranged it and we talked briefly. "I sat down and knew this was it," he said. He told me he was entranced by the teachings of a monk named Santikaro Bhikkhu, a student and translator for Buddhadasa Bhikkhu, who wrote *Mindfulness with Breathing, a Manual for Serious Beginners* I took notes and asked how to spell names. He was very affable, at ease. He explained that the teachings revisit, and comment on, the Buddha's original sermon, the *Anapanasati Sutra.* "That book is all I really need to do this thing," he said.

So now we meet again in the lobby by the meditation hall. I sink into a plush sofa; he sits on cushions on the floor. We sip tea and talk.

The guy is my age, but on a different path. He's got a wife, two kids, a two-million-dollar house. The son, whose interest in Asia led the guy to the focus on his breath, is hiking in Nepal. Within a few years, the guy

tells me, he intends to spend a great deal more of his time meditating
and focusing on his breath.

I'm both envious and irked. What about conservation and global
warming? What about all the air pollution out there?

He goes on, talking passionately about the *Anapanasati Sutra*. He sips
his tea and looks up at me through his glasses. "It's absolutely limitless,"
he says. "Everything's there. Absolutely everything. Impermanence is so
clear in the breath. There's this constant birth, death."

His wife, two kids, the two-million-dollar house—all are imperma-
nent. He's on the path. Christianity hasn't worked for him. He tried for
years to get some satisfaction from it. "It's a belief system. This stuff is so
experiential. All you need to do is go back to one breath and it's all
there." He snaps his fingers over his tea. "The trap is trying to reexperi-
ence things."

Admittedly still a beginner, he sits twice a day at home. He comes
here regularly.

And he has discovered there are three clear stages of progress.

Rapture. "Once you experience rapture, how can you ever let it go?"

Bliss.

And, as he puts it, "osmosing through the veil to where nothing does
arise and nothing passes away."

He smiles. "But once you hit rapture, your practice is cemented."

Later that day, as the sun dives, I sit in the slanting light that pierces
the windows of the meditation hall and get focused on my breath again.
It takes me a minute or two to get past feeling air brushing across my lip.

"You've been so busy thinking that you can't learn anything at all."

Who said that?

Chogyam Trungpa, in *The Myth of Freedom*. I, and my breath, are
walled in by "the human realm," a psychic landscape overshadowed by
the intellect, which has too much going on and can be likened to "a huge
traffic jam of discursive thought." In short, my mind is short-circuited
from an overload of inputs on everything from wondering what Paracelsus liked to eat to the health of a tree I planted in the Jizersky Mountains

to will Jay Leiter think my descriptions of the physiology of the breath are stupid?

Never mind! The important question is this: Is the breath really that important?

Of course it is. Without air, I'd die. Buddhist Center or no Buddhist Center. But there'll be air around for a while. A long while. Probably with too much carbon dioxide, maybe with extinction events going on left and right, and possibly with human civilization in complete and utter disarray. But there will be air. People will breathe whatever they're given. Or perish. Even the most catatonic yogi sitting for months in a forest in India three thousand years ago, lost in the labyrinth of his glucose-oxidizing maze of neurons and synapses, lacked the power to throttle his breathing sensors in the medulla so that he would asphyxiate himself. He passed out first. Involuntary actions kicked in. Respiratory muscles went to work. Partial-pressure gradients insured gas diffusion. The little furnaces in his cells continued to make ATP and create energy. When he came to in the dark, eye-to-eye with Blake's tiger, or a cobra, or with a bewildered and curious child, he was alive. Physiology trumped willpower.

In any event, I bring up thought after thought after thought of so much intellectual baggage, so much reading. I watch each stabilize, decay, and pass away. I enter a welcoming void, mindfully unmindful, counting my breaths. Suddenly an energetic shock leaps out from my shoulder joint. The jolt instantly puts my mind back into the driver's seat of my existence. It screws up my breath count. Then my mind drifts away toward Buddha-ville again, leaving me, for a blissful moment, empty of thought.

And full of air.

Notes

The following notes identify the sources of quotes, ideas, and facts. I identify which texts were most useful for each chapter. I refrain from notes for quotes from individuals I interviewed and who are identified in the text. For the work and lives of many scientists I often turned to *The Dictionary of Scientific Biography* (New York: Charles Scribners and Sons, 1972), which I abbreviate *DSB*.

A select bibliography follows the notes.

Chapter 1: Our Airy Bodies

For facts on human and animal respiration, I relied on input from three physicians, Jay Leiter, Hannah Kinney, and Louis Dandurand. The following texts were extremely helpful: Roger Eckert, *Animal Physiology: Mechanisms and Adaptations,* 3rd ed. (New York: W. H. Freeman, 1988); Jacopo P. Mortola, *Respiration Physiology of Newborn Mammals* (Baltimore: Johns Hopkins University Press, 2001); Emile M. Scarpelli, ed., *Pulmonary Physiology: Fetus, Newborn, Child, and Adolescent,* 2nd ed. (Philadelphia: Lea & Febiger, 1990); and Knut Schmidt-Nielsen, *Animal Physiology* (Cambridge: Cambridge University Press, 1975).

7   *"anticipation"*: Kenneth Franklin, *Joseph Barcroft* (Oxford: Oxford University Press, 1953), 255.

7   *Schultze's swing*: I happened across this long forgotten maneuver in the stacks at the Dartmouth Medical School library. An explanation and diagram are found in Joseph B. DeLee, *The Principles and Practice of Obstetrics* (Philadelphia: W. B. Saunders Company, 1914), 808.

10   *a speck of flesh*: Hannah Kinney, interviews.

10–11   *Galen and the gladiators*: John F. Perkins, Jr., *Handbook of Physiology*, ed. Wallace O. Fenn and Hermann Rahh (Washington, D.C.: American Physiological Society, 1964), chap. 1, "Historical Development of Respiratory Physiology," 1: 10–11.

11   *Legallois*: details from *DSB*.

11   *"By that, I mean"*: Franklin, 225.

12   *"carbonic acid spreads out"*: Perkins, 42.

13–14   *respiratory efficiency of birds*: For a complete comparison of respiratory systems in vertebrates, see Kenneth V. Kardong, *Vertebrates*, 2nd ed. (Boston: McGraw-Hill, 1998), chap.11.

14   *Richard Owen*: Owen's conflicts with Darwin are legend. Kardong, 11, explains some of Owen's thinking: "Owen was a difficult man from the accounts of those who worked or tangled with him."

16–17   *diving reflex*: Jay Leiter, interview with author. Schmidt-Nielsen also has a good section on diving mammals and birds, in *Animal Physiology*, 220–235.

17   *Argyroneta aquatica and lungfish*: The remarkable way both species deal with air came from *Encyclopedia Britannica Online*, q.v. *"Argyroneta aquatica"* and *"lungfish."*

20   *EPO and performance*: Jere Longman, "Performance-Enhancing Drugs Stir Debate," *New York Times*, July 29, 2001, B11.

20–21   *Lance Armstrong*: Quotes and Armstrong's regime from ibid.

21–22   *"Push, push, push"*: All quotes and details from my pulmonary function tests are from Roy Ward, technician at Dartmouth Medical Center, Dartmouth, Mass.

Chapter 2: A Whole Lot of Air

Particularly helpful for this chapter were Thomas E. Graedel and Paul J. Crutzen, *Atmosphere, Climate, and Change* (New York: Scientific American Library, 1997); and Lewis Thomas, *Lives of a Cell* (New York: Penguin Books, 1974), especially the final chapter, "The World's Biggest Membrane," 145–148.

25   *"thin seam of dark blue light"*: Quoted in Graedel and Crutzen, 5.

25   *Venus and Mars*: For Mars, see Oliver Morton, "Mars Revisited," *National Geographic*, January 2004; for Venus, see the magazine's Web site at www.nationalgeographic.com/ngm.

29   *"I was terrified"*: Graedel and Crutzen, 5.

29   *"as much a part of life as wine and bread"*: Thomas, 148.

30 *John Dalton*: For different perspectives on Dalton, I relied on Frank Greenaway, *John Dalton and the Atom* (Ithaca: Cornell University Press, 1966); Arnold Thackeray, *John Dalton: Critical Assessments of His Life and Science* (Cambridge: Harvard University Press, 1972); and the *DSB*.

31 *"You can sometimes"*: Greenaway, 108.

32 *Dalton conducted simple experiments*: Greenaway, 126.

33 *Democritus of Abera*: For a humorous description of the "laughing philosopher," see Leon Lederman, with Dick Teresi, *The God Particle* (Boston: Houghton-Mifflin, 1993), 1. Also see the *DSB*. For the survival of Democritisus's theories, a good source is Henry M. Leicester, *The Historical Background of Chemistry* (New York: Dover Publications, 1956), 110.

33 *Pierre Gassendi's universe*: Leicester, 112–114.

34 *"We might as well"*: Greenaway, 108.

34 *"form an almost infinite number"*: Perkins, 12.

35 *number of working scientists in the world*: Greenaway, 34.

36 *Etienne and Joseph Montgolfier*: Details on their lives are from *Encyclopedia Britannica Online*, q.v. "Montgolfier, Joseph-Michel and Jacques-Etienne."

36 *Speculation about the atmosphere*: For a sweeping view of origin myths and concepts of the sky, see Wendell C. Beane and William G. Doty, *Myths, Rites, Symbols: A Mircea Eliade Reader,* vol. 2 (New York: Harper & Row, 1976). Also see Leicester, chap. 2–8.

37–40 *Torricelli and Pascal experiments*: Both experiments are classics of scientific literature. Good descriptions are found in Hugh Kearney, *Science and Change: 1500–1700* (New York: McGraw-Hill, 1971); Louis B. Young, *Earth's Aura* (New York: Alfred A. Knopf, 1977); and J. H. Broome, *Pascal* (New York: Barnes & Noble, 1965). Galileo's house arrest is from Rom Harre, *Great Scientific Experiments: Twenty Experiments That Changed Our View of the World* (Oxford: Phaidon Press, 1981), 76–77. Malcolm Muggeridge, *A Third Testament* (Boston: Little, Brown, 1976), a book on religious thinkers and "God's spies," which was a companion volume to a TV series in the mid-1970s, discusses Pascal's conversion and his troubled relationship with science thereafter. Pascal's notion of deus absconditus is from Annie Dillard, *Pilgrim at Tinker Creek* (New York: HarperCollins, 1998), 9.

38 *"Nature would not, as a flirtatious"*: David I. Blumenstock, *The Ocean of Air* (New Brunswick: Rutgers University Press, 1959), 187–188.

40   *Otto von Guericke experiment*: Leonard C. Bruno, *The Tradition of Science* (Washington, D.C.: Library of Congress, 1987), 266–267.

41   *Joseph Guy-Lussac*: *Encyclopedia Britannica Online*, q.v. "Guy-Lussac, Joseph-Louis."

41   *the popular misconception*: Blumenstock, 163.

41   *Shelley's imagined balloon flight*: Paul Johnson, *Birth of the Modern: World Society 1815–1830* (New York: HarperCollins, 1991), 199; covered in larger context in Richard Holmes, *Shelley: The Pursuit* (New York: New York Review Books, 1994), 41.

42   *Philosophy of Storms*: *Encyclopedia Britannica Online*, q.v. "weather forecasting." In the twentieth century, meteorologists learned that in the Southern Hemisphere the rotation of the wind in storms was opposite that in the northern hemisphere; that is, they whirled clockwise.

42–47   *Coxwell and Glaisher's balloon flight*: Young, 13–16, gives a fine sketch of the infamous flight. For more detail, see Kurt R. Stehling and William Beller, *Skyhooks* (New York: Doubleday, 1961). For a personal account, see James Glaisher, "Coxwell and Glaisher's Dangerous Ascent," in *Voyages Aeriens* (Paris, 1870).

45   *High altitude sickness*: *Ch'ien Han Chu* and Father José de Acosta are mentioned by Frances Ashcroft, *Life at the Extremes: The Science of Survival* (Berkeley: University of California Press, 2000), 9–10.

46   *"Every 30 seconds I had to rest"*: Maurice Herzog, *Annapurna*, trans. Nea Morin and Janet Adam Smith (New York: E. P. Dutton & Company, 1952), 199.

47   *Tissandier's fatal flight*: Ashcroft, 13.

48–50   *sounding balloons*: The stratosphere experiments of Leon de Bort and Lawrence Rotch are from Seth Cagin, *Between Earth and Sky* (New York: Pantheon Books, 1993), 130–131.

49–50   *Absolute zero*: Charles Seife, *Zero* (New York, Penguin Books, 2000), 44.

51   *The cosmic rays are rattling down* (epigraph): Quote is from Cagin, 143. Much of my description of Piccard and Kipfer's journey to the region of "cosmic bullets" comes from Cagin, chap. 9, "Columbus of the Stratosphere."

52   *American Army Captain Hawthorne Gray*: Stehling and Beller, 202–210.

53   *The next time you fly*: The consequences of a popped cabin window are drawn from Ashcroft, 19–20.

54   *a popped window in the Concorde*: Ashcroft, 22.

55   *Piccard's premature obituary*: Cagin, 142.

56–57   *the furthest spheres*: Details drawn primarily from Joseph S. Weisberg, *Meteorology: The Earth and Its Weather*, 2nd ed. (Boston: Houghton-Mifflin, 1981), 24–32.

Chapter 3: The Origins of Air

59   *"out of blind confusion"*: Ovid, trans. Allen Mandelbaum, *Metamorphoses* (New York: Harvest Books, 1995), Book I: The Creation. The full stanza reads: "So things evolved, and out of blind confusion / Found each its place, bound in external order. / The force of fire, that weightless element, / Leaped up and claimed the highest place in heaven; / Below it, air; and under them the earthSank with its grosser portions, and the water, / Lowest of all, held up, held in, the land."

59   *"it takes a membrane"*: Thomas, 148.

60   *"one of the great true"*: Graedel and Crutzen, 59.

61–63   *the story of the creation of earth's atmosphere*: For greater detail of this fascinating tale, see Sidney Liebes, Elisabet Sahtouris, and Brian Swimme, *A Walk Through Time: From Stardust to Us* (New York: John Wiley & Sons, 1998); Carl Zimmer, *Evolution: The Triumph of an Idea* (New York: Harper Collins, 2001); and William K. Hartman and Ron Miller, *The History of Earth* (New York: Workman, 1991).

63   *"Taken all in all, the sky"*: Thomas, 148.

63   *"the imperishability of the human body"*: Jeffrey Burton Russell, *A History of Heaven: The Singing Silence* (Princeton, N.J.: Princeton University Press, 1997), 46–48.

64   *the archae*: Liebes, Sahtouris, and Swimme, 50–75.

67   *"a penal form"*: Jennifer Ackerman, *Chance in the House of Fate: A Natural History of Heredity* (Boston: Houghton-Mifflin, 2001), 20.

67   *"scientists are amazed"*: Liebes, Sahtouris, and Swimme, 66–67.

67   *"It is they who, in the last half-century"*: Ackerman, *Chance in the House of Fate*, 20.

70   *"the pounded powder of the moon"*: Thomas, 148.

73   *the cosmic fecundity of Big Bang debris*: Graedel and Crutzen, chap. 4, "Climates of the Past," was indispensable for me. Liebes, Sahtouris, and Swimme's *A Walk Through Time* also presents a graphic timeline of change from the earth's origins to the present.

74   *"Earth appears to have been especially favored"*: Graedel and Crutzen, 60–61.

75   *bacteria didn't really make love*: Liebes, Sahtouris, and Swimme, 94.

75   *"Every great scientific truth goes through three stages"*: Quoted in Hartman and Miller, 3.

76   *cartoon*: Arthur C. Clark, foreword, *Visions of Spaceflight: Images from the Ordway Collection,* by Frederick I. Ordway III (New York: Four Walls Eight Windows, 2000), 6.

77   *diffuse about 10,000 times faster*: Roger Eckert, *Animal Physiology: Mechanisms and Adaptations,* 3rd ed. (New York: W. H. Freeman, 1988); and Kenneth V. Kardong, *Vertebrates: Comparative Anatomy, Function, Evolution,* 2nd ed. (Boston: McGraw-Hill, 1998), were my guides to animal evolution and physiology, from chordates to reptiles to *Homo sapiens*.

78   *"Ontogeny recapitulates phylogeny"*: from Ernst Haeckel biography available online at www.ucmp.berkeley.edu/history/haeckel.html.

79   *insect respiration systems*: Eckert, chap. 14, "Exchange of Gases," provides a thorough overview of how insects breathe.

80   *"heaped like corpses in a drawer"*: Dillard, 129.

80–81 *reptiles become birds*: Kardong, 20, sketches this evolutionary sequence.

82   *Sixty-five million years ago, the atmosphere was violated*: Edward O. Wilson, *The Diversity of Life* (Cambridge: Harvard University Press, 1992), chap. 3, "The Great Extinction," provides a rich look at the consequence of this bolide hit of Earth. So does Zimmer, chap. 7, "Extinction: How Life Ends and Begins Again."

82   *energy released by the bolide*: Graedel and Crutzen, 73.

84   *"Humanity has initiated"*: Wilson, 244.

84   *"In general, five million years"*: Wilson, 31.

85   *father-and-son team*: For much of the information about the Alvarezes and their theory of a huge asteroid hitting the earth 65 million years ago, I relied on Zimmer, 159–164.

86–87 *Krakatau*: Zimmer, 151–154; and Wilson, 26–22.

87   *"It was darker than any night"*: Quoted in Zimmer, 153.

88   *"If anything in the natural world merits"*: Stephen Jay Gould, *I Have Landed* (New York: Harmony Books, 2002), 14.

89   *"In reality, the human species is one of thousands"*: Kardong, 22.

89   *Hunting down of species*: Examples, quotes, are all from Wilson, 246–253.

91   *"For Darwin, extinction was simply the exit"*: Zimmer, 143.

92   *"Evolution loves death more than it loves you or me"*: Dillard, 178.

93–94 *the "walking whale"*: Zimmer, 135–141.

94–95   *"In Arabic, absurdity is not being able to hear"*: Diane Ackerman, *A Natural History of the Senses* (New York: Vintage Books, 1991), 175–177.

Chapter 4: Once Air Was Simply Marvelous

For the ancients' view of air, Henry M. Leicester, *The Historical Background of Chemistry* (New York: Dover Publications, 1956), was very helpful, as was Hugh Kearney, *Science and Change: 1500–1700* (New York: McGraw-Hill, 1971), with his overview of the natural sciences during two very fertile centuries of scientific change.

99   *speaking Enochian*: Mark Dery, *Escape Velocity: Cyberculture at the End of the Century* (New York: Grove Press, 1997), 62.

99   *"It might be said that the vastness"*: George Leonard, *The Silent Pulse* (New York: Bantam Books), 176.

101   *"the outer skin of the cosmos"*: Jeffrey Burton Russell, *A History of Heaven: The Singing Silence* (Princeton: Princeton University Press, 1997), 22.

101   *"we can know neither what God"*: Russell, 91.

102–104   *the ancients sky*: Leicester, chaps. 2, 3, and 4.

104   *Empedocles' water clock*: Leicester, 22–23.

105–107   *Plato and Aristotle*: DSB; Adler, *Aristotle for Everybody* (New York: Collier Books, 1991); and Robert Maynard Hutchins, ed., *The Works of Aristotle I* (Chicago: Encyclopedia Britannica, 1956).

108   *"I was strange"*: Henry M. Pachter, *Magic into Science: The Story of Paracelsus* (New York: Henry Schuman, 1951), 193. For a broad, deep view of the colorful, exasperating genius, I turned to Pachter frequently. Evan S. Connell, *The Alchymist's Journal* (New York: Penguin Books, 1992), a fictitious look at Paracelsus's life, imagined a litany of snarls directed at doctors: "Curs barking after genius . . . Sycophants vomiting yellowed lies, sons of cuckolds that grope toward paradise in a milk-maid's crotch. Indentured almond-pickers prescribing slough water and sow-piss . . . servile boasting quacks in shit-stained breeches, malsters, sodomites fornicating upon rear stoops with spaniels or kitchen help—how many mumble and glory at the title of Physician? Doctor Slop!" Jacobi Jolande, ed., *Paracelsus: Selected Writings* (New York: Pantheon Books, 1951), reveals that Paracelsus was as quotable as Connell imagined.

109   *"the universities do not teach all things"*: Encyclopedia Britannica Online, q.v. "Paracelsus."

110   *"All they can do is gaze at piss"*: Pachter, 40.

110   *the four humours*: Kearney, 116–125.

110–111   *the dirt pharmacy*: Pachter, 52–53.

112   *"a Tower of Babel run by a medley of maguses"*: Kearney, 125.

112   *Could you tease a beautiful sylph*: Paracelsus coined the word *sylph* and *gnome* also.

113   *"For want of a name, I have called that vapor, Gas"*: Oxford English Dictionary, q.v. "gas."

113–115   *Robert Boyle and Robert Hooke*: DSB was very helpful on the lives of both figures, along with Kearney, 171–188. Also see Harre; and Bruno, 264–271, for more of Boyle's experiments with illustrations.

115   *John Mayow*: DSB; and Perkins, 19–20, 46.

115   *ideas of Seton and Sendivogius*: Leicester, 120.

116–119   *Thomas Hales*: Leicester, 132–133; and DSB.

119–121   *Phlogiston*: Leicester, 121–124; and DSB on Stahl's career.

Chapter 5: Air, God, and the Guillotine

125   *Cessation of Lavoisier's respiratory functions*: Hannah Kinney interview.

125   *"It took them only an instant"*: available online at www.woodrow.org/teachers/ci/1992/Lavoisier.html, 5.

127   *"I have discovered an air"*: available online at www.woodrow.org/teachers/ci/1992/Priestley.html, 3.

127   *Dissenters*: See Johnson, 377–383, for Priestley's conflicts with Anglican Church; and Richard Holmes, *Coleridge: Early Visions* (London: Hodder & Stoughton, 1989), 44–75, for details on Priestley's being driven from England and Coleridge's dreaming of taking the same route. During the 1790s, an estimated two thousand Englishmen, many of them Quakers and Unitarians, left for the Susquehanna region. Many returned to England with bitter memories of their experiences. Rural life in America hadn't quite lived up to the ideals of the Romantic Age.

127–128   *The two men were born*: Details about Priestley and Lavoisier came mainly from the *DSB*, Web sites, and a visit to the Priestley House, a historic site in Northumberland. For a dramatic look at Lavoisier's life and death, see Andrew Susac, *The Clock, the Balance, and the Guillotine: The Life of Antoine Lavoisier* (Garden City: Doubleday, 1970).

129   *"The streets stank of manure"*: Patrick Suskind, *Perfume* (New York: Washington Square Press, 1999), 3.

129 *royals no better*: Joseph Amato, *Dust: A History of the Small and the Invisible* (Berkeley: University of California Press, 2000), 29.

132 *"injury which is continually done"*: available online at www.woodrow.org/teachers/ci/1992/Preistley.html.

132 *Preistley and Captain Cook*: Data Sober, *Longitude* (New York: Walker Publishing, 1995), 158–159, tells how sauerkraut kept away scurvy.

133 *"The feeling of it my lungs"*: American Chemical Society, *Joseph Preistley: Discoverer of Oxygen* (Washington, D.C.: ACS, 2001), 5–6.

134 *Phlogiston is imaginary*: Susac, 102.

135 *"a million-headed leech"*: Susac, 11.

136 *"possibly as a result"*: Thomas Kuhn, *The Structure of Scientific Revolutions*, 2nd ed. (Chicago: University of Chicago Press, 1970), 53–54.

138 *"strongest fire we yet know"*: Quoted in *DSB* from Albert Henry Smyth, ed., *Writings of Benjamin Franklin*, vol. 8,314.

139 *"If it is any trouble"*: Encyclopedia Britannica Online, q.v. "Henry Cavendish."

139–140 *Montgolfiers*: *DSB* on Lavoisier, 78–79, describes the famous flights and Lavoisier's response to them.

141 *"to have routed all earlier chemistry work"*: Ibid., 79.

142 *"toadies laud him"*: For the Marat and Lavoisier feud over the wall erected around Paris, see Susac, 31, 199–200.

143 *"Breathing, rusting, burning, it's all one"*: Susac, 8.

Chapter 6: Mythic Gods of Air

145 *sky gods*: Mircea Eliade, *Myths, Rites, Symbols: A Mircea Eliade Reader,* ed. Wendell C. Beane and William G. Doty (New York: Harper and Row, 1976), 2:352–354.

147 *Greek origin myth*: Felix Guirands, *Greek Mythology*, trans. Delano Ames (London: Paul Hamlyn Ltd., 1963), 11–12; see also Eliade, 358.

148 *"I am the Earth . . . Your wives are to you"*: Quoted from Eliade, 386.

148 *Rangi and Papa*: Joseph Campbell, *The Hero with a Thousand Faces* (Princeton: Princeton University Press, 1968), 282–283.

148 *the children "cut the cords"*: Eliade, 363–364.

149 *vagina dentata*: Ibid., 409.

149 *"It would not be too much to say"*: Campbell, 3.

150 *"Live as though the day were here"*: Nietzsche, quoted in Campbell, 391.

152 *Viking world's surreal birthing*: H. R. Ellis Davidson, *Gods and Myths of*

Northern Europe (New York: Penguin Books, 1964), 26–28; also Campbell, 284.

152   *A hammer-throwing, red-bearded, gluttonous pagan*: See Davidson, chap. 3, "The Thunder God," for a full portrait of Thor.

153   *four dwarves holding the sky*: Davidson, 27.

153   *ritualistic self slaughters*: Davidson, 51.

153   *account of Ibn Fablan*: Davidson, 52.

154   *Yahweh*: Eliade, 366–369.

154   *Indra*: Campbell, 328–329.

154   *"Smaller than the smallest atom"*: See Charles Seife, *Zero: The Biography of a Dangerous Idea* (New York: Viking Penguin, 2000), 64–66, for a brief account of the atman, "the Spirit, the Self" concealed in everyone.

Chapter 7: Is There Air in Heaven?

155   *Crossing the skies, Stormfield faints*: Mark Twain, *Extract from Captain Stormfield's Visit to Heaven* (New York: Oxford University Press, 1996), 14–15.

157   *the graviton*: Leon Lederman, *The God Particle* (Boston: Houghton-Mifflin, 1993), in list of "dramatis personae."

157   *hell is "oddly fleshy"*: Alice K. Turner, *The History of Hell* (New York: Harcourt Brace & Company, 1993), 3.

157   *Gallup poll*: Turner, 4.

158   *heaven with property lines*: Russell, 126.

158   *English names for heaven*: Russell, 104.

158   *climbers to heaven*: Davidson, 19–20.

159   *Enoch travels aloft*: Russell, 38.

160   *St. Augustine's version of heaven*: Ibid., 88–89.

160   *"Stars get all tangled up in their hair"*: Reverdy's lines are in the translator's preface of Arthur Rimbaud, *A Season in Hell and Illuminations*, trans. Bertrand Mathieu (Brockport, N.Y.: BOA Editions, 1991), 2.

161   *"a miraculous achievement" and "magnificence"*: Thomas, 148.

161   *"in the deepest sense"*: Russell, 41.

162   *Woe to you, wicked souls!*: Dante, *The Inferno of Dante*, trans. Robert Pinsky (New York: Farrar, Straus and Giroux, 1995), canto III, l. 69–72.

162–163   *The Vision of Tundal*: Turner, 97–100.

164   *Dreams of Hell, an allegory*: Turner, 104.

164   *stabbing of Johannes Scotus Erigena*: Turner, 89.

Chapter 8: The Little Atmospheric Disruptions of Man

167   *Centralia*: The coal fires began in 1961. A burning dump managed to touch a vein of coal and set it on fire. Smoke, fumes, and toxic gases were soon rising through people's lawns and filling their cellars. The federal government bought homes and businesses, and relocated the owners. By 1999, when the last people moved out of Centralia, more than $40 million had been spent trying to stop the fires. None of the attempts worked. The latest proposal is to dig a 500-foot trench around all the burning coal beneath the abandoned town. When the fires reach the trench, they'll have no more fuel and will die out. For more information on a town not even on maps anymore, see www.offroaders.com/album/centralia.

170   *Leonardo de Vinci compared the blood*: Gould, 257.

170   *brown cloud over Asia*: Marianne Bray, CNN online, available at www.cnn.com/2002/WORLD/asiapcf/south/08/12/asia.haze.

172   *lungs of Stone Age people*: J. R. McNeill, *Something New Under the Sun: An Environmental History of the Twentieth-Century World* (New York: W. W. Norton & Co. 2000), 55.

172   *oldest known smelter*: Thomas H. Maugh II, "Ancient Copper Factory Discovered in Jordan," *Burlington Free Press*, July 8, 2002, 5A.

173   *"Otherwise, the noxious and deadly vapor"*: "Lead Pollution Fouled Ancient Skies," *National Geographic* (Earth Almanac), June 1995, 137.

173   *Song dynasty in China*: McNeill, 56.

174   *"stagnant, turbid, thick"*: McNeill, 56.

174–175 *first anti-smoke firebrand*: One of the planet's first environmentalists. Quotes are from John Evelyn, *Fumiugium, or the Inconvenience of the Air and Smoake of London dissipated, together with some Remedies humbly proposed by John Evelyn, Esq., to his Sacred Majesty's Command*, published in 1661. Evelyn's walk by Scotland yard was described by Peter Brimblecombe, *The Big Smoke: A History of Air Pollution in London Since Medieval Times* (London: Methuen & Co., 1987), 47.

175   *amount of carbon deposited*: Emsley, 176–177.

175   *Archbishop William Laud*: Brimblecombe, 49–50.

176   *"Great Sinking Fogs"*: The Gadbury, Arbuthnot, and Walker books are mentioned by Brimblecombe, 59, 74.

176   *"smoke is irritating to the air-passages"*: Committee on the Epidemiology of Air Pollutants, *Epidemiology and Air Pollution* (Washington, D.C.: National Academy Press, 1985), 21.

178   *Lavoisier's final experiments with air*: DSB on Lavoisier.

178   *"common people who for a modest sum"*: Perkins, 31.

178–179   *John Hutchinson*: Perkins, 47–48; and *DSB* on Hutchinson.

180   *Manchester*: Mark Girouard, *Cities and People* (New Haven: Yale University Press, 1985), chap. 12, "Manchester and the Industrial City," 258–259.

181   *atmospheric signatures of cities*: Amato, 8.

182   *"Here were a few, perhaps a hundred"*: Quoted in B. W. Clapp, *An Environmental History of Britain Since the Industrial Revolution* (London: Longman House, 1994), 33.

183   *Jerusalems versus Babylons*: Girouard, chap.17, "Babylon or Jerusalem," 343–376.

183   *Manchester Association for the Prevention of Smoke*: Clapp, 48.

183   *"these smoke-producing monopolists"*: Clapp, 33.

183–184   *Lord Palmerston and Earl of Derby alliance*: Johnson, 372, 901; Clapp, 24–26.

184   *Alkali Act of 1863*: Clapp, 23–26, 225–228.

184   *Saint Rollox soda works*: Johnson, 280.

186   *Gossage towers*: Clapp, 24, 228.

186   *Royal Committee of Noxious Vapors*: Clapp, 26.

186   *"a vast heap of reeking filth"*: Clapp, 30.

187   *ammonia and Ammonians*: P. W. Atkins, *Molecules* (New York: W. H. Freeman and Company, 2000), 26.

187   *suburbanization began in earnest*: Girouard, 269–284.

188   *"It's amazing"*: D. H. Lawrence, *Lady Chatterly's Lover* (New York: Signet Books, 1959), 86.

189   *"Ours is essentially a tragic age"*: Lawrence, 5.

189   *"Hell is a city much like London"*: Percy Bysshe Shelley, *Peter Bell the Third*, pt. 2, stanza 1.

189   *"a compound of fen fog"*: Southey, quoted in McNeill, 57.

189   *coal consumption had increased*: McNeill, 31.

190   *"Smoke is the incense"*: Christine Meisner Rosen, "Businessmen Against Pollution in Nineteenth-Century Chicago," *Business History Review* 69 (1995): 385.

190   *W. P. Rend*: Rosen, 385–386. Rend's speech to the Union League Club claimed no one could stop smoke and that it might be make people disease resistant.

191   *fog's almost mythical status*: McNeill, 61, describes people tumbling into the Thames; Stephen J. Gould, preface *Evolution: The Triumph of an*

Idea, by Carl Zimmer (New York: Harper Collins, 2001), x, mentions the darkening of the wings of London moths during the nineteenth century because of soot; David Stradling and Peter Thorsheim, "The Smoke of Great Cities: British and American Efforts to Control Air Pollution, 1860–1914," *Environmental History* 4 (June 1999): 11, mentions fruitless attempts by Parliament to keep smoke out of its chambers.

191　*"Hell with the lid taken off"*: Quoted in McNeill, 68.

191　*early smoke ordinances*: For a thorough history of the first antismoke ordinances, see Stradling and Thorsheim, 6–16. Also see David Stradling, *Smokestacks and Progressives: Environmentalists, Engineers, and Air Quality in America, 1881–1951* (Baltimore: Johns Hopkins University Press, 1999).

192　*emancipated women in America*: Stradling, 42–44.

193　*Pittsburgh meeting to discuss smoke*: Rosen, 385–386; Stradling, 13.

194–195　*"the secrets of metabolism unroll"*: Perkins, 35–36.

195　*Pflüger*: Perkins, 33.

195　*"Carbon dioxide traveling in the blood"*: Perkins, 42.

195　*August and Marie Krogh*: Perkins, 53–55.

197　*Semmelweis*: see www.uh.edu/engines/epi622.htm.

199　*a coal miner knew he was in trouble*: The pathology of coal dust is from Alfred P. Fishman, *Pulmonary Diseases and Disorders* (New York: McGraw-Hill, 1980), 675, 685.

200　*nuisance law*: Stradling, 61.

200　*"lungs of a city"*: Stradling and Thorsheim, 10.

201　*You can't stop it!*: Rosen, 386.

201–203　*America's First Air Fiasco*: For this section on Chicago's attempt to control smoke at the Columbian Exposition, I relied on Rosen. The context of the fair itself, and facts about the Ferris wheel, are from Erik Larson, *The Devil in the White City* (New York: Crown Publishers, 2003).

203　*"It is pretty generally agreed"*: Rosen, 353.

204　*"If the voice of the people"*: Rosen, 381.

204–205　*the sensational new Ferris Wheel*: Larson, 185, 193–194, 206, 279–281.

207　*costs of smoke*: Rosen, 354.

207　*"In the end, however,"* Stradling and Thorsheim, 23.

208　*"its thick dim distances"*: Quoted in Girouard, 347–348.

209　*"The Doom of London"*: The early sci-fi tale is sketched by Brimblecombe, 127–128. Details of Barr's life are from Bernard Benstock and

Thomas F. Staley, eds., *The Dictionary of Literary Biography: British Mystery Writers 1860–1919* (Detroit: Gale Research Company, 1968).

210   *"We saw the greasy heavy brown swirl drifting past us"*: Quoted in Brimblecombe, 126, from Doyle's Sherlock Holmes tale, "The Adventure of the Bruce-Partington Plans."

210   *"cavorite"*: Frederick Ordway III, *Visions of Spaceflight: Images from the Ordway Collection* (New York: Four Walls Eight Windows, 2000), 102.

210   *"repulsite"*: Ordway, 103.

210   *Making Mars a lot friendlier*: Ordway, 107.

210   *cartoons*: Ordway, 111, 113. *Amazing Stories* and *Science Wonder Stories* had characters blasting through space uninhibitedly, traveling in buslike vehicles with fish-eye headlights and wearing bulbs over their heads to supply them with oxygen.

210–211   *Lord Brabazon and the Coal Smoke Abatement Society*: Stradling and Thorsheim, 15–16.

212   *American anti-smoke revival*: Stradling and Thorsheim, 18; Stradling, 66–71.

213   *"the ethics of air"*: Stradling, 59.

214   *"To breathe pure air must be reckoned"*: Stradling and Thorsheim, 16.

214   *rise of engineers as combustion experts*: Stradling, 74–77; Stradling and Thorsheim, 21.

215   *Mellon Institute of Smoke Investigations*: Stradling and Thorsheim, 19–20.

216   *"Whether one believed that the war"*: Stradling and Thorsheim, 23.

Chapter 9: A Few Twentieth-Century Air Nightmares

219   *Ypres*: I relied on Franklin, 96–97, for a physician's point of view. Also see *Encyclopedia Britannica Online,* q.v."chemical warfare," and Rhodes, *The Making of the Atomic Bomb,* (New York: Simon & Schuster, 1987), pp. 90–91.

221   *The first toxic weapons*: John Noble Wilford, "From Hydra Venom to Anthrax Myth," *New York Times*, October 7, 2003. For this and other *New York Times* citations see www.nytimes.com. An interesting weapon I didn't mention was Greek fire, a mix of sulfur, quicklime, and naphtha shot by siphon pumps. Greek fire was introduced in 673 by Constantine V, the Byzantine emperor, in a defense of Constantinople against a Muslim attack. The mix was fired at the Muslim vessels and engulfed many of them in flames.

222   *At a hospital in Boulogne*: All details about Barcroft's stint treating the injured soldiers after the chlorine gas attack at Ypres are from Franklin, 96–100.

222   *an invitation from Nathan Zuntz*: Franklin, 69–74. En route to Tenerife aboard the *Konig Friedrich August*, Barcroft wrote home to his wife about an unusual experience involving airborne communications: "Last night we spent a good while in the wireless telegraphy room which is one of the most fascinating places in the ship, we were somewhere a little north of Gibraltar and I heard with my ears messages from Holland, from Madeira and from Marseilles within a few minutes of one another. They were none of them meant for this ship, but fancy the whole air alive all day with messages going to and from one part of the world to another."

222   *"dead space"*: Perkins, 39; and Jay Leiter, interview with author.

223   *Marie and August Krogh*: Perkins, 53–55.

224   *"It is surprising how little is known"*: Franklin, 97.

224   *Chlorine gas, it would be learned later*: Fishman, 794–795.

224   *Working for the Trench Warfare Department*: Franklin, 102–107.

224   *Phosgene*: Fishman, 793.

225–226   *Klop, mustard gas*: Rhodes, *Atomic Bomb*, 92–94.

226   *"So it came to pass"*: Rhodes, *Atomic Bomb*, 94.

227   *Promoting his so-called smokeless stove*: Count Rumford was born Benjamin Thomson in America. A royalist, he fled to England as a safe haven during the Revolutionary War and there developed his famous stove. See *DSB* for an account of Thompson's life.

227   *Geneva Protocol of 1925*: *Encyclopedia Britannica 2003*, *q.v.* "chemical warfare." The protocol was superseded by the 1972 Biological Weapons Convention, which was signed by one hundred nations, including the United States and Russia. The 1972 protocol outlawed biological and chemical weapons. The United States subsequently destroyed large stockpiles of Q fever, VEE, and other chemical weapons. According to protocol conventions, the United States kept small amounts of germs and toxins for use as defensive weapons, and also conducted experiments for new weapons that were neither manufactured in large quantities nor stockpiled. As Judith Miller, Stephen Engelberg, and William Broad wrote in *Germs* (New York: Simon and Schuster, 2001), 288, "The treaty was maddeningly vague, and government lawyers had spent years trying to translate its provisions into practical rules.... Some

experts believed such experiments were acceptable, as long as they were not intended for war. Other government officials contended that a weapon was, by definition, meant to inflict harm and was therefore out of bounds, even for defensive studies. A bomb was a bomb was a bomb, they would say."

228–231   *Smog Man*: The *L.A. Times* documented Tucker's visit. Reporter Ed Ainsworth wrote the most lively coverage in "'Times' Bringing Smog Expert Here," *L.A. Times*, December 1, 1946, part 1, p. 2; "Smog Hunted by Expert," *L.A. Times*, December 21, 1946, part 2, p. 1; and "Smog Fighter Ends Survey," *L.A. Times*, December 23, 1946, part 2, p. 1.

229   *In 1946 Los Angeles contained*: "Yes, Smog Is Still with Us," editorial, *L.A. Times*, August 10, 1947, part 2, p. 4.

229   *"eye-smarting tolerances"*: Ainsworth, "Smog Hunted by Expert."

229   *"unceasing fight and aroused citizenry"*: From a speech by Raymond Tucker to Smog Advisory Committee, December 19, 1946.

230   *a kind of invisible ceiling*: See Joe Sherman, *Charging Ahead* (New York: Oxford University Press, 1998), 39.

230   *A culprit was needed*: Bill Kelly, ed., *Southland's War on Smog* (Los Angeles: South Coast Air Quality Management District, 2002), 2.

230   *"I can tell you one thing"*: Ainsworth, "Smog Fighter Ends Survey."

231   *"Prof. Tucker waved no wands"*: "This Smog Fight Is No Picnic," editorial, *L.A. Times*, December 24, 1946, part 2, p. 4.

232   *A Smog Czar*: Ed Ainsworth, "Supervisors Pick Chief for Smog Fight," *L.A. Times*, August 9, 1947, part 1, p. 4.

232   *Many read like pages torn*: Kelly, 13.

233   *"an immensely complex scientific task"*: See "Deutch Tells Smog Problem," *L.A. Examiner*, July 27, 1947, 8.

234   *"Wide snaking rivers of concrete"*: K. T. Berger, *Where the Road and the Sky Collide* (New York: Henry Holt and Company, 1993), 4.

235   *When General Motors brought*: Jim Klein and Martha Olson, *Taken for a Ride,* documentary presented on *P.O.V.* (Public Broadcasting Service, 1996).

236   *The first guy to call this freedom*: Sherman, 38.

236–237   *Donora*: There are many accounts of the Donora tragedy. See Devra Davis, *When Smoke Ran Like Water* (New York: Basic Books, 2002). Also see Edwin Kiester, Jr., "A Darkness in Donora," *Smithsonian*, November 1999, available at www.smithsonianmag.si.edu/smithsonian/issues99/nov99/phenom_nov99.html. Chris Bryson, "The Donora Fluoride Fog:

A Secret History of America's Worst Air Pollution Disaster,"*Earth Island Journal* (fall 1998), argues persuasively that the Donora Zinc Works discharged deadly fluoride gas and that the facts were subsequently covered up by U.S. Steel and the U.S. Public Health Service, available at www.earthisland.org/eijournal.

237 *highest ozone level and "smog complex"*: Kelly, 8, 17.

238 *identifying ozone and measuring its concentrations*: Kelly, 5.

239 *"Since even small fluctuations may grow"*: Berger, 23.

244 *"It would simply lift a chunk"*: Richard Rhodes. *Dark Sun: The Making of the Hydrogen Bomb* (New York: Simon and Schuster, 1995), 402.

244 *cosmic rays*: Louise B. Young, *Earth's Aura* (New York: Alfred A. Knopf, 1977). For a brief history of rays from space, chap. 13, "Rain from the Universe," is excellent.

244 *Traces of C-14*: Richard P. Turco, *Earth Under Siege: From Air Pollution to Global Climate Change* (New York: Oxford University Press, 1997), 205.

245 *the harm from radon*: Young, 195. Radon consists of alpha particles, the energetic nuclei of helium atoms. These particles can't pierce skin, but once inhaled are biologically dangerous because inside the body they can cause mutations.

245 *an ironic and tragic disconnect*: Bruno, 290. Turco, 210–211, wrote about the number of scientists who died from exposure. Marie Curie's almost defiant nonchalance about exposure to radiation is well documented; I relied on Sarah Dry, *Curie* (London: Haus Publishing, 2003), 103.

246 *radioactive elements possessed*: Turco, 203.

246 *when you smashed deuterium*: Turco, 204.

247 *first atoms of radioactive plutonium*: Emsley, 227–229.

247 *"Unprecedented, magnificent, beautiful"*: Quoted in Young, 89–90.

248 *might ignite the atmosphere*: Rhodes, *Atomic Bomb*, 418.

248–249 *"Sinews of Peace" speech*: Rhodes, *Dark Sun*, 236–243. Rhodes provides the context and explains the politics behind Truman's decision to ignore the advice of the Acheson-Lilienthal report and to embrace Churchill's tougher stand and to keep the atomic bomb a secret. At least for a couple years, until spies and leaks resulted in Russia's first atomic blast in 1949.

249 The section of my book "Under the Vault of the Desert Sky," owes a debt of gratitude to Carole Gallagher's *American Ground Zero: The Secret Nuclear War* (Boston: MIT Press, 1993), from which I borrowed liberally and which I highly recommend to any reader interested in this dark

chapter of American history. All the anecdotes and quotes, except those below, were drawn from her work.

251 *"most prodigiously reckless program"*: Keith Schneider, forward to Gallagher, *American Ground Zero*, xv.

251 *code named Project Greek Island*: Located beneath the Greenbrier Resort, the congressional bomb shelter included eighteen dorms, a small hospital, and cooking and communication facilities. It could hold eleven hundred people in a pinch. Project Greek Island was made public only in 1992, once Russia collapsed. By then the shelter was not much protection anyway, given the size of the latest atomic bombs and their pinpoint accuracy. All from Brad Lemley, "To the End of the World: Bomb Shelters Never Go Out of Style," *Discover*, October 2003, 66–73.

252 *the most toxic element known*: There is no permissible dose for plutonium. A very little bit (one pound, according to Gallagher) could kill everybody, but only if the pound were finely ground and evenly distributed and inhaled by everyone—a virtual impossibility. Still, the image of a golf-ball-size sphere being able to wipe out humanity has become ingrained in the public imagination because the example has been endlessly repeated in the media. In fact, plutonium probably is an element from hell. But not as capable of making the human race extinct as is often suggested. Its dangers arise as much from how it interacts with human biology as from its distribution in the atmosphere. Emsley, 228, writes, "Plutonium is dangerous because it tends to concentrate on the surface of our bones rather than being uniformly distributed throughout the bone mass like other heavy metals." As you might guess, some radiation physicists take issue with plutonium's bad reputation and go a little ballistic when they hear the contention that a one-pounder of the stuff can kill everybody. "There are many people who have inhaled measurable quantities of plutonium many years ago and have suffered no ill effects," Richard J. Burke, Jr., executive secretary of the Health Physics Society, claimed in 1993. Explaining a dose of plutonium in "A Perspective on Dangers of Plutonium" (April 1995), W. G. Sutcliffe said, "People very near the disposal site could experience serious acute health effects or significant increased cancer risks, but it is inconceivable that large numbers of people would suffer grave health effects, as implied by the news media." See www.llnl.gov/csts/publications/sutcliffe/ for a defense of plutonium.

252 *fallout map*: Frontispiece in Gallagher, *American Ground Zero*. The map is

from Richard L. Miller, *Under the Cloud: The Decades of Nuclear Testing* (Two-Sixty Press, 1999).

252 *"Enough plutonium was scattered"*: Emsley, 228.

252 *"delayed mutation effect"*: Schneider, forward to Gallagher, *American Ground Zero*, xxvii.

254 *Alpha rays . . . Beta rays . . . gamma rays*: Encyclopedia Britannica.

255 *"The outcome of these assaults"*: Rhodes, *Atomic Bomb*, 731–732.

255–256 *So it came to pass*: Judith Miller, Stephen Engelberg, and William Broad, *Germs* (New York: Simon and Schuster, 2001), 34–65. The authors give the history of Fort Detrick, which grew "from a rural outpost in farm country" to "a dense metropolis of 250 buildings and living quarters for five thousand people" virtually overnight in 1943. They describe the Q-fever experiments conducted on the Old Testament guinea pigs, the Seventh-Day Adventists.

257 *"the devil's own laboratory"*: John Sutherland, "'Dead Cities:' Jeremiah with a MacArthur Grant," a review in *New York Times*, November 3, 2002, mentions not only Mike Davis's distaste for Dugway Proving Ground, where "The American government has conducted tests on a barrage of deadly biological, chemical and nuclear weapons, sometimes on live volunteers but more frequently (in the case of Army personnel and local inhabitants) on less-than-volunteers," but also H. G. Wells's *The War in the Air*. Written in 1908, the novel opens with the line "Lower Manhattan was soon a furnace of crimson flames, from which there was no escape."

257 *Wendover Range in Toole County*: Michael Janofsky, "Utah County's Toxic Tradition Is under Threat," *New York Times*, October 20, 2002, 16A.

257 *Adventists were led*: Miller, Engelberg, and Broad, 47–48.

259 *America expanded its germ warfare program*: Miller, Engelberg, and Broad, 48–49.

260 *"just makes you think you want to die"*: Miller, Engelberg, and Broad, 50.

260 *a very special time-release germ weapon*: Miller, Engelberg, and Broad, 53–57.

Chapter 10: Clean Air Dreams, and Denials

265 *It justified its ambition*: For an excellent summary of the history of air pollution law in America, see Arthur C. Stern, "History of Air Pollution Legislation in the United States, " *Journal of the Air Pollution Control*

Association, January 1982, 44–61. For a more detailed look at the original
federal act in 1963, see Randall B. Ripley, "Congress and Clean Air: The
Issue of Enforcement, 1963," in *Congress and Urban Problems* (Washington, D.C.: Brookings Institution, 1969). See also the excellent study by
James E. Krier and Edmund Ursin, *Pollution and Policy: A Case Essay on
California and Federal Experience with Motor Vehicle Air Pollution, 1940–1975*
(Los Angeles: University of California Press, 1977).

265 *All future air quality legislation:* Stern, 61.

265 *"sinister and little-recognized partners":* Rachel Carson, *Silent Spring* (New
York: Houghton Mifflin Company, 1962), 6. The new chemicals were
the "sinister and little-recognized partners in changing the very nature
of the world—the very nature of its life." See also Carson, chapter 4,
"Elixirs of Death," 15–37, which discusses the chemistry of pesticides.

266 *AEC regulated radioactive materials:* Walker, *Permissible Dose,* chap. 2,
"The Debate Over Nuclear Power and Regulation," 29–66, discusses all
the key issues and the conflicts that assailed the original Atomic Energy
Commission and its multiple, conflicting responsibilities.

267–268 *"Nobody in Washington thought":* Dick Valentinetti interview.

269 *"The first responsibility of Congress":* Quoted in Paul J. Miller, *Technology-
Following, Technology-Forcing, and Collusion in the Auto Industry* (W. Alton
Jones Foundation, 1995), 10.

271 *"The highway lobby had convinced":* Deborah Gordon, *Steering a New
Course: Transportation, Energy, and the Environment* (Cambridge, Mass.:
Union of Concerned Scientists, 1991).

271 *"Detroit absolutely refused to change":* Dick Valentinetti interview.

271 *"Old Gerstenberg the bookkeeper":* Quoted in James J. Flink, *The Automobile Age* (Cambridge, Mass.: MIT Press, 1992), 249.

272 *Old G. "was obsessed":* Quoted in Flink, 249.

273 *"More than 100 of our cities":* Gordon, 65, for figures on smog, deaths,
and costs.

274 *catalytic converters:* For the chemistry of the catalysts used in the converters, see Graedel and Crutzen, 37; for skepticism about the technology, see Flink, 388.

275 *"slightly hysterical" mood:* John B. Rae, *The American Automobile Industry*
(Boston: Twayne Publishers, 1984), 140.

276 *Ford's quirks:* Ford himself wrote about his quirks in his autobiography,
A Time to Heal. For his explanation of his tumble from *Air Force One,* see
289.

276   *listened to the car guys*: Letters from Henry Ford II, chairman of the Ford Motor Company, and Lee Iacocca, its president, are from letters to the president files at the Gerald R. Ford Library. Both executives urged the president to favor extensions for the meeting of emission standards. Iacocca encouraged the president to support "a voluntary compliance program—rather than a mandated legislative route with its go/no-go flexibility, costly and frustrating administrative burdens and the inherent adversary relationship between government and the private sector."

276–277   *The car ran beautifully*: Ford, 49.

278   *"society desires pristine air, but"*: Lester B. Lave and Gilbert S. Omenn, *Clearing the Air: Reforming the Clean Air Act* (Washington, D.C.: The Brookings Institution, 1981), 5.

279   *"You could get the science"*: Dick Valentinetti interview.

280   *most Americans wanted a car*: Berger, 152–155.

280   *Reagan's moratorium on new regulations*: Henry A. Waxman, "An Overview of the Clean Air Act of 1990," *Environmental Law* 21 (1991): 1725; also Seth Cagin, *Between Earth and Sky* (New York: Pantheon Books, 1993), 227–242, painted a portrait of Reagan's relentless fights with environmentalists. Cagin said Reagan put some "cowboy soul" in government and had a nostalgic love for "rugged individualism." Ironically, cowboy independence in the American West was "heavily dependent on minimally restricted use of public lands."

281   *James Watt*: Cagin, 233–238.

281   *"I was flown twice"*: Cagin, 234–235.

282   *Bhopal*: Waxman, 1729–1730 for statistics; 1774 for EPA's failure to list hazards.

282   *A staggering 2.7 billion pounds*: David Durenberger, "Air Toxics: The Problem," *EPA Journal* 17, 30.

282–283   *"Are We Poisoning Our Air?"*: Noel Grove, "Air: An Atmosphere of Uncertainty," *National Geographic*, April 1987, 502–537.

283–284   *Clean Air Act amendments of 1990*: See Waxman, 1723–1816, for detailed particulars. A general look appears in U.S. Environmental Protection Agency, *The EPA's Plain English Guide to the Clean Air Act* (EPA, 1993). For a critical look, along with pros and cons of the revised act, see "The New Clean Air Act and What It Means to You," *EPA Journal* 17 (January/February 1991).

284   *"Many clean-coal technologies"*: Kenneth Leung, "Opportunities in the Clean Up," *EPA Journal* 17 (January/February 1991): 52.

285   *"The goal of the act"*: William G. Rosenberg, *EPA Journal* 17
      (January / February 1991): 5.

285   *"the detail and complexity"*: Craig N. Oren, "The Clean Air Act Amend-
      ments of 1990: A Bridge to the Future?" *Environmental Law* 21 (1991):
      1817.

286   *air improvements*: William K. Reilly, "The New Clean Air Act: An Envi-
      ronmental Milestone," *EPA Journal* 17 (January / February 1991): 3–4.

287   *"Air pollution is modern man's wolf at the door"*: Grove, 000.

288   *L.A. air remained the filthiest*: Berger, 322.

288   *often alluded to study*: Shari Roan, "Air Sickness," *Los Angeles Times*, April
      3, 1990, E1.

288–289 *ISTEA*: Available online at www.transact.org. Berger, 354–356, also gives
         an overview of the act and its expenditures.

289   *Houston's "new think versus old think"*: Jim Yardley, "Legal Fight Stalls a
      City's Plans for Light-Rail Relief," *New York Times*, February 13, 2001, A14.

290   *supporters could rightfully brag*: "EPA Air Trend Highlights" (2002), avail-
      able at www.epa.gov.

290   *"The present administration is trying"*: Dick Valentinetti interview.

292   *"These guys are smart"*: Dick Valentinetti interview.

292   *eco-friendly jingoism*: Michael Janofsky, "Nominee for EPA Defends His
      Job As Utah Governor," *New York Times*, August 14, 2003.

292   *"All government does regulations"*: Harold Garabedian interview.

293   *"Reagan's anti-environmental tactics"*: Dick Baldwin interview.

293   *"The public's interest in air"*: Barbara Page, telephone conversation with
      author.

293   *The EPA's shoddy handling*: Kirk Johnson, "Study Says Ground Zero Soot
      Lingered," *New York Times*, September 11, 2003. Also see Kirk Johnson
      and Jennifer Lee, "When Breathing Is Believing," *New York Times*,
      November 30, 2003.

297   *necrology of air figures*: McNeill, 103.

297   *"the health consequences of air pollution"*: McNeill, 103.

299   *"the central species"*: Graedel and Crutzen, 36.

299   *"In the bloodstream it can act"*: conversation with Dr. David Brown.

299   *Health-conscious Victorians*: Emsley, 157–158, including stanza from
      poem.

299–300 *ozone damages*: For tree damages, see *Audubon Reports* (July / August,
         1994), 18. For lower crop yields and damages to plankton and reptiles,
         see Graedel and Crutzen, 108–109.

301–302   *ozone discoveries*: Levy's establishing a baseline is from Graedel and Crutzen, 102–103; other milestones in understanding of ozone are presented in a time line at www.beyond/beyonddiscovery.org.

303   *James Lovelock's electron capture detector*: See www.beyond discovery.org, for the history of the detector. Lovelock's biographical statement is online at www.ecobooks.com/authors/lovelock.html. Cagin, 170–171; 201–202, links Lovelock's work on CFCs with the ozone hole theory. Said Lovelock: "This is one of the more plausible of the doomsday theories, but it needs to be proved."

304   *Midgley "had more impact...."*: McNeill, 111–113.

304   *hydroxl radical*: Graedel and Crutzen, 44.

305   *"I was just looking"*: F. Sherwood Rowland, quoted in www.nrdc.org/reference. For thorough look at the origin of the ozone depletion theory, see Cagin, chaps. 15–18.

306   *Vienna Convention*: See www.unep.org.

307   *damning evidence continued*: Cagin, 277–301.

307   *"a hole in our atmosphere"*: Cagin, 291.

307–308   *Tom DeLay and ozone*: Bill Dawson, *Houston Chronicle* online (November 3, 1995), includes details of Arizona decision to ignore protocol.

308   *ozone hole and cancer rates*: Graedel and Crutzen, 108–109.

308   *Hodel suggested that Americans buy sunglasses*: Cagin, 332–334.

309   *an exemption for methyl bromide*: Andrew C. Revkin, "At Meetings, U.S. to Seek Support for Broad Ozone Exemption," *New York Times,* November 10, 2003.

310   *"an inescapable truth,"* Darcy Frey, "George Divoky's Planet," *New York Times Magazine,* January 6, 2002, 24–25.

311   *"Whether the glaciers of Greenland"*: Frey, 28.

311–312   *Top of the world changes*: For Ellesmere Island glacier loss, see Maggie Fox, "Arctic Ice Shelf Breakup Reported," Reuters, September 23, 2003. The Kenai peninsula devastation is described by Tim Appenzeller, "The Case of the Missing Carbon," *National Geographic,* February 2004, 105–106. In the early twenty-first century the average summer temperature in Alaska was up five degrees Fahrenheit, and the winter temperature was up ten degrees Fahrenheit, with the increases occurring since the 1970s. See Timothy Egan, "Alaska, No Longer Frigid, Starts to Crack, Burn, and Sag," *New York Times,* June 16, 2002.

313   *carbon dioxide is not classified as a pollutant*: This is an interesting point. Without carbon dioxide, life would perish, yet with too much of it, life

would fry. Whether it's a pollutant or not really depends on its concentrations and their impacts.

313 *30 percent more carbon dioxide*: Graedel and Crutzen, 97, 105.

313 *Earth hadn't seen, or had to adapt to*: Yann Arthus-Bertrand, *Earth from Above* (New York: Abrams, 2002), 292.

313–314 *there were multiple factors at work*: Graedel and Crutzen, 111, 121. All the greenhouse gases stay aloft for varying lengths of times. Methane stays for eight to ten years, carbon dioxide for up to a hundred. CFCs float unperturbed for decades, unless they gravitate up to the stratosphere. Without the gases in its atmosphere, the planet would be about fifty-seven degrees Fahrenheit colder, putting its average temperature at zero and making water ice everywhere. Greenhouse gases primarily warm the atmosphere not by warming up from sunlight coming through the atmosphere but by absorbing outgoing radiation. Heat emitted from the planet is long-wave radiation, which lacks the intense energy of incoming sunlight. The outgoing radiation cannot break molecules apart, but it is easily absorbed by large, blocking molecules—again mostly carbon dioxide, along with water vapor. So, the more greenhouse gases in the air absorbing outgoing radiation, the less of it that makes it into space.

314 *overall global temperatures are projected to rise*: Graedel and Crutzen, 135, 143–153; Kenneth Chang, "As Earth Warms, the Hottest Issue Is Energy," *New York Times*, November 4, 2003, gives a summary of various temperature scenarios of the near future.

315 *Chalcaltaya*: Juan Forero, "As Andian Glaciers Shrink, Water Worries Grow," *New York Times*, November 24, 2002.

315 *glacial accident in the Caucasus Mountains*: James Wines, "Rising Star Lost in Russia's Latest Disaster," *New York Times*, September 24, 2002, A11.

316 *carbon hunger of young forests*: Appenzeller, 88–117, discusses the carbon cycle, and the roles of forests and oceans as the world's biggest carbon sinks; he also explains new technologies such as carbon fertilization and geological sequestering.

317 *"Geritol solution"*: Appenzeller, 116.

317 *Oceanographer at Princeton*: Appenzeller, 113.

318 *Kyoto Accord . . . unfair to the United States*: "Bush Cites 'Energy Crisis' for Shift on Emissions," *Portland Press Herald*, March 15, 2001, 4A.

319 *Competitive Enterprise Institute*: Andrew C. Revkin, "Suit Challenges Climate Report by U.S.," *New York Times*, August 8, 2003.

319   *irate lawmakers slashed funds*: H. Josef Hebert, "Legislation Would Cut Global Warming Education," *Burlington Free Press,* July 7, 1998, 3A.

320–321   *making India and Brazil pale as carbon producers*: For growth of China's auto industry, see Keith Bradsher, "China's Factories Aim to Fill the World's Garages," *New York Times,* November 11, 2003. For change in the typical Chinese family, see Keith Bradsher, "China's Boom Adds to Global Warming Problem," *New York Times,* November 22, 2003.

321   *"Maybe six or seven of them"*: The super project approach to wrestling with global warming is reported by Chang.

322   *present energy use by percentages*: Chang, .

Chapter 11: Black Triangle

332   *The razing of Most*: From documentary film I watched in Liberec Library.

333   *The once great spa town of Teplice*: Rob Humphreys, *Czech and Slovak Republics: The Rough Guide* (London: Penguin Books, 1996), 212–214.

333   *a popular local legend*: Zuzana Smekalova interview.

337   *"without alarm bells ringing"*: Jon Thompson, "East Europe's Dark Dawn," *National Geographic,* June 1991, 38–69.

343   *the black sheep of energy resources*: John Crane, interview with author; Jan Pinoš, interview with author.

346–347   *coal production in the twentieth century and now*: For twentieth-century figures, see McNeill, 31–32; for China's coal consumption in the 1990s, see Patrick E. Tyler, "China's Inevitable Dilemma: Coal Equals Growth," *New York Times,* November 29, 1995, A1; Andrew C. Revkin, "Sunken Fires Menace Land and Climate," *New York Times,* January 15, 2002, F1; and Jasper Becker, "China's Growing Pains," *National Geographic,* March 2004, 68–95.

344   *Libkovice*: For the full story, see Susan Gockeler and Hester Reeve, *Libkovice* (Prague: British Council, 1997).

346–347   *so-called reborn landscapes*: Stanislav Stys, *The Region of Most: A New Born Landscape* (Most: Town of Most, 1995).

347   *Karel Krejčí*: Gockeler and Reeve, 140–142.

348   *"Today the Czech government"*: John Crane interview.

349–353   *NEMAK and Jan Rajter*: Input came from NEMAK European chief Jorge Rada and corporate spokespersons. Pavel Franc, Ibra Ibrahimovič, Petr Pakosta, and Jan Pinoš. See www.nemak.org, and www.bankwatch.org/press/2003/press7.html for more details about ongoing conflict over the siting of the aluminum plant.

353 *dioxin*: For an education on the toxicity of dioxin, see www. ejnet.org/dioxin.

353 *came down on the tundra*: Barry Commoner, et al., *Long Range Air Transport of Dioxin from North American Sources to Ecologically Vulnerable Receptors in Nunavut*, Arctic Canada (Montreal: North American Commission for Environmental Cooperation, 2000); and Sheila Watt-Cloutier, interview with author.

Chapter 12: One Clean Breath

363 *between the idea and the reality*: Reference to T. S. Eliot, *The Hollow Men*, vol. 5.

363 *thousands of trace elements*: Graedel and Crutzen, 36.

364 *"that seemingly empty cube"*: David Bodanis, "Pulling What Out of the Air," *Smithsonian*, April 1995, 76.

366 *French duo called Air*: Mark White, "Vive la Difference," *HB* (June/July, 2001), 28.

367 *The object of ibex desire*: Virginia Morell, "Animal Attraction: It's his show, it's her choice," *National Geographic*, July 2003, 29–53.

367 *"about being a pompous, jumped-up rock god"*: Steve Kurtz, *Details*, August 2002, 80.

368 *"life is lost in the years"*: Loren Eiseley, *The Night Country* (New York: Charles Scribner's Sons, 1971), 229.

368 *Cleopatra's nose*: Oddly enough, the lineage of this anonymous quote can be traced back to Pascal, who wrote in the *Pensées*: "Cleopatra's nose, had it been shorter, the whole face of the world would have been changed."

368 *"Give me a reason . . . "*: Lyric from Tupac's "Breathin'" in *Until the End of Time* CD.

368 *You must take the most strenuous efforts*: Huang Po, quoted by Jim Murphy, Jr., in conversation.

Afterword: My Short Stay at the Buddhist Breath Center

369 *"stank of lies"*: Herman Hesse, *Siddhartha* (New York: Bantam, 1981) 14.

369–370 *the Buddha's bio*: Tilman Vetter, *The Ideas and Meditative Practices of Early Buddhism* (Leiden: E. J. Brill, 1988).

372 *"a class of holy men"*: A. L. Bashan, *The Wonder That Was India* (New York: Grove Press, 1959), 243–245.

373 *"Breathing in, I know I am breathing in"*: I relied on Thich Nhat Hanh, *Breathe! You Are Alive*, rev. ed. (Berkeley, Calif.: Parallax Press, 1996), chap. 3, "Analysis of the Sutra's Content," to acquaint myself with the Buddha's *Anapanasati Sutra*. Another excellent and accessible guide through the sutra is by Larry Rosenberg, *Breath by Breath: The Liberating Practice of Insight Meditation* (Boston: Shambhala, 1999).

374 *"Practicing awareness in this way"*: Hanh, 23.

375 *the words of Shunryu Suzuki*: Shunryu Suzuki, *Zen Mind, Beginner's Mind*, ed. Trudy Dixon (New York: John Weatherhill, 1970).

375 *"exists freely in the fulness"*: Suzuki, 18.

377–378 *"I sat down and knew this was it"*: The guy who was on the road by following his breath requested he remain anonymous. I thank him for his openness and wish him more rapture. The book he used is Buddhadasa Bhikkhu, *Mindfulness with Breathing: A Manual for Serious Beginners*, trans. Santikaro Bhikkhu (Bangkok: Dhamma Study & Practice Group, 1989).

378 *"You've been so busy thinking"*: Chogyam Trungpa, *The Myth of Freedom* (Boston: Shambhala, 2002), 31.

Select Bibliography

Ackerman, Diane. *A Natural History of the Senses*. New York: Vintage Books, 1991.

Ackerman, Jennifer. *Chance in the House of Fate: A Natural History of Heredity*. Boston: Houghton Mifflin, 2001.

Adler, Mortimer J. *Aristotle for Everybody*. New York: Collier Books, 1991.

Alberts, Bruce, Dennis Bray, Julian Lewis, and others. *Molecular Biology of the Cell*. New York: Garland Publishing, 1983.

Amato, Joseph A. *Dust: A History of the Small and the Invisible*. Berkeley: University of California Press, 2000.

Appenzeller, Tim. "The Case of the Missing Carbon," *National Geographic*, February 2004, 88–117.

Ashcroft, Frances. *Life at the Extremes: The Science of Survival*. Berkeley: University of California Press, 2000.

Atkins, P. W. *Molecules*. New York: W. H. Freeman and Company, 2000.

Berger, K. T. *Where the Road and the Sky Collide*. New York: Henry Holt and Company, 1993.

Blumenstock, David I. *The Ocean of Air*. New Brunswick: Rutgers University Press, 1959.

Brimblecombe, Peter. *The Big Smoke: A History of Air Pollution in London Since Medieval Times*. London: Methuen & Co., 1987.

Broome, J. H. *Pascal*. New York: Barnes & Noble, 1965.

Bruno, Leonard C. *The Tradition of Science: Landmarks of Western Science in the Collections of the Library of Congress*. Washington, D.C.: Library of Congress, 1987.

Buddhadasa, Ajahn. *Mindfulness with Breathing*. Translated by Santikaro
    Bhikkhu. Boston: Wisdom Publications, 1997.

Cagin, Seth, and Philip Dray. *Between Earth and Sky: How CFCs Changed Our
    World and Endangered the Ozone Layer*. New York: Pantheon Books, 1993.

Campbell, Joseph. *The Hero with a Thousand Faces*. Princeton: Princeton Univer-
    sity Press, 1968.

Cane, Philip. *Giants of Science*. New York: Grosset & Dunlap, 1959.

Capra, Fritjof. *The Tao of Physics*. Boulder, Colo.: Shambhala Publications, 1975.

Carson, Rachel L. *Silent Spring*. Boston: Houghton Mifflin, 1994.

Clapp, B. W. *An Environmental History of Britain Since the Industrial Revolution*.
    London: Longman House, 1994.

Connell, Even S. *The Alchymist's Journal*. New York: Penguin Books, 1992.

Crossley-Holland, Kevin. *The Norse Myths*. New York: Pantheon Books, 1980.

Davidson, H. R. Ellis. *Gods and Myths of Northern Europe*. New York: Penquin
    Books, 1964.

DeLee, Joseph B. *The Principles and Practice of Obstetrics*. Philadelphia: W. B.
    Saunders, 1914.

Dickens, Charles. *Bleak House*. Boston: Houghton Mifflin, 1956.

Dillard, Annie. *Pilgrim at Tinker Creek*. New York: Harper's Magazine Press, 1974.

Dry, Sarah. *Curie*. London: Haus Publishing, 2003.

Eckert, Roger. *Animal Physiology: Mechanisms and Adaptations*. New York: W. H.
    Freeman and Company, 1988.

Eiseley, Loren. *The Night Country*. New York: Charles Scribner's Sons, 1971.

Eliade, Mircea. *Myth and Reality*. New York: Harper & Row, 1975.

———. *Myths, Dreams, and Mysteries*. Translated by Philip Mairet. New York:
    Harper & Row, Publishers, 1975.

———. *Myths, Rites, Symbols: A Mircea Eliade Reader*. Edited by Wendell C. Beane
    and William G. Doty. New York: Harper & Row, 1975.

Elliot, Gil. *Twentieth Century Book of the Dead*. New York: Charles Scribner's
    Sons, 1972.

Emsley, John. *Molecules at an Exhibition*. New York: Oxford University Press,
    1998.

Finlayson-Pitts, Barbara J., and James N. Pitts, Jr. *Atmospheric Chemistry*. New
    York: John Wiley & Sons, 1986.

Fishman, Alfred P. *Pulmonary Diseases and Disorders*. New York: McGraw-Hill, 1980.

Flink, James J. *The Automobile Age*. Cambridge, Mass.: MIT Press, 1992.

Franklin, Kenneth J. *Joseph Barcroft*. Oxford: Blackwell Scientific Publications,
    1953.

Frey, Darcy. "Watching the World Melt Away," *New York Times Magazine*, January 6, 2002, 22–33, 47.

Gallagher, Carole. *American Ground Zero: The Secret Nuclear War*. Boston: MIT Press, 1993.

Gillispie, C. C., ed. *The Dictionary of Scientific Biography*. New York: Charles Scribners's Sons, 1975.

Girouard, Mark. *Cities & People*. New Haven: Yale University Press, 1985.

Gordon, Deborah. *Steering a New Course: Transportation, Energy, and the Environment*. Cambridge, Mass.: Union of Concerned Scientists, 1991.

Gould, Stephen J. *I Have Landed*. New York: Harmony Books, 2002.

———. "Writing in the Margins." *Natural History,* November 1998, 16–20.

Graedel, Thomas E., and Paul J. Crutzen. *Atmosphere, Climate, and Change*. New York: Scientific American Library, 1997.

Greenaway, Frank. *John Dalton and the Atom*. Ithaca: Cornell University Press, 1966.

Grove, Noel. "Air: An Atmosphere of Uncertainty." *National Geographic,* April 1987, 502–537.

Hammond, J. L., and Barbara Hammond. *The Rise of Modern Industry*. London: Methuen & Co., 1966.

Harre, Rom. *Great Scientific Experiments: Twenty Experiments That Changed Our View of the World*. Oxford: Phaidon Press, 1981.

Hartmann, William K., and Ron Miller. *The History of Earth*. New York: Workman Publishing, 1991.

Hobsbawn, Eric. *The Age of Empire: 1875–1914*. London: Abacus, 1994.

Hutchins, Robert Maynard, ed. *The Works of Aristotle I*. Chicago: Encyclopedia Britannica, 1956.

Janofsky, Michael. "Utah County's Toxic Tradition Is Under Threat," *New York Times*, October 20, 2002, 16A.

Johnson, Paul. *Birth of the Modern: World Society 1815–1830*. New York: HarperCollins, 1991.

Kardong, Kenneth V. *Vertebrates: Comparative Anatomy, Function, Evolution*. 2nd ed. Boston: McGraw-Hill, 1998.

Kearney, Hugh F. *Science and Change, 1500–1700*. New York: McGraw-Hill, 1971.

Kelly, Bill, ed. *Southland's War on Smog*. Los Angeles: South Coast Air Quality Management District, 2002.

Kingsley, Ayala, ed. *The New Century: A Changing World 1900–1914*. London: Chancellor Press, 1993.

Kuhn, Thomas S. *The Structure of Scientific Revolutions*. Chicago: University of Chicago Press, 1970.

Larson, Erik. *The Devil in the White City*. New York: Crown Publishers, 2003.

Lederman, Leon, with Dick Teresi. *The God Particle*. Boston: Houghton Mifflin, 1993.

Leicester, Henry M. *The Historical Background of Chemistry*. New York: Dover Publications, 1956.

Leopold, Aldo. *A Sand County Almanac*. Oxford: Oxford University Press, 1949.

Levi-Strauss, Claude. *Tristes Tropiques*. New York: Penquin Books, 1992.

Liebes, Sidney, Elisabet Sahtouris, and Brian Swimme. *A Walk Through Time: From Stardust to Us*. New York: John Wiley & Sons, 1998.

Lockard, Duane. *Coal: A Memoir and Critique*. Charlottesville: University Press of Virginia, 1998.

McKie, Douglas. *Antoine Lavoisier: Scientist, Economist, Social Reformer*. New York: Henry Schuman, 1952.

McNeill. J. R. *Something New Under the Sun: An Environmental History of the Twentieth-Century World*. New York: W. W. Norton & Co., 2000.

Metford, J. C. J. *Dictionary of Christian Lore and Legend*. London: Thames and Hudson, 1983.

Miller, Judith, Stephen Engelberg, and William Broad. *Germs: Biological Weapons and America's Secret War*. New York: Simon & Schuster, 2001.

Mortola, Jacopo P. *Respiration Physiology of Newborn Mammals*. Baltimore: Johns Hopkins University Press, 2001.

Muggeridge, Malcolm. *A Third Testament*. Boston: Little, Brown and Company, 1976.

Nhat Hanh, Thich. *Breathe! You Are Alive*. Rev. ed. Berkeley, Calif.: Parallax Press, 1996.

Ordway, Frederick I., III. *Visions of Spaceflight: Images from the Ordway Collection*. New York: Four Walls Eight Windows, 2000.

Pachter, Henry M. *Magic into Science: The Story of Paracelsus*. New York: Henry Schuman, 1951.

Perkins, John F., Jr. "Historical Development of Respiratory Physiology." Chapter 1 in *Handbook of Physiology*, edited by Wallace O. Fenn and Hermann Rahh. Vol. 1. Washington: American Physiological Society, 1964.

Rhodes, Richard. *The Making of the Atomic Bomb*. New York: Simon & Schuster, 1986.

———. *Dark Sun: The Making of the Hydrogen Bomb*. New York: Simon & Schuster, 1995.

Rimbaud, Arthur. *A Season in Hell and Illuminations*. Translated by Bertrand Mathieu. Brocksport, N.Y.: BOA Editions, 1991.

Rosen, Christine Meisner. "Businessmen Against Pollution in Late Nineteenth Century Chicago." *Business History Review* 69 (autumn 1995): 351–397.

Rosenberg, Larry. *Breath by Breath*. Boston: Shambhala, 1998.

Russell, Jeffrey Burton. *A History of Heaven: The Singing Silence*. Princeton: Princeton University Press, 1997.

Sawday, Jonathan. *The Body Emblazoned*. London: Routledge, 1996.

Scarpelli, Emile M., ed. *Pulmonary Physiology: Fetus, Newborn, Child, Adolescent*. Philadelphia: Lea & Febiger, 1990.

Schlosser, Eric. *Fast Food Nation*. Boston: Houghton Mifflin, 2001.

Schmidt-Nielsen, Knut. *Animal Physiology*. Cambridge: Cambridge University Press, 1975.

Schofield, Robert E. *A Scientific Autobiography of Joseph Priestley (1733–1804)*. Cambridge, Mass.: MIT Press, 1966.

Seife, Charles. *Zero*. New York: Penguin Books, 2000.

Seinfeld, John H., and Spyros N. Pandis. *Atmospheric Chemistry and Physics*. New York: John Wiley & Sons, 1998.

Sobel, Dava. *Longitude*. New York: Walker Publishing, 1995.

Soeng, Mu. *Heart Sutra*. (Cumberland, R.I.: Primary Point Press, 1991).

Sperling, Daniel. *Future Drive*. Washington, D.C.: Island Press, 1995.

Stradling, David. *Smokestacks and Progressives: Environmentalist, Engineers, and Air Quality in America, 1881–1951*. Baltimore: Johns Hopkins University Press, 1999.

Stradling, David, and Peter Thorsheim. "The Smoke of Great Cities: British and American Efforts to Control Air Pollution, 1860–1914." *Environmental History* 4, no. 1 (June 1999): 6–31.

Thackeray, Arnold. *John Dalton: Critical Assessments of His Life and Science*. Cambridge, Mass.: Harvard University Press, 1972.

Thomas, Lewis. *The Lives of a Cell*. New York: Penguin Books, 1978.

Thompson, Jon. "East Europe's Dark Dawn." *National Geographic*, June 1991, 37–69.

Trungpa, Chogyam. *The Myth of Freedom*. Boston: Shambhala, 1988.

Turco, Richard P. *Earth Under Siege: From Air Pollution to Global Change*. Oxford: Oxford University Press, 1997.

Turner, Alice K. *The History of Hell*. New York: Harcourt Brace & Company, 1993.

Walker, J. Samuel. *Permissible Dose: A History of Radiation Protection in the Twentieth Century*. Berkeley: University of California Press, 2000.

Weisberg, Joseph S. *Meteorology: The Earth and Its Weather*. Boston: Houghton Mifflin, 1981.

Wilford, John Noble. "From Hydra Venom to Anthrax Myth." *New York Times*, October 7, 2003. Available at <www.nytimes.com/2002/10/07/science/ 07WAR.html>.

Wilson, Edward O. *The Diversity of Life*. Cambridge, Mass.: Harvard University Press, 1992.

Young, Louise B. *Earth's Aura*. New York: Alfred A. Knopf, 1977.

Acknowledgments

For their indispensable help with this book I'd like to thank the following people:

Jay Leiter, Hannah Kinney, David Brown, and Louis Dandurand for input on medical matters; Dick Valentinetti and Harold Garabedian for advice on the politics of air in the United States; and Blažena Hušková, Tony Šinágl, and Jan Pinős for assistance in the Black Triangle of the Czech Republic. I am much indebted to Sarah Anderson and Norm Pangrass for their help with historical research. And to Ludek Rathousky for artwork and Ibra Ibrahimovič for photographs of the Black Triangle. Ibra's partner, Lenka Zemanova, provided translation services and was always encouraging. Laura Tatro provided generous assistance with the illustrations.

The Gerald Ford Library awarded me a research fellowship, and the Center for Buddhist Studies in Barre, Massachusetts, provided me with accommodations and access to its facilities and library. To both organizations, I give my heartfelt thanks.

I also want to thank the editor Erika Goldman for her interest and encouragement during the early stages of this project, my agent, Sally Brady, for hanging in there throughout various contractual complications, and publishers Jack Shoemaker and Trish Hoard for making publication a collaborative enterprise.

The book's epigraph by Michael Longley was provided courtesy of the poet.

Illustration Credits

Chapter 1. Painting by Ludek Rathousky

Chapter 2. From *Travels in the Air*, J. Glaisher, C. Flammarion, E. de Fontvielle, and G. Tissandier (Paris, 1870).

Chapter 3 and frontispiece. From John Flamsteed, *British History of the Heavens*, 1776.

Chapter 4. Reprinted by permission from Frederick I. Ordway III collection.

Chapter 5. Painting by Jacques-Louis David, Rockefeller University, New York: courtesy of the Library of Congress.

Chapter 6. By Gustav Dore. Reprinted by permission from Frederick I. Ordway III collection.

Chapter 7. Painting by Ludek Rathousky.

Chapter 8. *Manchester*, William Wylde, 1851. Courtesy of Her Majesty the Queen.

Chapter 9. From *Los Angeles Times* Photographic Archives. Department of Special Collections. Charles E. Young Research Library, UCLA.

Chapter 10. Painting by Ludek Rathousky.

Chapter 11. Photographs throughout chapter by Ibra Ibrahimovič

Chapter 12. Photograph courtesy of CSIRO Atmospheric Research, Australia.